Sample Preparation with Nanomaterials

Sample Preparation with Nanomaterials

Next Generation Techniques and Applications

Dr. Chaudhery Mustansar Hussain
Dr. Rüstem Keçili
Dr. Chaudhery Ghazanfar Hussain

WILEY-VCH

Authors

Dr. Chaudhery Mustansar Hussain
New Jersey Institute of Technology
Department of Chemistry &
Environmental Sciences
University Heights, Newark
New Jersey
United States of America

Dr. Rüstem Keçili
Anadolu University
Yunus Emre Vocational School of Health Services
Yeşiltepe, Çamlıdere Sk. No. 6
26210 Tepebaşı/Eskişehir
Turkey

Dr. Chaudhery Ghazanfar Hussain
Department of Computer Science and
Information Technology
University of Lahore
Defence Road
Lahore
Pakistan

Cover Courtesy of Chaudhery Mustansar Hussain

Library of Congress Card No.:
applied for

British Library Cataloguing-in-Publication Data
A catalogue record for this book is available from the British Library.

Bibliographic information published by the Deutsche Nationalbibliothek
The Deutsche Nationalbibliothek lists this publication in the Deutsche Nationalbibliografie; detailed bibliographic data are available on the Internet at <http://dnb.d-nb.de>.

Print ISBN: 978-3-527-33817-7
ePDF ISBN: 978-3-527-68560-8
ePub ISBN: 978-3-527-68561-5
oBook ISBN: 978-3-527-68562-2

Typesetting SPi Global, Chennai, India
Printing and Binding CPI Group (UK) Ltd, Croydon, CR0 4YY

Printed on acid-free paper

C094265_260321

Contents

1

Nanomaterials (NMs) in Analytical Sciences

1.1 Introduction

Analytical science is the branch of science that finds solutions and improvement of classical methods by fulfilling the demands for analytical information modeled by modern economic and scientific society [1]. Analytical information is about describing chemical systems and determination of their components. Technically, analytical information consists of two major components, i.e. analytical capabilities and analytical properties of the system to be analyzed. The measurement of upgrading of an analytical method, characterization, and its results is called analytical capabilities, while analytical properties can be of three different types, i.e. basic, principal, and socio-environmental. In the past, most of the developments are generally focused on techniques and solving instrumental problems, and there was no scope of socio-environmental aspects.

The rapid progresses of the microelectronic industry since 1960s established semiconductor preparation techniques to enhance the density of transistors in integrated circuits. In 1980s, these techniques led to the first design and fabrication of the microelectromechanical systems (MEMSs) [2]. With the introduction of microelectronics and computer technologies, analytical science has undergone remarkable expansions in terms of automation, miniaturization, higher sensitivities, and greater precision and complete analytical effectiveness.

Nanotechnology is defined as the technology that appreciates and controls the matter at dimensions between 1 and 100 nm, called nanomaterials (NMs). NMs have received much attention in the past decade due to the novel physical and chemical properties associated with their size and shape [3–12]. Widespread applications and outstanding performance of NMs not only have accelerated the development of materials science but also offer many opportunities in other related disciplines and have a significant impact on many fields of science, including chemistry, electronics, optics, medicine, and bioanalysis. NMs possess different properties compared to the same material in their coarser or bulk form. NM structures can be in the form of particles, pores, wires, or tubes or combination of these. Unusual properties such as high conductivity, greater heat transfer, higher melting temperature, exceptional optical and magnetic properties, and super adsorbent, etc., provide NMs for a wide range of novel applications [13–20]. The main objective of nanotechnology is to use

Sample Preparation with Nanomaterials: Next Generation Techniques and Applications,
First Edition. Chaudhery Mustansar Hussain, Rüstem Keçili, and Chaudhery Ghazanfar Hussain.

these exceptional properties of NMs and develop new products, tools, and methods. Nanotechnology can play an important role in analytical sciences. Principal design and fabrication of NMs with the incorporation of interfacial elements would be of paramount importance for the whole process of molecular analysis at present times and near future. The objective of this chapter is to focus on recent developments with different types of NMs in analytical sciences, i.e. sample preparation, separation, extraction, and identification techniques.

NMs can be highly selective materials when it comes to purification techniques. During last few years, these have been intensively researched in all types of chromatographic techniques such as gas chromatography [21], liquid chromatography [22], capillary chromatography [23], and membrane technology [24]. Overall, NMs are proven extremely important and exhibited effective improvements over present systems. However, obstacles to overcome them in separation sciences are the aggregations, stability, safety, and economic issues.

This chapter provides an overview of NMs in analytical sciences. The chapter starts with the description of the types of NMs including carbon nanotubes (CNTs), fullerenes (FULs), graphene, inorganic nanoparticles, and magnetic nanoparticles. Then, various examples on the application of NMs are demonstrated.

1.2 Types of NMs

Unique properties of NMs and new methods for the analysis of NMs are strong areas for research these days. NMs have initiated the development of new ways of performing target concentration and detection and new analytical methods and instrumentation for measuring the properties of NMs. The optical and electronic properties of NMs are often dominated by their surface chemistry, and this makes the task of analyzing NMs immensely more challenging than bulk materials or homogeneous solutions. A particularly intriguing incentive for developing new approaches to NMs fabrication is the potential for creating new constructs with nanoscale order and unique functional properties. NMs for analysis can be constructed with the manipulation of one or more components.

1.2.1 Graphene

Graphene was discovered in 2004 by Professor Andre Geim and Professor Kostya Novoselov from the University of Manchester [25]. Since then, many efforts were put into the design and development of new graphene-based functional materials. In addition, the Nobel Prize for Physics in 2010 was awarded to Geim and Novoselov in the University of Manchester for their research on graphene [26]. The excellent features of graphene make it the starting point of new technologies in many application areas. Although graphene has a very low thickness, it is more robust than the diamond and 300 times stronger than steel. It conducts electricity better than copper, passes light, and bends without any deformation and can be put into any desired shape. It also exhibits great features such as large surface area, high stability, and layered structure. These superior features of graphene make it as an excellent NM in many applications such as separation processes, sensors, bioimaging, and drug delivery [27–30].

Due to the large requirement for research and applications, there has been much investment put into the development of methods to prepare the high-quality graphene in bulk. The preparation methods can be classified into two categories, one is "top-down" and the other is "bottom-up." The former one is to break graphite into graphene through approaches like mechanical cleavage, liquid exfoliation, thermal expansion, and electrochemical exfoliation. The latter one is to synthesize graphene by the techniques such as chemical vapor deposition (CVD), arc discharge, epitaxial growth on SiC, and unzipping CNTs [31–34].

1.2.2 Carbon Nanotubes (CNTs)

CNTs were first discovered by Ijima [35] in 1991 and were successfully employed for different purposes in analytical sciences due to their mechanical, electric, optical, and magnetic properties as well as their extremely large surface area [36, 37]. CNTs are hollow graphitic materials composed of one or multiple layers of graphene sheets: single-walled carbon nanotubes (SWCNTs) and multiwalled carbon nanotubes (MWCNTs), respectively. The schematic depiction of the formation of SWCNTs and MWCNTs by rolling of graphite layer is shown in Figure 1.1.

The synthesis of CNTs can be carried out by means of three main techniques: CVD [39], laser ablation (LA) [40], and catalytic arc discharge (CAD) [41]. CVD seems to be the most efficient approach for the synthesis of CNTs for analytical applications due to high purity and desirable tuning at low temperature. However, for all the synthesis methods, the presence of different undesired by-products (such as

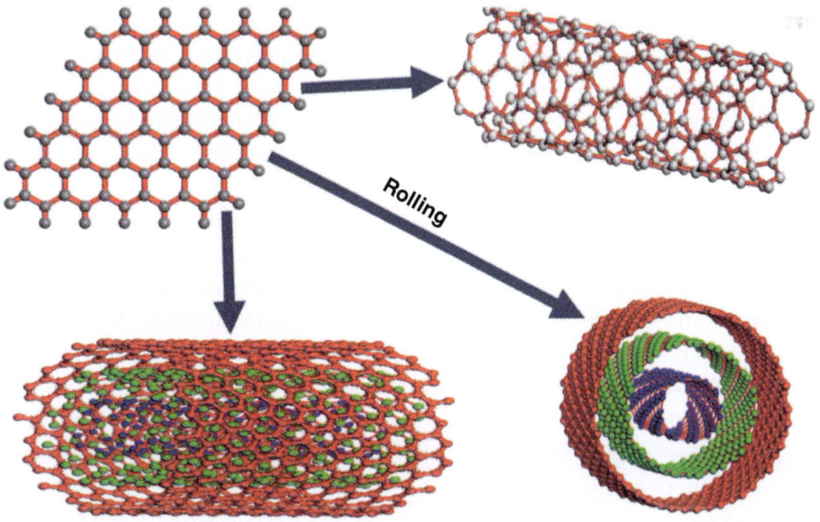

Figure 1.1 The schematic depiction of the formation of SWCNTs and MWCNTs by rolling of graphite layer. Source: Khan et al. [38]. © 2017, Elsevier.

carbonaceous residues, amorphous carbon, metal impurities, and others) makes it necessary to purify CNTs. To purify the synthesized nanotubes, different strategies including chemical oxidation, physical separation, or combination of both chemical and physical techniques have been employed so far. Chemical oxidation is a purification system based on the fact that carbonaceous impurity residues are oxidized sooner than CNTs. The main advantage is its easy use, but it should be noted that the oxidation process affects the structure of the nanotube introducing functional groups (hydroxyl, carbonyl, and carboxyl) and defects in the side walls. Physical purification procedures are based on the different physical properties (such as size, ratio, weight, electrical and magnetic characteristics, etc.) between impurities and CNTs. Filtration, centrifugation, chromatography, and electrophoresis are the commonly employed techniques. The disadvantages of these procedures are: first, the elimination of certain impurities is inefficient; second, a high dispersion of CNTs is required; and third, only a low quantity of CNT can be purified.

The adsorption sites on CNTs are on the wall and in the interstitial spaces between tubes. These sites are easily accessed for both adsorption and rapid desorption. The impurity coverage on the CNT reduces their availability because the sorbate has to diffuse through these impurities to reach the CNT. Moreover, the porous structure of impurities introduces mass transfer limitations, slowing both adsorption and desorption. Understanding these characteristics is important for their application separation media. The excellent features of CNTs, along with their nanoscale features, make them ideal candidates for microscale devices for gas as well as liquid phase analysis. In gas phase analysis, one can implement a micro-concentrator or a micro-sorbent trap, which has been used in a variety of online chromatography and sensing applications. The purpose of such devices is usually to act as a fast pre-concentrator or to modulate the concentration of a stream for real-time monitoring. They have been fabricated in small capillary tubing and also by micromachining silicon and other substrates. The interesting liquid phase pre-concentrating applications are microscale solid-phase extraction (μ-SPE) and solid-phase microextraction (SPME).

Functionalization of CNTs offers a unique opportunity of altering the physical and chemical characteristics of the CNTs [42, 43]. The presence of a covalently attached functional group can alter the retention/affinity of the CNT surface and important properties such as polarity, hydrophilicity, and other specific interactions. The functional groups may also alter diffusional resistance and reduce the accessibility and affinity of CNT surfaces for certain analytes. Functionalization also enhances interaction with polymers and other materials, thus facilitating the formation of composites that can be used as pre-concentrating substrates. These include polymers and sol–gel-type immobilization.

1.2.3 Fullerenes (FULs)

FULs have attracted considerable attention in different fields of science since their discovery in 1985 [44]. Physical, chemical, and biological properties of

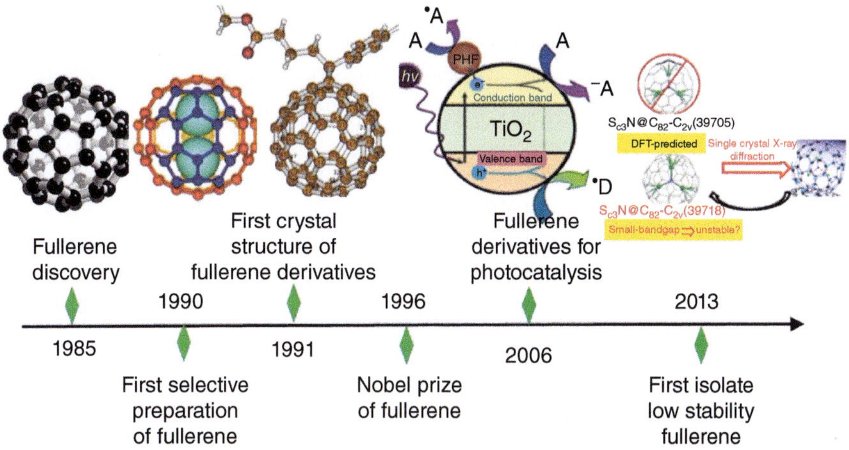

Figure 1.2 The historical development of fullerenes and their derivatives since their discovery. Source: Pan et al. [45]. © 2020, Elsevier.

fullerenes have shown their promising applications for future. It is inferred that size, hydrophobicity, three-dimensionality, and electronic configurations make them an appealing subject in chemistry. The historical development of fullerenes and their derivatives since their discovery is illustrated in Figure 1.2.

FULs are often accompanied with large amounts of soot; therefore, purification is a necessary step before their real applications are realized. Purification of FULs normally involves extraction from soot and then separation by liquid chromatography. Additionally, other purification techniques, like elution on molecular sieves or activated charcoal, gel permeation, and supercritical fluids purification, have also been used.

FULs are practically insoluble in water and generally tend to form aggregates in aqueous media [45]. To disperse these in water, prolonged stirring for days is needed. Addition of additives, e.g. surfactants, cyclodextrins, or sugar polymers, offer a possibility to disperse these in water more efficiently. FULs can also be dispersed in water by first dissolving it in an organic solvent, then adding the solution to water and evaporating-off the organic solvent.

Incorporation of FULs into polymeric materials to improve their processability has become a common practice in applications. The unique properties of FULs, such as their spherical shape, conjugated three-dimensional π-electronic system, and the ability to exhibit donor–acceptor or π–π interactions, make them attractive candidates for analytical applications [46]. As a result, these unique NMs can be successfully employed as chromatographic stationary nanophases in a wide temperature range (80–360 °C), with good thermal stability and high selectivity for aromatic compounds. Chemical functionalization of FULs is often carried out to alter their chemical properties and/or improve miscibility with host polymers.

1.2.4 Inorganic Nanoparticles

Inorganic NMs have received interest in analytical sciences because of the large specific surface area, lower weight, excellent mechanical strength, and magnetism [47–49]. However, in the field of sample preparation, only a few are prominent, while others remain unexplored yet. The prominent ones are gold, silver, titanium, and silica nanoparticles.

1.2.4.1 Gold and Silver Nanoparticles

Recent interest in the development of novel strategies for the generation of gold nanoparticles stems from their potential applications in the fields of physics, chemistry, biology, medicine, and materials science. Gold nanoparticles coupled with tunable surface properties in the form of inorganic or inorganic–organic hybrid are excellent platform for developing a broad range of analytical methods. Therefore, depending on their size, shape, degree of aggregation, and nature of the protecting organic shells on their surface, these can be engineered for individual and multifold applications, including chemical sensing and imaging [50, 51]. First, the gold core is essentially inert, and monodispersed nanoparticles can be synthesized with core sizes ranging from 1 to 150 nm. Further versatility is imparted by relatively simple functionalization via thiol linkages. Additionally, gold nanoparticles have been shown to exhibit size-dependent optical properties, which make them candidates for spectroscopic techniques such as surface-enhanced Raman spectroscopy (SERS). Gold nanoparticle-based fluorescent probes have also been used in various applications [52, 53]. The advantage of using gold nanoparticles in biological labeling is that visible light can be used to observe a color shift from red to blue when it forms aggregates. The ability to synthesize Au nanoparticles of different size is a great asset, for example, very small nanoparticles (4–6 nm) are essential for biological labeling, while SERS is typically carried out using relatively larger (around 35 nm) particles. Because of their unique optical features, gold nanoparticles find applications in many research fields. The coating of materials with gold nanoparticles is extensively applied for the sampling of scanning electron microscope (SEM) to enhance the electronic stream, which helps to achieve great SEM images with high quality.

Silver nanoparticles have attracted a great deal of attention in recent years due to their optical, physical, and chemical properties that differentiate them from corresponding bulk properties [54]. Hence, they have found many applications in catalysis, photonics, optoelectronics, information storage, and antibacterial materials. Silver powders, having ultrafine and uniformly distributed particle size, are of considerable use in the electronics industry as thick film conductors in integrated circuits due to their unique properties such as high electrical and thermal conductivity and high resistance to oxidation. Silver nanoparticles are also important SERS-active substrates that significantly enhance Raman intensities [55]. Currently, there are many kinds of silver-based SERS substrates, such as charged silver colloidal solution, acid-etching silver foil, metal island film formed through thermal evaporation, laser-ablated silver plate, etc.

1.2.4.2 Titanium Nanoparticles

Titanium dioxide (TiO_2) is one of the most widely used of the NMs and has found wide ranging applications such as heterogeneous photocatalysis and energy conversion, etc. [56, 57]. TiO_2 naoparticles (NPs) are normally synthesized by sol–gel process, hydrothermal methods, and also high-energy methods such as laser pyrolysis/ablation, CVD, and plasma methods. Interest in Titania in analytical chemistry arises from its high chemical stability and cost-effectiveness. Strong oxidizing power of the holes, redox selectivity, high photo-stability, and easy preparation make TiO_2 a powerful candidate as an efficient adsorbent.

1.2.4.3 Silica Nanoparticles

Mesoporous silicas are composed of a honeycomb-like porous structure with hundreds of empty channels (mesopores) that are able to absorb/encapsulate relatively large amounts of analytes. Unique properties, such as high surface area (>900 $m^2\ g^{-1}$), large pore volume (>0.9 $cm^3\ g^{-1}$), tunable pore size with a narrow distribution (2–10 nm), and good chemical and thermal stability, make them attractive for various sample preparation applications. Silica nanoparticles have gained great interest for use as sorbents, separation media in chromatography, and membrane separation [58, 59]. For mesoporous silica materials, it is difficult to separate the small particles from liquid. Therefore, these normally incorporate some functionalization, which provides nanosilica immobility, mechanical stability, and water insolubility.

1.2.5 Magnetic Nanoparticles

Magnetic nanoparticles are interesting advanced materials and have gained great attention among researchers from a wide range of disciplines in science [60–63]. Magnetic nanoparticles were successfully applied for different applications such as magnetic fluids [64], catalysis [65], magnetic resonance imaging [66], and environmental remediation [67, 68]. Magnetic nanoparticles exhibit superparamagnetic feature which can be attracted to a magnetic field, but retain no residual magnetism after the field is removed. Thus, it is very easy to separate these magnetic nanoparticles adhered with target compound(s) from complex matrices such as environmental, food, and biological samples by using an external magnetic field. There is no centrifugation or filtration step needed during this process. However, an unavoidable drawback associated with these nanoparticles in this range of their size is instability of these magnetic nanoparticles that tends to cause formation of agglomerates in the sample solution. Moreover, naked metallic nanoparticles are highly chemically active and can easily be oxidized in air that leads to loss of their dispersibility and magnetism feature. Therefore, it is important to design and develop protection approaches for the chemical stabilization of the naked magnetic nanoparticles. These approaches for the modification of the surfaces of magnetic nanoparticle comprise grafting technique or coating with an inorganic layer such as carbon or silica or coating with organic compounds including polymers and surfactants [69].

1.3 Applications of NMs

1.3.1 NMs in Separation Processes

NMs can be successfully employed in separation processes. For example, gold nanoparticles were successfully applied in the capillary electrophoresis (CE)-based analysis due to the improved separation resolutions [20]. These nanoparticles also behave as a pseudo-stationary phase in the existence of poly(ethylene oxide) (PEO) for the efficient separation of double-stranded DNA. The interaction between PEO and DNA molecule is increased in the existence of nanoparticles thereby enhancing the sieving ability of PEO without any change in its viscosity. The separation of acidic and basic proteins using a capillary filled with surfactant-capped gold nanoparticles was also reported [70] along with great separation efficiencies toward various target compounds [71]. A facile approach was developed and demonstrated for the fabrication of highly effective separation columns coated with octadecylamine-capped gold nanoparticles for open tubular capillary gas chromatography [72].

Monodispersed silica nanotubes having desired size and shape were designed and successfully prepared by using template-assisted synthesis approach. These developed nanotubes have inner voids that can be filled with various compounds varying from large protein molecules to small molecules, and the inner and outer surfaces of the nanotube can be efficiently functionalized. One of the interesting applications of the nanotubes is the nanophase extractor for the efficient removal of small compounds from the solution. Because the outer surface of the nanotubes exhibits hydrophilic feature and inner surface shows hydrophobic behavior, these NMs are promising candidates for the efficient extraction of the lipophilic compounds from the aqueous samples. The nanotube functionalized with antibody enables an ultimate extraction selectivity. The antibody produced against the drug, 4-[3-(4-fluorophenyl)-2-hydroxy-1-[1,2,4]-triazol-1-yl-propyl]-benzonitrile (FTB), selectively binds to the RS enantiomer and the Fab fragments of the antibody are immobilized on the inner and outer surfaces of the silica nanotubes [73] (Figure 1.3).

1.3.2 NMs in Biomedical Applications

Nanotechnology addresses these issues by designing and applying NM-based drug delivery systems that are capable of achieving enhanced biopharmaceutical features by altering the biopharmaceutical and pharmacokinetics features of the target pharmaceutical compound. Various NMs such as polymeric nanoparticles, liposomes, nanoemulsions, micelles, and dendrimers can be successfully used as nanodrug delivery systems (Figure 1.4). With the innovative nanoscience- and nanotechnology-based approaches, NMs having different functionalities, shapes, and distinct physical, chemical, and biological features were designed and developed for the successful applications in the pharma industry.

The functionalized NMs exhibit biomimetic features since they can change their unique features that enable them to be applied in drug delivery and self-healing

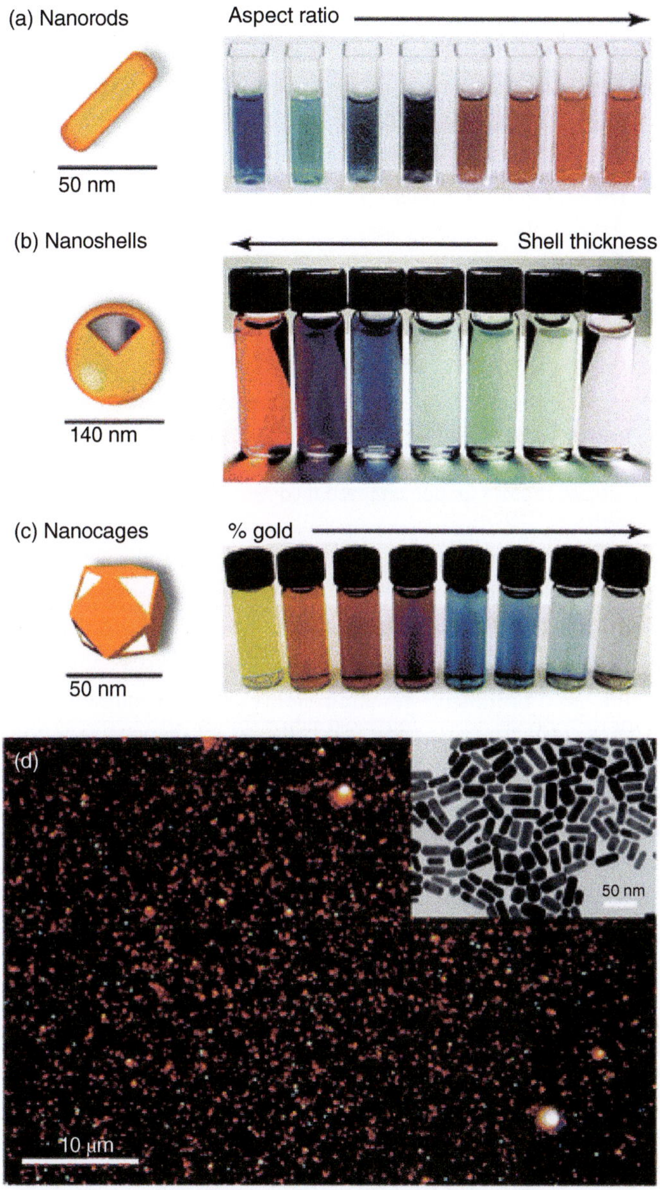

Figure 1.3 Color dependence of various gold-based nanomaterials on their shape and size. (a) Gold nanorods, (b) silica–gold core–shell nanoparticles, and (c) gold nanocages. The intense color of these nanoparticles arises from the collective excitation of their conduction electrons or surface plasmon resonance modes. (d) Optical dark-field scattering micrograph of gold nanorods. Source: Dreaden et al. [74]. © 2020, Royal Society of Chemistry.

Figure 1.4 Nanotechnology-based tools for personalized medicine. Source: Sharma and Hussain [75]. © 2020, Elsevier.

materials [76]. The examples include the application of polymeric NMs as artificial muscles that can contract and return to original shape when short-circuited. These can replicate muscular action and can have strong visual effects.

One of the interesting and promising research fields is the designing and development of NMs whose structure and chemical features are responsive to the biocatalytic action of natural enzymes. Enzymes play a crucial role in all metabolic and biological processes, and their dysregulation is a feature of many diseases. Therefore, the NMs can efficiently sense these enzymes in the physiological environment and behave as an effective and promising tool for therapeutic and diagnostic applications. Various examples of enzyme-responsive NMs are polymeric NMs [77], phospholipids [78], and inorganic NMs [79]. Functional mesoporous silica nanoparticles (MSNs) were successfully designed and prepared by using alkoxysilane tether, α-cyclodextrin, and multifunctional peptides to target tumor cells and decrease the side effects of doxorubicin (DOX), which is an antitumor drug. The multifunctional peptides were composed of the cell-penetrating peptide of seven arginine (R7) sequences, an enzyme-cleavable peptide of Gly-Phe-Leu-Gly (GFLG peptide), and a tumor-targeting peptide of Arg-Gly-Asp-Ser (RGDS peptide). When MSNs loaded with DOX are incubated with tumor cells and normal cells, the nanoparticles could efficiently target tumor cells through the specific interactions between RGDS and integrin receptor $\alpha_v\beta_3$ overexpressed on tumor cells, followed by penetrating cell membrane with the aid of R7 sequence. Once DOX-loaded silica nanoparticles penetrate into the cellular membrane, the antitumor drug DOX is quickly released because of the breakage of GFLG peptide cleaved by cathepsin B, resulting in increased antitumor activity [80]. This effective enzyme-responsive nanoparticle-based drug delivery system exhibits an excellent potential in the area of nanomedicine (Figure 1.5).

Among various applications of NMs, the most promising applications are in the area of biomedical sciences. These interesting applications have crucial effects on

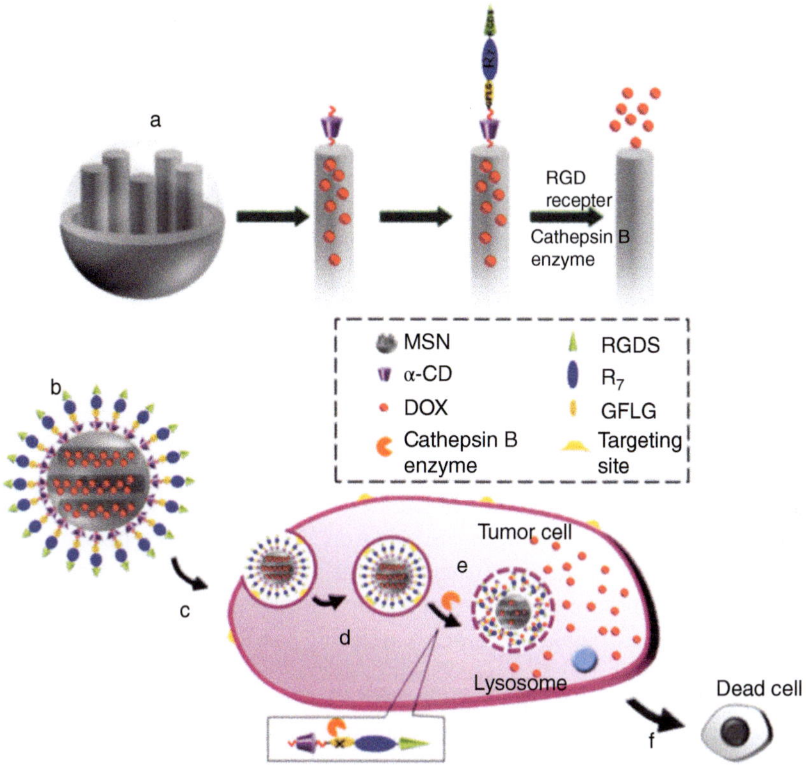

Figure 1.5 The schematic representation of the (a) functionalization procedure of the MSNs, (b) MSNs loaded with DOX under physiological conditions, (c) RGDS targeted to the tumor cell, (d) endocytosis into a specific tumor cell, (e) Cathepsin Benzyme-triggered drug release in the cytoplasm, (f) apoptosis of the tumor cell. Source: Cheng et al. [80]. © 2015, American Chemical Society.

the human health since these NMs are targeted toward the diagnostic techniques for disease which are fast and low cost. To decrease the side effects and increase the therapeutic effects, the pharmaceutical compounds should be selectively targeted at the disease site and accumulated for a prolonged time in the body. Drug release refers to the design and development of efficient approaches, techniques, and formulations required to transport any pharmaceutical compound safely within the body to obtain the efficient therapeutic effects [81]. In this regard, NM-based drug release systems have opened new opportunities in the area of medicine. NM-based drug release systems accumulate and bind specifically to the disease target with controlled release behavior. The careful design and preparation of these NMs as effective drug carriers should address various key issues [82] such as biodegradability and biocompatibility, high stability at physiological pH values, excellent drug-loading capacity without any toxicity, and industrial production of these NMs in large scale for clinical applications. The stimuli-responsive drug release systems deliver the drug at specific tissues in the systematic administration. These do not freely extravasate during the blood circulation and are released at the targets where the nanocarriers accumulate by active or passive targeting strategy [83].

An immunoassay is a biochemical technique that is efficiently applied in the detection of the concentration of a small molecule or a macromolecule in a sample solution by using an antigen or an antibody. In this approach, the key issue is to obtain a measurable signal in response to a binding effect. Screen-printed electrodes have attracted great attention as immunosensors because of their unique features such as miniaturized dimension, low-cost and facile fabrication, and large-scale production. A single drop of sample, i.e. only a few microliters of a sample solution, is used to conduct all the immunological stages, thereby decreasing the consumption of the reagents. Screen-printed electrodes can be successfully coated with nanoparticles that increase the sensitivity of these electrodes [84]. In an interesting study, screen-printed electrode was modified with silver nanoparticles and effectively applied for the preliminary screening of cystic fibrosis that is a common genetic disease caused by the autosomal recessive gene known as the cystic fibrosis transmembrane conductance regulator gene [85]. It affects multiple organs such as lungs and intestines, which causes the irregular transport of chloride and sodium ions across the epithelial cells [86]. It has been reported that the oxidation of silver is sensitive to concentration of chloride ions [87]. Sweat chloride levels are important for the effective diagnosis of cystic fibrosis. A layer of silver nanoparticles is deposited on the working electrode and anodically stripped off to generate silver cations. The anodic stripping voltammetry of silver nanoparticles yields a single silver oxidation peak in the absence of chloride ions, whereas two voltammetry stripping peaks (one for AgCl and other for the oxidation of Ag to Ag(I) ions) are obtained observed in the existence of chloride ions. The silver chloride peak is used for the quantitative determination of chloride concentration [88].

On the other hand, Parkinson's disease and Alzheimer's disease are the most common neurodegenerative diseases that affected 1.6% of the world population and around 26 million people, respectively [89, 90]. This has posed a growing challenge for the patients, clinicians, care givers, and society. The current treatment approaches rely on the clinical symptoms of the diseases, and therefore, the efficient diagnostic techniques depend on the proficiency of the treating physician. Nanoparticle-based sensor systems are successfully applied to distinguish groups of Parkinson's disease and Alzheimer's disease patients from a healthy control group. The diagnostic approach relies on the identification of patterns of volatile organic compounds in the exhaled breath [91]. The disease-related changes in the blood chemistry may be transmitted to the alveolar exhaled breath through the lungs even in the initial stages of the disease. The nanoparticle-based sensor undergoes fast and reversible changes in electrical resistance on being exposed to characteristic volatile organic compounds. Thus, breath prints can be the basis for the design and fabrication of facile, low-cost, non-invasive biomarker using nanoparticle-based sensor systems [92].

1.3.3 NMs in Sensor Platforms

Graphene and graphene-based nanocomposite materials were successfully applied in sensor applications. For example, Kong et al. reported the preparation of a nanocomposite material composed of reduced GO and Au/Ag bimetallic nanoparticles for the sensitive detection of daunorubicin in human serum samples [93]. In this study, the prepared nanosensor showed high recognition behavior toward the target compound daunorubicin in a linear concentration range between 0.01 and 15 μg ml^{-1}. The detection limit was obtained as 3 ng ml^{-1}.

In another important study conducted by Arumugam and Kim [94], the sensitive recognition of ascorbic acid in aqueous samples was successfully carried out by using the prepared GO modified with CdSe and ZnS quantum dots. In their study, the prepared GO-based nanosensor displayed great affinity and sensitivity for the ascorbic acid, and the detection limit was achieved as 568 pM.

GO and GO-based NMs were also successfully applied for the selective recognition and removal of various pollutants from environmental samples. For example, Ghorbani et al. prepared GO functionalized with ethylene diamine for the removal of Cd(II) and Pb(II) ions from wastewater samples [95]. The effects of various factors such as the amount of the adsorbent, shaking rate, pH, and time on the extraction behavior of the prepared ethylene diamine-functionalized GO toward the target ions were investigated. The obtained results confirmed that the prepared ethylene diamine-functionalized GO successfully removed the target Cd(II) and Pb(II) ions from wastewater samples. The extraction percentages for Cd(II) and Pb(II) ions were achieved as 99.4% and 99.6%, respectively.

In another important study conducted by Keshvardoostchokami et al. [96], GO/Ag nanocomposite was developed for the efficient removal of an insecticide pollutant imidacloprid from water samples. The obtained results showed that approximately 63% of the target insecticide imidacloprid was removed from 50 ml water sample containing 10 mg l^{-1} by using 30 mg GO/Ag nanocomposite.

On the other hand, molecularly imprinted polymers (MIPs) are highly selective polymeric materials that mimic the shape-specific binding mechanism which is found naturally occurring in the biological structures (i.e. 3-D binding site of an enzyme or the glycoprotein recognition sites in antibodies). However, since they are designed and prepared from artificial polymers, MIPs display the same robustness and adaptability found in sensors constructed from artificial materials. MIPs as biomimetic recognition components efficiently combine the accuracy and highly specific binding affinity of a sensor system toward the target compound(s), with the great reusability and robustness [97–105].

The chemiluminescence technique is extensively applied in a number of areas because of its great sensitivity toward the target compound(s). However, it has some drawbacks such as low selectivity. Thus, one of the most crucial and urgent fields of the researches is to design and develop new systems having improved selectivity. Many efforts were put to overcome this significant problem. The combination of the chemiluminescence technique and molecular imprinting technology is a powerful approach that can be successfully applied for the development of molecular imprinting-based spectroscopic sensor systems toward pharmaceuticals. For example, Zor et al. prepared a photoluminescent sensor system composed of magnetic silica, photoluminescent graphene quantum dots (GQDs) and molecularly imprinted polypyrrole for the sensitive recognition of pesticide tributyltin that exhibits genotoxic feature and lead to increase in endocrine disruptions [106]. In their research, tributyltin was chosen as the template compound for the synthesis of selective MIP. The developed photoluminescent composite nanosensor based on molecular imprinting was effectively employed for the selective and sensitive recognition of tributyltin in sea water and water samples. The achieved limit of detection values for tributyltin in sea water and water samples were 42.56 and 12.78, respectively.

On the other hand, in the fluorescence-based detection of the target compound(s) by using molecular imprinting-based nanosensors, appropriate functional monomers having fluorescence feature are chosen for the development of molecular imprinting-based fluorescence sensor systems [107]. Once the target compound(s) selectively binds to recognition sites of the molecular imprinting-based sensor, intensity of the fluorescence rises or reduces depending on the design of the imprinted nanosensor.

In another interesting research carried out by Zhang et al. [108], a fluorescence-based nanosensor composed of ZnS/CdSe quantum dots has a selective MIP layer toward carbaryl which is an insecticide. The researchers successfully applied this nanosensor for the effective recognition of carbaryl in food samples including rice and cabbage. The achieved outputs from the experiments for the fluorescence-based detection of the target insecticide carbaryl displayed that the fluorescence-based imprinted nanosensor showed excellent recognition behavior and selectivity for carbaryl in the existence of its anologues such as isocarb and metolcarb.

In a research reported by Masteri-Farahani et al. [109], a molecular imprinting-based fluorescence nanosensor was developed for the efficient detection of methamphetamine, which is a quite addictive and illegal psychostimulant drug. For this purpose, they embedded GQDs into the MIP particles as schematically illustrated in Figure 1.3a. The obtained outputs showed that the developed molecular imprinting-based fluorescence nanosensor can sensitively and selectively recognize methamphetamine even in the existence of other similar compounds (i.e. ibuprofen, amphetamine, morphine, and codeine). The detection limit was achieved as $1.7\,\mu g\,l^{-1}$.

Liu et al. prepared a sensitive molecular imprinting-based fluorescence nanosensor for the recognition of metronidazole in serum samples [110]. In this research, first, carbon dots (CDs) were synthesized using longan peels by applying high-pressure microwave approach. Then, the preparation of CDs-embedded MIPs was carried out by using the functional monomers acrylamide (AM) and glycidyl methacrylate (GMA), cross-linker ethylene glycol dimethylacrylate (EGDMA), and initiator 2,2′-azobisisobutyronitrile (AIBN) as schematically illustrated in Figure 1.4.

The developed molecular imprinting-based fluorescence nanosensor was employed for the sensitive and selective recognition of metronidazole in serum samples. The achieved sensor response toward metronidazole was linear in the range from 50 to $1200\,ng\,ml^{-1}$, and the limit of detection was obtained as $17.4\,ng\,ml^{-1}$.

In a work performed by Jalili et al. [111], penicillin G in milk samples was successfully detected by using the molecular imprinting-based fluorescence nanosensor having CDs. In their work, first, the synthesis of silica nanoparticles

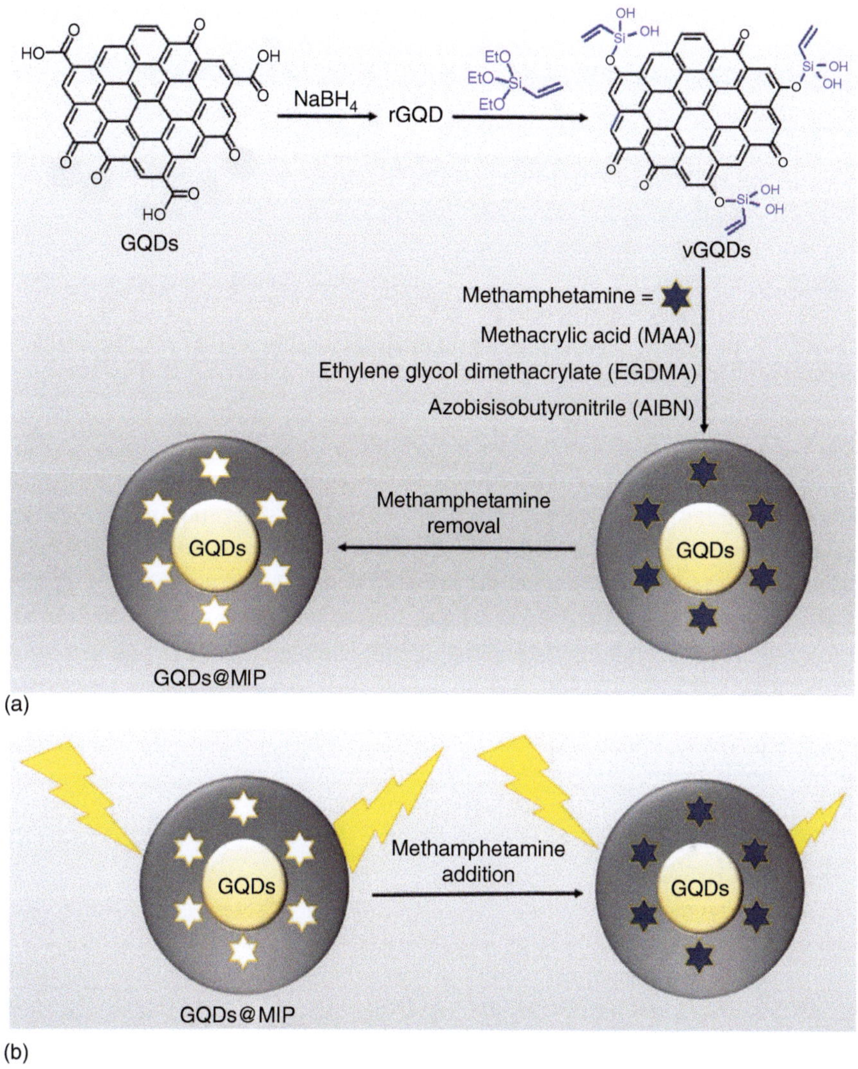

Figure 1.6 (a) The scheme of the preparation of molecular imprinting-based fluorescence nanosensor for methamphetamine. (b) Fluorescence quenching of molecular imprinting-based nanosensor in the existence of methamphetamine. Source: Masteri-Farahani et al. [109]. © 2020, Elsevier.

embedded with blue-emissive CDs was performed. Then, MIP film was formed on the surface of the prepared nanocomposite (CDs@silica nanoparticles) by using 3-aminopropyltriethoxysilane (APTES) and tetraethoxysilanz (TEOS) as functional monomers. The prepared molecular imprinting-based fluorescence nanosensor having CDs displayed excellent sensitivity toward penicillin G. The obtained sensor response was linear in the range between 1 and 32 nM. The limit of detection was found as 0.34 nM (Figure 1.6).

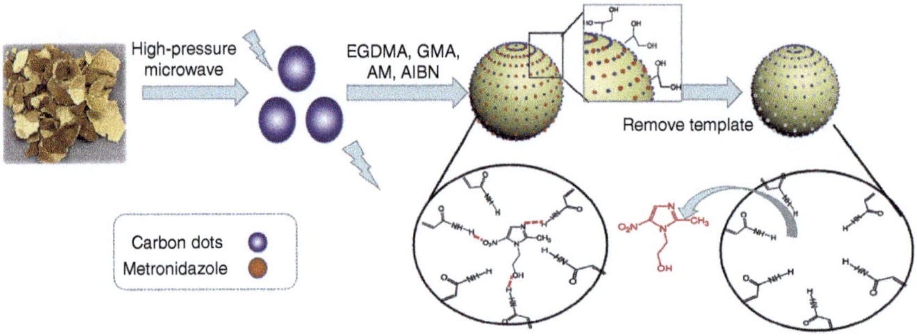

Figure 1.7 The scheme of the preparation of molecular imprinting-based fluorescence nanosensor for the recognition of metronidazole. Source: Liu et al. [110]. © 2019, Elsevier.

Stoian et al. developed a molecular imprinting-based electrochemical sensor for the selective and sensitive recognition of an antibiotic azithromycin in biological samples including tears, plasma, and urine samples [112]. For this purpose, the electrode position of the selective MIP layer was carried out on the surface of a glassy carbon electrode as schematically demonstrated in Figure 1.7. The functional monomer 3-thienyl boronic acid and cross-linker 2,2′-bithiophene were used for the formation of MIP layer on the electrode surface. The fabricated electrochemical sensor displayed excellent response toward the target antibiotic azithromycin in a linear range from 13.33 nM to 66.67 μM. The limit of detection was obtained as 0.85 nM (Figure 1.8).

1.4 Conclusions

NMs offer unique opportunities for designing ultrasensitive (bio)sensors and analytical assays. The broad potential of multifunctional nanoparticles for (bio)molecular recognition events and separations are required.

The remarkable sensitivity of the new multifunctional nanoparticle-based sensing protocols opens up the possibility for detecting disease markers, biothreat agents, chemicals, or infectious agents that cannot be measured by conventional methods. The use of multifunctional nanoparticles for detecting proteins is still in its infancy (most of the work has been focused on separations), but the experience gained in ultrasensitive DNA detection may provide a useful starting point. Despite analytical research in the field is active, there is still much space to explore and to explain tremendous opportunities and significant challenges to be afforded. For example, whereas present work is more concentrated on multifunctional magnetic nanoparticles, very few studies on multifunctional gold nanoparticles or quantum dots (QDs) are still appearing. Probably, the most prominent advantage of multifunctional magnetic nanoparticles over others lies in the fact that the particles can be easily functionalized and magnetically manipulated using permanent magnets independent of normal chemical or biological analysis.

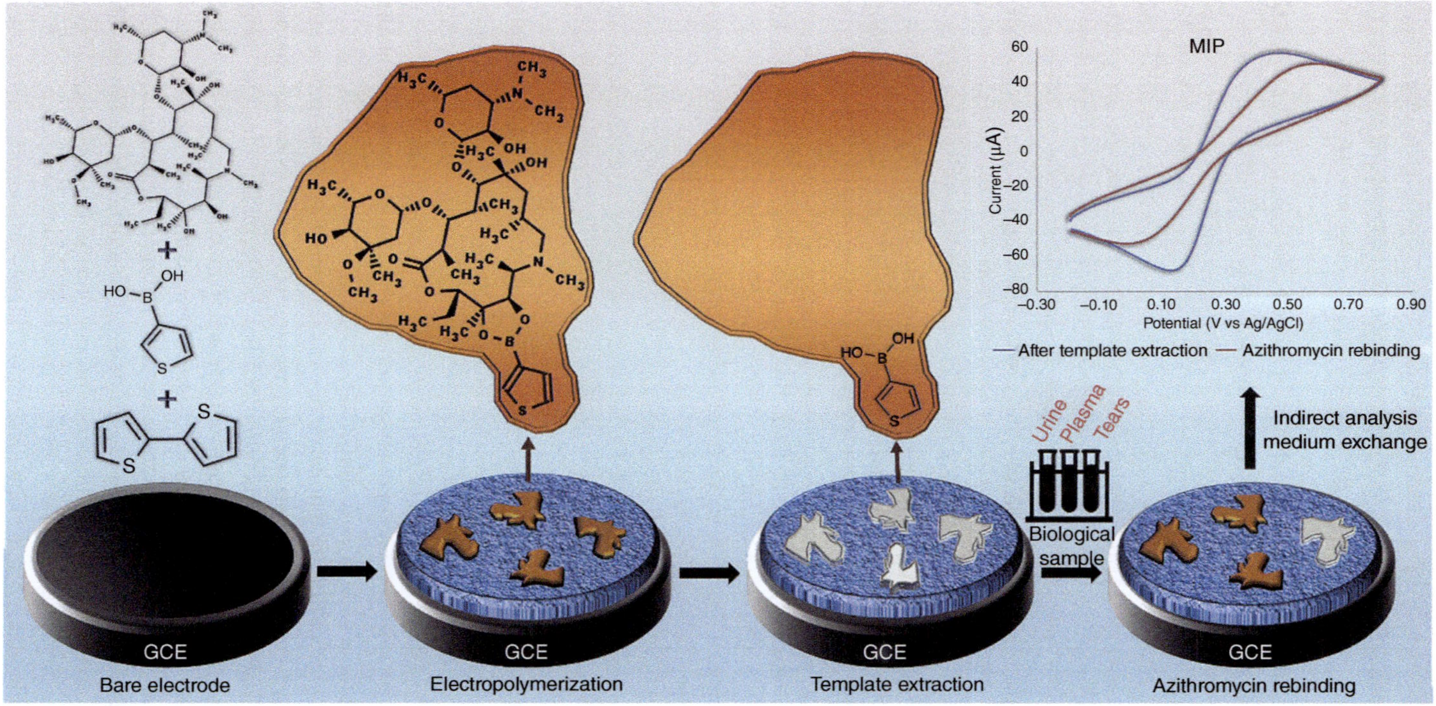

Figure 1.8 The illustration of the fabrication of molecular imprinting-based electrochemical sensor for azithromycin. Source: Stoian et al. [112]. © 2020, Elsevier.

The challenges of processing nanoparticles easily and their multifunctionalization with (bio) molecules or other nanoparticles in a biological and chemical friendly manner are just beginning to be addressed. Just starting with the base nanoparticle, one of the most significant barriers along the road is the poor reproducibility and the wide range of size distribution of nanoparticle product synthesized from wet chemical methods. Improved synthetic methodologies that can lead to true monodispersity and satisfactory reproducibility of nanoparticles with well-controlled shape remain to be developed. On the other hand, nanoparticle surface modification is also a critical step for protecting the nanoparticle core, improving aqueous dispersibility and obtaining appropriate surface multifunctionality for the development of highly sensitive and selective "nanosensors." Even, from a physicochemical point of view, clear and quantitative characterizations of the surface chemistry of the nanoparticles should be provided in terms of the number of surface reactive functions that are effectively present at the surface of the particles.

The rapid advancement of imaging technology over the past decade has opened opportunities for the design of specific and multifunctional nanoprobes for medical and pharmaceutical applications such as molecular imaging, drug delivery, targeting, and cancer research, some of which can be considered analytical. The use of multifunctional nanoparticles would have advantage for preoperative tumor assessment and intraoperative surgical guidance for the tumor tissue resection. Whereas successful experiments were conducted in vitro, progress in practical applications of these multifunctional nanoparticles may be slow, particularly due to the unknown effects of nanoparticles in human health.

The development of novel multifunctional NMs and the exact investigation of their performance characteristics open the way to new analytical devices in the future. It is a highly multidisciplinary area, requiring a range of scientific knowledge, from inorganic/organic chemistry involved in the preparation of different types of nanoparticles, through chemistry and biological sciences to allow for their functionalization and, of course, the basic physics of optics and magnetic materials. It is now generally recognized that nanotechnologies and biosciences will be one of the leading and most promising areas of research and development in the twenty-first century. We believe that multifunctional nanoparticles could play a very important role in these developments.

A wide number of novel applications of carbon-based NMs have been recently reported. Besides chemical composition, size and shape are probably the most important variables identified. In addition, the hydrophobicity of the surface (affected by the number of functional groups) plays a key role in the optical, electrochemical, and adsorptive properties of carbon-based NMs and should be carefully evaluated. Typically, hydrophobic particles can be used to enhance nonspecific interactions (via increases in surface area) with organic molecules. On the other hand, highly derivatized carbon-based surfaces provide excellent platforms to develop applications based on electrostatic and specific interactions.

Although there is an incredible volume of literature supporting the use of carbon-based NMs, researchers should carefully assess the properties of such materials before claiming exceptional behavior. Additionally, while the main focus of reports published in past years has been the use of CNTs, this trend may change as researchers develop new types of carbon-based materials. Among them, graphene was identified as one emerging allotrope that could play a fundamental role in the preparation of future sensors. All things considered, the use of carbon-based nanoparticles in analytical chemistry has been obviously advantageous and has enabled the integration of analytical chemistry with a large number of fields. Although today the use of nanotechnology in analytical chemistry has a fairly young approach that mixes art, intuition, and science, many researchers around the world have recognized the utility of NMs. We believe that the analytical applications of carbon-based NMs will continue to rise and will soon develop into a mature and independent field. Materials science has always been an exciting research area. There are two capabilities of advanced materials science and technology that we found fascinating always. The physico-chemical properties of materials can be controlled via structural designs or manipulation of physical dimensions, as well as through incorporation of suitable components within the materials or modification of their surfaces. When incorporated appropriately into analytical chemistry, advanced materials ought to enhance performance levels of many existing analytical techniques or create new exciting techniques derived from novel applications and exploitation of unique properties of these advanced materials. Analytical chemistry is progressing toward analysis at the molecular level, whereby interfacing materials that interact directly with analytes will play increasingly important roles. In the same way, extensive research that interfaces between nanomaterials and analytical techniques will be very important in near future.

References

1 Adams, F. and Barbante, C. (2015). Nanotechnology and analytical chemistry, Chapter 4. In: *Comprehensive Analytical Chemistry*, vol. 69 (eds. F. Adams and C. Barbante), 125–157. Elsevier.

2 Bao, M. and Wang, W. (1996). Future of microelectromechanical systems (MEMS). *Sensor. Actuat. A* 56: 135–141.

3 Hussain, C.M. and Keçili, R. (2020). *Modern Environmental Analysis Techniques for Pollutants*. The Netherlands: Elsevier.

4 Büyüktiryaki, S., Keçili, R., and Hussain, C.M. (2020). Functionalized nanomaterials in dispersive solid phase extraction: advances and prospects. *Trends Anal. Chem.* https://doi.org/10.1016/j.trac.2020.115893.

5 Keçili, R., Büyüktiryaki, S., and Hussain, C.M. (2019). Advancement in bioanalytical science through nanotechnology: past, present and future. *Trends Anal. Chem.* 110: 259–276.

6 Hussain, C.M. and Kharisov, B. (2017). *Advanced Environmental Analysis – Application of Nanomaterials*. The Royal Society of Chemistry.

7 Hussain, C.M. (2017). Magnetic nanomaterials for environmental analysis. In: *Advanced Environmental Analysis-Application of Nanomaterials* (eds. C.M. Hussain and B. Kharisov). The Royal Society of Chemistry.

8 Hussain, C.M. (2019). *Handbook of Nanomaterials in Analytical Chemistry: Modern Trends in Analysis*. Elsevier.

9 Büyüktiryaki, S., Sümbelli, Y., Keçili, R., and Hussain, C.M. (2019). Lab-on-chip platforms for environmental analysis. In: *Encyclopedia of Analytical Science*, 3e (eds. P. Worsfold, C. Poole, A. Townshend and M. Miró), 267–273. Academic Press.

10 Keçili, R. and Hussain, C.M. (2018). Recent progress of imprinted nanomaterials in analytical chemistry. *Int J. Anal. Chem.*: 8503853.

11 Keçili, R. and Hussain, C.M. (2018). Mechanism of adsorption on nanomaterials, Chapter 4. In: *Nanomaterials Chromatography* (ed. C.M. Hussain), 89–115. Elsevier.

12 Keçili, R., Büyüktiryaki, S., and Hussain, C.M. (2018). Engineered nanosensors based on molecular imprinting technology, Chapter 57. In: *Micro and Nano Technologies, Handbook of Nanomaterials for Industrial Applications* (ed. C.M. Hussain), 1031–1046. Elsevier.

13 Zhi, D., Yang, T., Yang, J. et al. (2020). Targeting strategies for superparamagnetic iron oxide nanoparticles in cancer therapy. *Acta Biomater.* 102: 13–34.

14 Alphandéry, E. (2020). Iron oxide nanoparticles for therapeutic applications. *Drug Discovery Today* 25 (1): 141–149.

15 Anzar, N., Hasan, R., Tyagi, M. et al. (2020). Carbon nanotube – a review on synthesis, properties and plethora of applications in the field of biomedical science. *Sensor. Int.* 1: 100003.

16 Verma, B. and Balomajumder, C. (2020). Surface modification of one-dimensional carbon nanotubes: a review for the management of heavy metals in wastewater. *Environ. Technol. Inno.* 17: 100596.

17 Tiwari, S., Purabgola, A., and Kandasubramanian, B. (2020). Functionalised graphene as flexible electrodes for polymer photovoltaics. *J. Alloys Compd.* 825: 153954.

18 Nandanapalli, K.R., Mudusu, D., and Lee, S. (2019). Functionalization of graphene layers and advancements in device applications. *Carbon* 152: 954–985.

19 Korkmaz, S. and Kariper, İ.A. (2020). Graphene and graphene oxide based aerogels: synthesis, characteristics and supercapacitor applications. *J. Energy Storage* 27: 101038.

20 Xu, T. and Lee, J.-K. (2014). Nanomaterials: electrical, magnetic, and photonic applications. *JOM* 66 (4): 654–654.

21 Talebpour, Z., Haghighi, F., and Sanati-Nezhad, A. (2018). On-a-chip gas chromatography with nanomaterials (NMs), Chapter 12. In: *Nanomaterials in Chromatography* (ed. C.M. Hussain), 343–358. Elsevier.

22 Zhang, M. and Qiu, H. (2015). Progress in stationary phases modified with carbonaceous nanomaterials for high-performance liquid chromatography. *TrAC Trends Anal. Chem.* 65: 107–121.

23 Hu, W., Hong, T., Gao, X., and Ji, Y. (2014). Applications of nanoparticle-modified stationary phases in capillary electrochromatography. *TrAC Trends Anal. Chem.* 61: 29–39.

24 Keçili, R., Büyüktiryaki, S., and Hussain, C.M. (2020). Membrane applications of nanomaterials, Chapter 7. In: *Handbook of Nanomaterials in Analytical Chemistry* (ed. C.M. Hussain), 159–182. Elsevier.

25 Geim, A.K. and Novoselov, K.S. (2007). The rise of graphene. *Nat. Mater.* 6 (3): 183–191.

26 Gerstner, E.D. (2010). Nobel Prize 2010: Andre Geim and Konstantin Novoselov. *Nat. Phys.* 6: 836–837.

27 Liu, Z., Wang, W., Ju, X. et al. (2017). Graphene-based membranes for molecular and ionic separations in aqueous environments. *Chin. J. Chem. Eng.* 25: 1598–1605.

28 Nag, A., Mitra, A., and Mukhopadhyay, S.C. (2018). Graphene and its sensor-based applications: a review. *Sensor. Actuat. A Phys.* 270: 177–194.

29 Lin, J., Chen, X., and Huang, P. (2016). Graphene-based nanomaterials for bioimaging. *Adv. Drug Delivery Rev.* 105: 242–254.

30 Liu, J., Dong, J., Zhang, T., and Peng, Q. (2018). Graphene-based nanomaterials and their potentials in advanced drug delivery and cancer therapy. *J. Controlled Release* 286: 64–73.

31 Kim, H., Abdala, A.A., and Macosko, C.W. (2010). Graphene/polymer nanocomposites. *Macromolecules* 43 (16): 6515–6530.

32 Young, R.J., Kinloch, I.A., Gong, L., and Novoselov, K.S. (2012). The mechanics of graphene nanocomposites: a review. *Compos. Sci. Technol.* 72 (12): 1459–1476.

33 Sutter, P.W., Flege, J.I., and Sutter, E.A. (2008). Epitaxial graphene on ruthenium. *Nat. Mater.* 7 (5): 406–411.

34 Sutter, P., Hybertsen, M.S., Sadowski, J.T., and Sutter, E. (2009). Electronic structure of few-layer epitaxial graphene on Ru(0001). *Nano Lett.* 9 (7): 2654–2660.

35 Iijima, S. (1991). Helical microtubules of graphitic carbon. *Lett. Nat.* 354: 56–58.

36 Mohamed, A.E.-M.A. and Mohamed, M.A. (2020). Carbon nanotubes: synthesis, characterization, and applications, Chapter 2. In: *Micro and Nano Technologies, Carbon Nanomaterials for Agri-Food and Environmental Applications* (ed. K.A. Abd-Elsalam), 21–32. Elsevier.

37 Indarto, A., Ikhsan, N.A., and Wibowo, I. (2020). Applications of carbon nanotubes for controlling waterborne pathogens, Chapter 20. In: *Waterborne Pathogens* (eds. M.N.V. Prasad and A. Grobelak), 433–461. Butterworth-Heinemann.

38 Khan, I., Saeed, K., and Khan, I. (2019). Nanoparticles: properties, applications and toxicities. *Arabian J. Chem.* 12: 908–931.

39 Sharma, A., Patwardhan, A., Dasgupta, K., and Joshi, J.B. (2019). Kinetic study of boron doped carbon nanotubes synthesized using chemical vapour deposition. *Chem. Eng. Sci.* 207: 1341–1352.

40 Mwafy, E.A. and Mostafa, A.M. (2019). Multi walled carbon nanotube decorated cadmium oxide nanoparticles via pulsed laser ablation in liquid media. *Opt. Laser Technol.* 111: 249–254.

41 Mohammad, M.I., Moosa, A.A., Potgieter, H., and Ismael, M.K. (2013). Carbon nanotubes synthesis via arc discharge with a Yttria catalyst. *ISRN Nanomater.*: 785160.

42 Abousalman-Rezvani, Z., Eskandari, P., Roghani-Mamaqani, H., and Salami-Kalajahi, M. (2020). Functionalization of carbon nanotubes by combination of controlled radical polymerization and "grafting to" method. *Adv. Colloid Interface Sci.* 278: 102126.

43 Merum, S., Veluru, J.B., and Seeram, R. (2017). Functionalized carbon nanotubes in bio-world: applications, limitations and future directions. *Mater. Sci. Eng., B* 223: 43–63.

44 Kroto, H.W., Heath, J.R., O'brien, S.C. et al. (1985). C60: buckminsterfullerene. *Nature* 318: 162–163.

45 Pan, Y., Liu, X., Zhang, W. et al. (2020). Advances in photocatalysis based on fullerene C60 and its derivatives: properties, mechanism, synthesis, and applications. *Appl. Catal., B* 265: 118579.

46 Yan, L., Zhao, F., Li, S. et al. (2011). Low-toxic and safe nanomaterials by surface-chemical design, carbon nanotubes, fullerenes, metallofullerenes, and graphenes. *Nanoscale* 3 (2): 362–382.

47 Malmsten, M. (2013). Inorganic nanomaterials as delivery systems for proteins, peptides, DNA, and siRNA. *Curr. Opin. Colloid Interface Sci.* 18 (5): 468–480.

48 Ariga, K., Ji, Q., McShane, M.J. et al. (2012). Inorganic nanoarchitectonics for biological applications. *Chem. Mater.* 24 (5): 728–737.

49 Zinchenko, A. (2012). Templating of inorganic nanomaterials by biomacromolecules and their assemblies. *Polym. Sci. Ser. C* 54 (1): 80–87.

50 Peixoto de Almeida, M., Pereira, E., Baptista, P. et al. Gold nanoparticles as (bio)chemical sensors, Chapter 13. In: *Comprehensive Analytical Chemistry*, vol. 66 (eds. M. Valcárcel and Á.I. López-Lorente), 529–567. Elsevier.

51 McMahon, S.J. and Currell, F.J. (2013). Gold nanoparticles for imaging and radiotherapy, Chapter 3. In: *Frontiers of Nanoscience*, vol. 5 (ed. H. Summers), 65–93. Elsevier.

52 Xiao, Q., Gao, H., Lu, C., and Yuan, Q. (2012). Gold nanoparticle-based optical probes for sensing aminothiols. *TrAC Trends Anal. Chem.* 40: 64–76.

53 Chen, L. and Liang, J. (2020). An overview of functional nanoparticles as novel emerging antiviral therapeutic agents. *Mater. Sci. Eng., C* 112: 110924.

54 Deshmukh, S.P., Patil, S.M., Mullani, S.B., and Delekar, S.D. (2019). Silver nanoparticles as an effective disinfectant: a review. *Mater. Sci. Eng., C* 97: 954–965.

55 Tong, Q., Wang, W., Fan, Y., and Dong, L. (2018). Recent progressive preparations and applications of silver-based SERS substrates. *TrAC Trends Anal. Chem.* 106: 246–258.

56 Lan, Y., Lu, Y., and Ren, Z. (2013). Mini review on photocatalysis of titanium dioxide nanoparticles and their solar applications. *Nano Energy* 2: 1031–1045.

57 Mahmoud, W.M.M., Rastogi, T., and Kümmerer, K. (2017). Application of titanium dioxide nanoparticles as a photocatalyst for the removal of micropollutants such as pharmaceuticals from water. *Curr. Opin. Green Sustain. Chem.* 6: 1–10.

58 Sierra, I., Pérez-Quintanilla, D., Morante, S., and Gañán, J. (2014). Novel supports in chiral stationary phase development for liquid chromatography. Preparation, characterization and application of ordered mesoporous silica particles. *J. Chromatogr. A* 1363: 27–40.

59 Grazia De Angelis, M. and Sarti, G.C. (2012). Gas sorption and permeation in mixed matrix membranes based on glassy polymers and silica nanoparticles. *Curr. Opin. Chem. Eng.* 1: 148–155.

60 Rahim, S., Jan Iftikhar, F., and Malik, M.I. (2020). Biomedical applications of magnetic nanoparticles, Chapter 16. In: *Micro and Nano Technologies, Metal Nanoparticles for Drug Delivery and Diagnostic Applications* (eds. M.R. Shah, M. Imran and S. Ullah), 301–328. Elsevier.

61 Keçili, R., Büyüktiryaki, S., Dolak, İ., and Hussain, C.M. (2020). The use of magnetic nanoparticles in sample preparation devices and tools, Chapter 5. In: *Handbook of Nanomaterials in Analytical Chemistry* (ed. C.M. Hussain), 75–95. Elsevier.

62 Hu, B., He, M., and Chen, B. (2020). Magnetic nanoparticle sorbents, Chapter 9. In: *Handbooks in Separation Science, Solid-Phase Extraction* (ed. C.F. Poole), 235–284. Elsevier.

63 Mirza, S., Ahmad, M.S., Shah, M.I.A., and Ateeq, M. (2020). Magnetic nanoparticles: drug delivery and bioimaging applications, Chapter 11. In: *Micro and Nano Technologies, Metal Nanoparticles for Drug Delivery and Diagnostic Applications* (eds. M.R. Shah, M. Imran and S. Ullah), 189–213. Elsevier.

64 Chikazumi, S., Taketomi, S., Ukita, M. et al. (1987). Physics of magnetic fluids. *J. Magn. Magn. Mater.* 65: 245–251.

65 Lu, A.H., Schmidt, W., Matoussevitch, N. et al. (2004). Nanoengineering of a magnetically separable hydrogenation catalyst. *Angew. Chem. Int. Ed.* 43: 4303–4306.

66 Li, Z., Wei, L., Gao, M.Y., and Lei, H. (2005). One-pot reaction to synthesize biocompatible magnetite nanoparticles. *Adv. Mater.* 17: 1001–1005.

67 Farrukh, A., Akram, A., Ghaffar, A. et al. (2013). Design of polymer-brush-grafted magnetic nanoparticles for highly efficient water remediation. *ACS Appl. Mater. Interfaces* 5: 3784–3793.

68 Lin, P.C., Yu, C.C., Wu, H.T. et al. (2013). A chemically functionalized magnetic nanoplatform for rapid and specific biomolecular recognition and separation. *Biomacromolecules* 14: 160–168.

69 Lu, A.H., Salabas, E.L., and Schuth, F. (2007). Magnetic nanoparticles: synthesis, protection, functionalization, and application. *Angew. Chem. Int. Ed.* 46: 1222–1244.

70 Yu, C.J., Su, C.L., and Tseng, W.L. (2006). Separation of acidic and basic proteins by nanoparticle-filled capillary electrophoresis. *Anal. Chem.* 78 (23): 8004–8010.

71 Liu, F.-K. (2009). Analysis and applications of nanoparticles in the separation sciences: a case of gold nanoparticles. *J. Chromatogr. A* 1216 (52): 9034–9047.
72 Qu, Q.S., Fei, S., Shen, M. et al. (2008). Open-tubular gas chromatography using capillary coated with octadecylamine-capped gold nanoparticles. *Anal. Chim. Acta* 609 (1): 76–81.
73 Mitchell, D.T., Lee, S.B., Trofin, L. et al. (2002). Smart nanotubes for bioseparations and biocatalysis. *J. Am. Chem. Soc.* 124 (40): 11864–11865.
74 Dreaden, E.C., Alkilany, A.M., Huang, X. et al. (2012). The golden age: gold nanoparticles for biomedicine. *Chem. Soc. Rev.* 41: 2740–2779.
75 Sharma, D. and Hussain, C.H.M. (2020). Smart nanomaterials in pharmaceutical analysis. *Arabian J. Chem.* 13: 3319–3343.
76 Yoshida, M. and Lahann, J. (2008). İntelligent nanomaterials. *ACS Nano* 2 (6): 1101–1107.
77 Wang, C., Chen, Q., Wang, Z. et al. (2010). An enzyme-responsive polymeric superamphiphile. *Angew. Chem. Int. Ed.* 122 (46): 8794–8797.
78 Wang, H., Thorling, C.A., Liang, X. et al. (2015). Diagnostic imaging and therapeutic application of nanoparticles targeting the liver. *J. Mater. Chem. B* 3 (6): 939–958.
79 Popat, A., Ross, B.P., Liu, J. et al. (2012). Enzyme-responsive controlled release of covalently bound prodrug from functional mesoporous silica nanospheres. *Angew. Chem. Int. Ed.* 51 (50): 12486–12489.
80 Cheng, Y.-J., Luo, G.-F., Zhu, J.-Y. et al. (2015). Enzyme-induced and tumor-targeted drug delivery system based on multifunctional mesoporous silica nanoparticles. *ACS Appl. Mater. Interfaces* 7 (17): 9078–9087.
81 Hrubý, M., Filippov, S.K., and Štěpánek, P. (2015). İntelligent polymers in drug delivery systems on crossroads: which way deserves following? *Eur. Polym. J.* 65: 82–97.
82 Liu, D. et al. (2016). The intelligent drug delivery system and its clinical potential. *Theranostics* 6 (9): 1306–1323.
83 Mura, S., Nicolas, J., and Couvreur, P. (2013). Stimuli-responsive nanocarriers for drug delivery. *Nat. Mater.* 12: 991.
84 Arduini, F., Micheli, L., Moscone, D. et al. (2016). Electrochemical biosensors based on nanomodified screen-printed electrodes: recent applications in clinical analysis. *TrAC Trends Anal. Chem.* 79: 114–126.
85 Kerem, B., Rommens, J.M., Buchanan, J.A. et al. (1989). Identification of the cystic fibrosis gene: genetic analysis. *Science* 245 (4922): 1073–1080.
86 Rowe, S.M., Miller, S., and Sorscher, E.J. (2005). Cystic fibrosis. *N. Engl. J. Med.* 352 (19): 1992–2001.
87 Brolo, A.G. and Sharma, S.D. (2003). Using probe beam deflection (PBD) to investigate the electrochemical oxidation of silver in perchlorate media in the presence and absence of chloride ions. *Electrochim. Acta* 48 (10): 1375–1384.
88 Toh, H.S., Batchelor-McAuley, C., Tschulik, K. et al. (2013). Electrochemical detection of chloride levels in sweat using silver nanoparticles: a basis for the preliminary screening for cystic fibrosis. *Analyst* 138 (15): 4292–4297.

89 Brookmeyer, R., Johnson, E., Ziegler-Graham, K. et al. (2007). Forecasting the global burden of Alzheimer's disease. *Alzheimers Dement.* 3 (3): 186–191.
90 de Rijk, M.C., Launer, L.J., Berger, K. et al. (2000). Prevalence of Parkinson's disease in Europe: a collaborative study of population-based cohorts. Neurologic Diseases in the Elderly Research Group. *Neurology* 54 (11 Suppl 5): S21–S23.
91 Song, G., Qin, T., Liu, H. et al. (2010). Quantitative breath analysis of volatile organic compounds of lung cancer patients. *Lung Cancer* 67 (2): 227–231.
92 Tisch, U., Schlesinger, I., Ionescu, R. et al. (2013). Detection of Alzheimer's and Parkinson's disease from exhaled breath using nanomaterial-based sensors. *Nanomedicine (London)* 8 (1): 43–56.
93 Kong, F.Y., Yao, L., Li, R.F. et al. (2019). Synthesis of nitrogen-doped reduced graphene oxide loading with Au–Ag bimetallic nanoparticles for electrochemical detection of daunorubicin. *J. Alloys Compd.* 797: 413–420.
94 Arumugam, N. and Kim, J.S. (2018). Quantum dots attached to graphene oxide for sensitive detection of ascorbic acid in aqueous solutions. *Mater. Sci. Eng., C* 92: 720–725.
95 Ghorbani, M., Shams, A., Seyedin, O., and Lahoori, N.A. (2018). Magnetic ethylene diamine-functionalized graphene oxide as novel sorbent for removal of lead and cadmium ions from wastewater samples. *Environ. Sci. Pollut. Res.* 25: 5655–5667.
96 Keshvardoostchokami, M., Bigverdi, P., Zamani, A. et al. (2018). Silver@graphene oxide nanocomposite: synthesize and application in removal of imidacloprid from contaminated waters. *Environ. Sci. Pollut. Res.* 25: 6751–6761.
97 Sellergren, B. (2001). *Molecularly Imprinted Polymers: Man-made Mimics of Antibodies and Their Application in Analytical Chemistry: Techniques and Instrumentation in Analytical Chemistry*. Amsterdam, The Netherlands: Elsevier.
98 Kryscioa, D.R. and Peppas, N.A. (2012). Critical review and perspective of macromolecularly imprinted polymers. *Acta Biomater.* 8: 461–473.
99 Cheong, W.J., Yang, S.H., and Ali, F. (2013). Molecular imprinted polymers for separation science: a review of reviews. *J. Sep. Sci.* 36: 609–628.
100 Piletsky, S., Canfarotta, F., Poma, A. et al. (2020). Molecularly imprinted polymers for cell recognition. *Trends Biotechnol.* 38: 368–387.
101 Kupai, J., Razali, M., Büyüktiryaki, S. et al. (2017). Long-term stability and reusability of molecularly imprinted polymers. *Polym. Chem.* 8 (4): 666–673.
102 Dolak, İ., Keçili, R., Hür, D. et al. (2015). Ion-imprinted polymers for selective recognition of neodymium(III) in environmental samples. *Ind. Eng. Chem. Res.* 54: 5328–5335.
103 Keçili, R., Say, R., Ersöz, A. et al. (2012). Investigation of synthetic lipase and its use in transesterification reactions. *Polymer* 53 (10): 1981–1984.
104 Keçili, R., Özcan Atılır, A., Ersöz, A. et al. (2011). Superparamagnetic nanotraps containing MIP based mimic lipase for biotransformations uses. *J. Nanopart. Res.* 13 (5): 2073–2079.
105 Keçili, R. (2018). Selective recognition of myoglobin in biological samples using molecularly ımprinted polymer-based affinity traps. *Int J. Anal. Chem.* 2018: 1–9.

106 Zor, E., Morales-Narváez, E., Zamora-Gálvez, A. et al. (2015). Graphene quantum dots-based photoluminescent sensor: a multifunctional composite for pesticide detection. *ACS Appl. Mater. Interfaces* 7: 20272–20279.

107 Tan, J., Wang, H.-F., and Yan, X.-P. (2009). Discrimination of saccharides with a fluorescent molecular imprinting sensor array based on phenylboronic acid functionalized mesoporous silica. *Anal. Chem.* 81 (13): 5273–5280.

108 Zhang, C., Cui, H., Cai, J. et al. (2015). Development of fluorescence sensing material based on CdSe/ZnS quantum dots and molecularly imprinted polymer for the detection of carbaryl in rice and chinese cabbage. *J. Agric. Food. Chem.* 63: 4966–4972.

109 Masteri-Farahani, M., Mashhadi-Ramezani, S., and Mosleh, N. (2020). Molecularly imprinted polymer containing fluorescent graphene quantum dots as a new fluorescent nanosensor for detection of methamphetamine. *Spectrochim. Acta A* 229: 118021.

110 Liu, H., Ding, J., Zhang, K., and Ding, L. (2020). Fabrication of carbon dots@restricted access molecularly imprinted polymers for selective detection of metronidazole in serum. *Talanta* 209: 120508.

111 Jalili, R., Khataee, A., Rashidi, M.R., and Razmjou, A. (2020). Detection of penicillin G residues in milk based on dual-emission carbon dots and molecularly imprinted polymers. *Food Chem.* 314: 126172.

112 Stoian, I.-A., Iacob, B.-C., Dudaş, C.-L. et al. (2020). Biomimetic electrochemical sensor for the highly selective detection of azithromycin in biological samples. *Biosens. Bioelectron.* 155: 112098.

2

Special Properties of Nanomaterials (NMs) for Sample Preparation

2.1 Introduction

Recent advancement in technology allows the synthesis of well-defined nanoscale objects, solids, and films with multiple subunits with high reproducibility. Nanomaterials (NMs) can be metals, ceramics, polymeric materials, or composite materials. Their most defining characteristic is a very small feature size in the range of 1–100 nm. NMs are not simply another step in miniaturization, but a different arena entirely; the nanoworld placed itself halfway between the scale of atomic and quantum phenomena and the scale of bulk materials. At the NMs level, some material properties are affected by the laws of atomic physics, rather than behaving as traditional bulk materials do [1–7].

NMs have great variety in morphologies and range of their properties and possible applications appear to be gigantic, from significantly small electronic devices, including miniature batteries, to biomedical devices, and as packaging films, superabsorbent, components of armor, and parts of automobiles. First vehicle to use the NMs for exterior automotive applications in running boards is claimed by General Motors. NMs' extremely small feature size is of the same scale as the critical size for physical phenomena, for example, the radius of the tip of a crack in a material may be in the range 1–100 nm. The way a crack grows in a larger-scale bulk material is likely to be different from crack propagation in a NM where crack and particle size are comparable. Fundamental electronic, magnetic, optical, chemical, and biological processes are also different at this level [8–15].

By controlling the structures of NMs at nanoscale dimensions, the properties of the nanostructures can be tailored in a very predictable manner to meet the needs for a variety of applications. Examples of the engineered nanostructures include metallic and nonmetallic nanoparticles, nanotubes, quantum dots and super lattices, thin films, nanocomposites, and nanoelectronic and optoelectronic devices, which utilize the superior properties of the NMs to fulfill the applications.

In this chapter, the properties of NMs will be shortly outlined from their possible applications perspective of nanomaterials. NMs are already in commercial use, with some having been available for several years or decades. The range of commercial products available today is very broad, including stain-resistant and wrinkle-free textiles, cosmetics, sunscreens, electronics, paints, and varnishes.

Sample Preparation with Nanomaterials: Next Generation Techniques and Applications,
First Edition. Chaudhery Mustansar Hussain, Rüstem Keçili, and Chaudhery Ghazanfar Hussain.

Nanocoatings and nanocomposites are finding uses in diverse consumer products, such as windows, sports equipment, bicycles, and automobiles. There are novel UV-blocking coatings on glass bottles, which protect beverages from damage by sunlight, and longer-lasting tennis balls using butyl rubber/nano-clay composites. Nanoscale titanium dioxide, for instance, is finding applications in cosmetics, sunblock creams, and self-cleaning windows, and nanoscale silica is being used as filler in a range of products, including cosmetics and dental fillings.

2.2 Mechanical Properties of NMs

As a primary microstructural parameter, grain size imparts significant influence on the mechanical behavior of materials. The drive for refining grain size comes from the demand for increased strength in many applications. A number of mechanisms have been proposed to explain this phenomenon: pileup of dislocations against the grain boundaries (GBs), GBs acting as dislocation sources and sinks, and presence of geometrically necessary dislocations in the vicinity of the GBs to account for the deformation compatibility of polycrystalline metals, etc. Certainly, the grain size, or average grain diameter, influences the mechanical properties of a material.

Some appealing characteristics of metals and alloys with potential significance for engineering applications include ultrahigh yield and strengths, decreased elongation and toughness, superior wear resistance, and the promise of enhanced super plastic formability at lower temperatures and faster strain rates compared to their microstructured counterparts. The need for scratch, mar, and/or abrasion resistance is well established in various markets, including fingernail polishes, flooring, plastic glazing, headlamp covers and other automotive parts, transportation windows, and optical lenses, where clear scratch-resistant coatings are used. Because of the nanosize, many of the mechanical properties of the materials are modified, among others, hardness and elastic modulus, fracture toughness, scratch resistance, fatigue strength, and hardness. Energy dissipation, mechanical coupling within arrays of components, and mechanical nonlinearities are influenced by structuring components at the nanometer scale. This includes also the interpretation of unusual mechanical behavior (e.g. strengths approaching the theoretical limit) and the exploration of new ways to integrate diverse classes of mechanically functional materials on the nanosize.

2.2.1 Hardness and Strength

Adjacent grains normally have different crystallographic orientations and common GB. In coarse-grained materials, the GB acts as a barrier to dislocation motion because of necessity in change of direction of motion in GB regions, and pileup of dislocations at GBs is envisioned as a key mechanistic process underlying an enhanced resistance to plastic flow from grain refinement. As the microstructure is refined from the micro and ultrafine-grained regime into the NC regime, this process invariably breaks down, and the flow stress vs. grain size relationship departs markedly from that seen at higher grain sizes. With further grain refinement, the yield stress peaks in many cases at an average grain size value on the order of 10 nm or so. Further decrease in grain size can cause weakening of the metal. Experimental evidence indicates that below a grain size of ~10 nm, strength decreases with further grain refinement (the so-called "inverse Hall–Petch-type" relationship). Although there is a growing body of experimental evidence pointing to such unusual deformation responses in NC materials, the underlying mechanisms are not well understood yet. The mechanical properties of fully dense face-centered cubic metals (Cu, Ni, and Pd) with grain size less than 100 nm have been primarily derived from uniaxial tension/compression tests and micro- or nanoindentation. Typically, these nanocrystalline metals exhibit significantly higher yield strength and reduced tensile elongation relative to their microcrystalline counterparts. Furthermore, hardness and yield strength have been found to increase with decreasing grain size in this regime (<100 nm) down to at least 15 nm. The reasons for this behavior are still under debate as dislocation sources within grains are not expected to operate at these grain sizes. In addition, there is no documented evidence of dislocation pileups in deformed specimens, and any dislocation activity is primarily believed to originate and terminate at GBs. Among many excellent mechanical properties of NMs [16–23], high hardness of nanocomposite systems is one of the most intriguing. A variety of superhard nanocomposites can be made of nitrides, borides, and carbides by plasma-induced chemical and physical vapor deposition. In the appropriately synthesized binary systems, the hardness of the nanocomposite exceeds significantly that given by the rule of mixtures in bulk. Since their discovery, carbon nanotubes have stimulated intensive research interests because of excellent mechanical properties [24–31]. The strength of the carbon fibers increases with graphitization along the fibber axis. Carbon nanotubes, which are formed of seamless cylindrical graphene layers, represent the ideal carbon fiber and presumably have the best mechanical properties in the carbon fibers species, showing a high Young's modulus and high tensile strength.

Improvement in the mechanical features of materials is a crucial issue when we incorporate NMs such as carbon nanotubes, nanoparticles, nanowires, nanocrystals, etc. Improvement is expected in various features including tensile strength, Young's modulus, bending strength, fracture features, and impact resistance. When taking into account the detailed analysis of the mechanical features of NMs, there are various parameters to consider, such as surface structure, porosity, functionalization, preparation methods, chemical treatments, etc., all of which have a direct impact on the mechanical features of the NM. Mechanical features also vary with the base material. Carbon-based NMs possess very high mechanical strength [32, 33]. Table 2.1 shows the comparison of Young's Modulus values for the carbon-based NMs with the other materials.

Atomic force microscope (AFM) has also been employed to measure the modulus of the carbon nanotubes [35]. This is realized by bending the anchored carbon nanotubes with AFM tip while simultaneously recording the force by the tube as a function of the displacement from its equilibrium position. The resultant Young's modulus was 1.28 ± 0.5 TPa. The values of Young's module measured from different

Table 2.1 Comparison of Young's modulus values for the carbon-based nanomaterials with the other materials.

Material	Young's modulus (GPa)
Rubber	0.1
Al	70
Fe	200
SiC	440
Fe nanoparticles (100 nm)	800
C nanotubes	1000
Diamond	1200

Source: Reghunadhan et al. [34]. © 2018, Elsevier.

ways were all in the range by theoretical prediction, proving the existence of the high elastic modulus of the carbon nanotubes. The tensile strength of carbon nanotubes has also been under consideration. An individual multiwalled carbon nanotube was mounted between two AFM tips, one on rigid cantilever and the other on soft cantilever. By recording the whole tensile loading experiment, both the deflection of the soft cantilever from which the force applied on the nanotube and the length change of the nanotube were simultaneously obtained. The carbon nanotubes broke in the outermost layer ("sword-in-sheath" failure), and the tensile strength of this layer ranged from 11 to 63 GPa, and the measured strain at failure can be as high as 12%. For comparison, the tensile strength of high-strength steel is 1–2 GPa. Enhancement of mechanical properties of polymeric materials by nanofillers is one of the very active applications of NMs. Micrometer-size fillers used in traditional polymer composites show improvements in their mechanical properties such as the modulus, yield strength, and glass transition temperature. However, these enhancements sacrifice the ductility and toughness of the materials. Comparably, polymer nanocomposites from nanosize fillers result in unique mechanical properties at very low filler weight fractions. The dramatic improvements in the yield stress (30%) and Young's modulus (170%) in polypropylene filled with ultrafine SiO_2, as compared to micrometer filled polypropylene, have been reported. A rubbery polyurethane elastomer filled with 40 wt% of 12 nm silica particles exhibits six times increase in the elongation at break and three times increase in the modulus as them compared to a micrometer-sized filler-reinforced polymer.

2.2.2 Ductility

NMs often exhibit low tensile ductility, which essentially limits their practical use. There are three factors limiting the ductility: the presence of structural artifacts arising from processing, such as porosity and cracks; crack nucleation or propagation instability; plastic instability in tension. Defects obviously decrease fracture strength as they serve as crack initiation sites resulting in easy crack nucleation and growth, promoting brittle behavior in tension, and causing failure before yielding has chance to occur. Plastic instability originates from the lack of an effective hardening mechanism and/or internal flaws; this instability manifests itself as either shear bands or through "early" necking. During deformation, the dislocation storage and annihilation at GBs are two key competing mechanisms influencing the level of the flow stress. The absence of substantial macroscopic tensile ductility in nanocrystalline fcc metals together with the observation of dimpled rupture on fracture surfaces leads to the hypothesis that deformation is localized. Localized deformation is clearly manifested through the appearance of shear bands on deformed specimen surfaces. Shear bands in ultrafine-grained copper are produced by severe plastic deformation (SPD) method. Plastic flow localization in materials in the absence of strengthening leads to the macroscopic necking, followed by the stress concentration in the neck region. Importance of the presence of nanoscale voids in the structure prior to deformation is now well understood. For example, the magnetron-sputtered Ni, containing some GB porosity, failed in a brittle manner via rapid coalescence of intergranular cracks, whereas in the laser deposited film that contained no porosity, the crack propagated slowly and is accompanied by continuous film thinning. Fracture surfaces resulting from tensile tests have frequently shown dimpled rupture in NC metals. Further, the dimple size is significantly larger than the average grain size; in addition, a pair of mating fracture surfaces illustrates the presence a significant stretching of the ligaments between the dimples that indicates an appreciable local plasticity. When the grain size is reduced down to ~10 nm or less, the resulting fracture surface still shows the dimpled rupture. The only difference is that the dimple diameter is finer in size (but still much larger than the grain size).

Fracture surface of a 30 nm grain size electrodeposited Ni tensile specimen. Based on the observations of dimpled rupture, dislocation activity at the crack tip, and the formation of voids at GBs and triple junctions in the regions ahead of the advancing crack, a model for damage evolution and fracture was proposed. In the early stages of deformation, dislocations are emitted from GBs under the influence of an applied stress, when intragranular slip is coupled with unaccommodated GB sliding to facilitate void formation at the GBs. Such voids do not necessarily form at every boundary. Triple junction voids and wedge cracks can also result from GB sliding if resulting displacements at the boundary are not accommodated by diffusional or power law creep. These GB and triple junction voids then act as sites for nucleation of the dimples, which are significantly larger than the individual grains, and the rim of these dimples on the fracture surface do not necessarily coincide with GBs. The deformation and fracture processes are closely related to the coupling of dislocation-mediated plasticity and formation and growth of voids. The fundamental difference between the two approaches is that the atomistic simulations reveal intergranular crack propagation, where the GBs that are chosen by the crack path are determined by plastic deformation processes, stretching in concert and finally leading to transgranular fracture. Whatever the fracture mechanism, it is evident that the fracture will be heavily influenced by microstructural features such as the presence of nanoscale voids or even bubbles and the presence of grown-in twins. It is well known that bubbles filled with hydrogen (hydrogen bubbles) are often present in electrodeposited metals, and these could serve as nucleation sites for the dimples. To improve the ductility of nanomaterials, four basic concepts are proposed. The first option is to suppress plastic flow localization through fabrication of a bimodal single-phase structure composed of nanograins and large grains. The second strategy is to suppress plastic flow localization through fabrication of a composite structure consisting of ductile second-phase inclusions embedded into NC matrix. Localization of the plastic flow is effectively hampered due to the strengthening effects provided by large grains and inclusions, respectively. The next option is to deform materials at low temperatures. In this case, recovery or dislocation annihilation processes at GBs are suppressed and do not compensate the dislocation storage at the boundaries. As a result, an NC specimen shows a good ductility due to strengthening that prevents plastic flow localization. NC materials with high ductility include: (i) a bimodal single-phase structure composed of nanograins and large grains; and (ii) nanocomposite consisting of nanoscale grains and dendrite-like inclusions of the second phase. The fourth option is using a positive strain rate sensitivity of the flow stress. The sensitivity means that a local increase of the plastic strain rate in the neck region leads to a local increase in a flow stress.

2.2.3 Applications of Mechanical Properties

Enhanced mechanical properties of the NMs have many potential applications both in nanoscale such as mechanical nanoresonators, mass sensors, microscope probe tips, and nanotweezers for nanoscale object manipulation and in macroscale applications such as structural reinforcement of polymer materials, lightweight high-strength materials, flexible conductive coatings, wear-resistant coatings, tougher and harder cutting tools, etc. Cutting tools made of NMs, such as tungsten carbide, tantalum carbide, and titanium carbide, are much harder, much more wear-resistant, erosion-resistant, and last longer than their conventional (large-grained) counterparts. In the recent years, so much effort was also put on the miniaturization, which is widely applied in chemical and biological processes, offering various superiorities compared to the macroscopic systems [36–40].

In automobiles, NMs are envisioned to be used in spark plugs. Also, automobiles waste significant amounts of energy by losing the thermal energy generated by the engine. So, the engine cylinders are envisioned to be coated with nanocrystalline ceramics, such as zirconia and alumina, which retain heat much more efficiently, resulting in complete and efficient combustion of the fuel. One of the key properties required of the aircraft components is the fatigue strength, which decreases with the component's age. The fatigue strength increases with a reduction in the grain size of the material. NMs provide such a significant reduction in the grain size over conventional materials that the fatigue life is increased by an average of 200–300%. In spacecrafts, elevated-temperature strength of the material is crucial because the components (such as rocket engines, thrusters, and vectoring nozzles) operate at much higher temperatures compared to aircrafts and at higher speeds. NMs are perfect candidates for spacecraft applications, as well. Ceramics are very hard, brittle, and hard to machine even at high temperatures. However, with a reduction in grain size, their properties change drastically. Nanocrystalline ceramics can be pressed and sintered into various shapes at significantly lower temperatures. Zirconia, for example, is a hard, brittle ceramic, that has even been rendered superplastic, i.e. it can be deformed to great lengths (upto 300% of its original length). However, these ceramics must possess nanocrystalline grains to be superplastic. Ceramics based on silicon nitride (Si_3N_4) and silicon carbide (SiC) have been used in automotive applications as high-strength springs, ball bearings, and valve lifters and because they possess good formability and machinability combined with excellent physical, chemical, and mechanical properties. They are also used as components in high-temperature furnaces. Aerogels are nanocrystalline porous and extremely lightweight materials and can withstand 100 times their weight. They are currently being used for insulation in offices, homes, etc. They are also being used as materials for "smart" windows, which darken when the sun is too bright, and they lighten themselves otherwise.

2.3 Thermal Properties of NMs

Most of the properties of NMs such as optical [17, 41–47], electrical [48, 49], and magnetic properties [50, 51] have been well studied. However, the thermal properties [52] of NMs have been only recently given serious consideration. This is partially due to the difficulties of experimental measuring and controlling the thermal transport in nanoscale. AFM has been introduced to measure the thermal transport of nanostructures with nanometer-scale high spatial resolution, providing a promising way to probe the thermal properties. Moreover, the theoretical simulations and analysis of thermal transport in nanostructures are still in infancy.

Several approaches such as numerical solution by Fourier's law, computational calculation based on Boltzmann transport equation, and molecular dynamics (MD) simulation have been applied to study thermal properties of NMs. However, at nanoscale level, the definition of temperature is difficult to comprehend. The thermal energy is mainly carried by phonons, which have a wide variation in frequency and the mean free paths (mfp) in nonmetals. The heat-carrying photons often have large wave vectors and mfp in the order of nanometer range at room temperature, so that the dimensions of the nanostructures are comparable to the mfp and wavelengths of photons. However, the general definition of temperature is based on the average energy of a material system in equilibrium. But for NM systems, the dimensions are too small to define a local temperature. Additionally, it is also difficult to use the concept of temperature, which is defined in equilibrium conditions for the nonequilibrium processes of thermal transport in NMs posing difficulties for theoretical analysis of thermal transport in nanoscales.

Although there are the complications for experimental and theoretical characterization of thermal properties of NMs, yet in recent years new developments in experiments have shown that NMs have extraordinarily thermal properties. For example, silicon nanowires have much smaller thermal conductivities compared to bulk silicon; similarly, tubular structures of carbon nanotubes have extremely high thermal conductivity in axial directions and high anisotropy in the heat transport over the specimen.

To determine the thermal properties of NMs, the most important thing is to locate the interfaces. Usually, the internal interfaces obstruct the flow of heat due to photon and phonon scattering. At the interface boundary between similar materials, the interface disorder scatters phonons, as the differences in elastic properties and densities of vibration states affect the transfer of vibration energy across interfaces between dissimilar materials. Consequently, the NM structures with high interface densities reduce the thermal conductivity of the materials. These interconnected factors joined together to determine the special thermal properties of the NMs. A good example is carbon nanotubes (CNTs), which are from same family of diamond and graphite and are well known for their high thermal conductivities.

The stiff sp^3 bonds in diamond structure result in high phonon velocity and consequently high thermal conductivity. In carbon nanotubes, the carbon atoms are held together by the even stronger sp^2 bonds, so that the nanotube structures, consisting of seamlessly joined graphitic cylinders, are expected to have extraordinarily high thermal conductivity. The rigidity of these nanotubes, combined with virtual absence of atomic defects or coupling to soft photon modes of the embedding medium, should make isolated nanotubes very good candidates for efficient thermal conductors. On the other hand, one-dimensional nanowires may offer ultralow thermal conductivities, quite different from that of carbon nanotubes. In nanowires, phonons behave differently from those in the corresponding bulk materials due to the quantum confinement in the one-dimensional structures. The nanowires' surface can introduce surface phonon modes, resulting in many different phonon polarizations other than the two transverse and one longitudinal acoustic branches found in the bulk semiconductors. Those changes in the dispersion relation can modify the group velocity and the density of states of each branch. The phonon lifetime also changes due to the strong phonon–phonon interactions and the boundary scattering within the nanostructures. Thus, the phonon transports and the thermal properties of the nanowires will be significantly different from that of the bulk materials.

The phonon transport of the semiconducting nanowires has been studied experimentally and theoretically. Multilayers and superlattices are thin films consisting of alternating layers of two or more different materials stacked upon each other. In multilayer structure, the films can be either amorphous or polycrystalline while in superlattices, the films are single crystal. There are many effects in the multilayers or superlattice structures that affect the phonon transport properties. Multilayers and superlattice nanostructures have many potential applications. For example, multilayer thin films can be used as thermal barriers at high-temperature environments, such as in engines to improve their efficiencies; epitaxial superlattices of semiconductor films with low thermal conductivity can be used in thermoelectric power generation because of the ability to the control both electronic band structure and phonon transport simultaneously.

To enhance the thermal transport with nanofluids is another promising application of the thermal properties of NMs. Usually, nanofluids are formed by dispersing nanometer-sized particles (1–100 nm) or droplets into heat transfer fluids.

Compared with micro-fluids or milli-fluids, nanofluids stay more stable, so nanofluids are promising for practical applications. The thermal conductivity enhancement of nanofluids shows a temperature-dependent characteristic – increase of enhancement with rising temperature, which makes the nanofluids more suitable for applications at elevated temperatures. The research on nanofluids is only at its infant stage. On the other hand, the large interface areas between the nanoparticles and the base fluids increase the heat transfer rate, so nanosheets, nanorods, and nanowires, which have large surface areas, perhaps are favorable to more significantly increase the thermal conductivity of nanofluids. Recently, four possible mechanisms for nanofluids that may contribute to the thermal conduction are: (i) Brownian motion of nanoparticles; (ii) liquid layering at liquid/particle interface; (iii) ballistic nature of heat transport in nanoparticles; and (iv) nanoparticle clustering in nanofluids. The most important feature of nanofluids is the significant increase of thermal conductivity compared with liquids without NMs, which has been proved by many experimental works.

2.4 Electrical Properties of NMs

Understanding the electronic properties of NMs is a key factor in the fabrication of useful electronic devices [53]. However, the electrical conductivity of NMs is difficult to understand due to their smaller size and based on distinct mechanisms. In general, these mechanisms can be categorized as: surface scattering, quantized conduction including ballistic conduction, coulomb charging and tunneling, and widening and discrete band gap. In addition, increased perfection, such as reduced impurity, structural defects, and dislocations, would affect the electrical conductivity of nanostructures and NMs. NMs can hold considerably more energy than conventional ones because of their large GB area. They are materials in which an optical absorption band can be introduced, or an existing band can be altered by the passage of current through these materials, or by the application of an electric field.

Recently, nanoelectronics and molecular electronics have shown remarkable growth in the electronics industry. Single molecules are used to control electron transport in molecular electronics, which offers the potential of exploring the vast variety of molecular functions for electronic devices. On the other hand, the molecules that are biologically active give route to bioelectronic devices. In molecular electronics, control over the electronic energy levels at the surface of conventional semiconductors and metals is achieved by assembling on the solid surfaces, poorly organized, partial monolayers of molecules instead of the more commonly used ideal ones. Once those surfaces become interfaces, these layers exert electrostatic rather than electrodynamic control over the resulting devices, based on both electrical monopole and dipole effects of the molecules. Thus, electronic transport devices, incorporating organic molecules, can be constructed without current flow through the molecules.

Conventional and rechargeable batteries are used in almost all applications that require electric power. The energy density (storage capacity) of these batteries is quite low requiring frequent recharging. Nanocrystalline materials are good candidates for separator plates in batteries because they can hold considerably more energy than conventional ones. Nickel-metal hydride batteries made of nanocrystalline nickel and metal hydrides are envisioned to require far less frequent recharging and to last much longer.

An electrochromic device consists of materials in which an optical absorption band can be introduced, or an existing band can be altered by the passage of current through the materials or by the application of an electric field. They are similar to liquid-crystal displays (LCDs) commonly used in calculators and watches and are primarily used in public billboards and ticker boards to convey information. The resolution, brightness, and contrast of these devices depend on the tungsten acid gel's grain size. Hence, NMs, such as tungstic oxide gel, are being explored for this purpose.

2.5 Optical Properties of NMs

NMs have attracted much interest for their novel optical properties, which differ remarkably from bulk forms [54, 55]. Significant contributory factors include quantum confinement of electrical carriers within nanoparticles, efficient energy and charge transfer over nanoscale distances, and, in many systems, a highly enhanced role of interfaces. The linear and nonlinear optical properties of such materials can be finely tailored by controlling the NMs' dimensions and the chemistry of their surfaces; fabrication technology becomes a key factor for the applications. Surface plasmons, also known as surface plasmon polaritons, are surface electromagnetic waves that propagate parallel along a metal/dielectric (or metal/vacuum) interface. Since the wave is on the boundary of the metal and the external medium (air or water, for example), these oscillations are very sensitive to any change of this boundary, such as the adsorption of molecules to the metal surface.

On the other hand, surface plasmon resonance (SPR) is the coherent excitation of all the "free" electrons within the conduction band, leading to an in-phase oscillation [56]. The principle of SPR technique is summarized in three terms, which are "*surface*," suggesting that the SPR sensor is normally used for the study of phenomena or reactions occurring at the surface; "*plasmon*," indicating the movement of free electrons; and "*resonance*," which indicates the energy transfer from incident light to surface plasmons. SPR can be successfully employed to investigate surface interactions through surface plasmons caused by resonance with incident light. The incident light is usually p-polarized visible light or infrared light. p-Polarized light is used since its polarization is in the direction parallel to the incidence plane. Thus, electronic surface plasmons can be excited. In the SPR sensor systems, the detection technique is based on the detection of the any changes in refractive index of very thin film layer formed on the surface the metal such as Au or Ag.

Unique optical property of NMs may also arise from another quantum size effect. When the size of a nanocrystal (i.e. a single-crystal nanoparticle) is smaller than the de Broglie wavelength, electrons and holes are spatially confined and electric dipoles are formed, and discrete electronic energy level would be formed in all materials. Similar to a particle in a box, the energy separation between adjacent levels increases with decreasing dimensions. The electronic configurations of NMs are significantly are different from that of their bulk counterpart. These changes arise through systematic transformations in the density of electronic energy levels as function of the size, and these changes result in strong variations in the optical and electrical properties with size. Nanocrystals lie in between the atomic and molecular limit of discrete density of electronic states and the extended crystalline limit of continuous band. In any material, there will be a size below which there is substantial variation of fundamental electrical and optical properties with size, when energy level spacing exceeds the temperature. For a given temperature, this occurs at a very large size (in nanometers) in semiconductors as compared with metals and insulators. In the case of metals, where the Fermi level lies in the center of a band and the relevant energy level spacing is very small, the electronic and optical properties more closely resemble those of continuum, even in relatively small sizes (tens or hundreds of atoms). In semiconductors, the Fermi level lies between two bands, so that the edges of the bands are dominating the low-energy optical and electrical behavior. Optical excitations across the gap depend strongly on the size, even for crystallites as large as 10 000 atoms. For insulators, the band gap between two bands is already too big in the bulk form. The quantum size effect is most pronounced for semiconductor nanoparticles, where the band gap increases with a decreasing size, resulting in the interband transition shifting to higher frequencies. In a semiconductor, the energy separation, i.e. the energy difference between the completely filled valence band and the empty conduction band is of the order of a few electrovolts and increases rapidly with a decreasing size. The same quantum size effect is also known for metal nanoparticles; however, in order to observe the localization of the energy levels, the size must be well below 2 nm, as the level spacing has to exceed the thermal energy (–26 meV). In a metal, the conduction band is half-filled and the density of energy levels is so high that a noticeable separation in energy levels within the conduction band (intraband transition) is only observed when the nanoparticle is made up of 100 atoms. If the size of metal nanoparticle is made small enough, the continuous density of electronic states is broken up into discrete energy levels. Glues containing nanoparticles have optical properties that give rise to uses in optoelectronics. Casings containing nanoparticles used in electronic devices, such as computers, offer improved shielding against electromagnetic interference. Electrochromic devices similar to LCDs are being developed with NMs. The incorporation of NMs in surface coatings can provide long-term abrasion resistance without significantly affecting the optical clarity, gloss, color, or physical properties. The cosmetics industry has used NMs as UV absorbers or sunscreens.

2.6 Magnetic Properties of NMs

Magnetic nanomaterials are those that exist in a state of permanent magnetization without the need to apply a field [57, 58]. The strength of a magnet is measured in terms of saturation magnetization and coercivity (H_c is the field required to reduce the magnetization to zero from saturation and is applied in the opposite direction to the original saturating field) values. These values increase with a decrease in the grain size and an increase in the specific surface area (surface area per unit volume) of the grains. Therefore, NMs also present good properties in this field. There are three main categories of magnetism: diamagnetism, paramagnetism, and ferromagnetism. Diamagnetism is a fundamental property of all atoms and the magnetization is very small and opposed to the applied magnetic fields direction. However, many materials exhibit paramagnetism, where a magnetization develops parallel to the applied magnet field as the field is increased from zero, but the strength of the magnetization is small. Ferromagnetism is the property of those materials that are intrinsically magnetically ordered and that develop spontaneous magnetization without the need to apply a field. The ordering mechanism is the quantum mechanical exchange interaction. Magnets made of nanocrystalline yttrium–samarium–cobalt grains possess very unusual magnetic properties due to their extremely large surface area.

2.7 Adsorption Properties of NMs

Adsorption properties of NMs have received considerable attention in recent years due to their potential application in environmental pollution problems [59–62]. NMs are relatively new adsorbents and hold interesting positions in adsorptive materials for many reasons. On the one hand, they provide chemically inert surfaces for physical adsorption, and their high specific surface areas stand in comparison with those of activated carbons (ACs). On the other hand, CNTs are essentially different from ACs in the aspect that their structure at the atomic scale is far more well-defined and uniform. Parameters such as the pore diameter distribution and adsorption energy distribution are needed to quantify adsorption on ACs, while, as for CNTs, one can deal directly with various well-defined adsorption sites available to the adsorbed molecules. From the viewpoint of CNT structures, the relationship between CNTs and other carbonaceous adsorptive materials is almost the same as that between single crystals and polycrystalline materials [63].

Due to their highly porous and hollow structure, large specific surface area, light mass density, and strong interaction between CNTs and pollutant molecules [64], the applications of CNTs for the removal of hazardous pollutants from gas streams [65–67] from large volumes of aqueous solutions [68–70] have being studied extensively via theoretical calculations, experimental measurements, and molecular simulations. Large number of experimental studies have already been carried out on the adsorption of small molecules [71–73], heavy metal ions [74–76], radio nuclides [77, 78], and organic chemicals [79, 80] on various CNTs (single-walled or multiwalled, closed- or open-ended CNTs). The study of adsorption properties of CNTs plays an important role in both fundamental and practical point of view because it could shed new light on the mechanism of adsorption in complex systems.

Adsorption properties of CNTs depend on the following factors, the first of which is the contribution of individual adsorption sites [81]. Through the temperature programmed desorption studies of various alkanes on nanotube bundles, one can know that the different groups of adsorption sites are present in the bundles [82, 83]. There are four possible sites in CNT bundles for the adsorption of different pollutants: (i) "internal sites" – the hollow interior of individual nanotubes (accessible only if the caps are removed and the open ends are unblocked); (ii) "interstitial channels (ICs)" – the interstitial channels between individual nanotubes in the bundles; (iii) "grooves" – the grooves present on the periphery of a nanotube bundle and the exterior surface of the outermost nanotubes, where two adjacent parallel tubes meet; and (iv) "outside surface" – the curved surface of individual nanotubes on the outside of the nanotube bundles.

Considerable studies have been carried out to determine the occupied sites by adsorbate molecules after the adsorption [84–88]. A consensus is that the adsorption on close-ended CNTs takes place on the grooves, on the outer surface sites, and on large-diameter stacking-defect induced [89–93]. For as-produced CNTs, both ends are generally closed, the first step of adsorption occurs in the grooves between adjacent tubes on the perimeter of the bundles and the largest accessible interstitial channels, but bigger molecules such as Xe and SF6 only occur in the grooves. The second step can be assigned to the adsorption on the convex external walls [94]. For purified opened CNT bundles, adsorption proceeds firstly by the population of the walls inside the opened nanotubes and forms one-dimensional chains in the grooves at the outer surface of the bundles, followed by the filling of the remaining axial sites inside the nanotubes and the completion of a quasi-hexagonal monolayer on the outer surfaces of the bundles [95]. It is important to emphasize that the adsorption reaches equilibrium much faster on external sites (grooves and outer surfaces) than on the internal sites (interstitial channels and inside the tube) under the same pressure and temperature conditions. The external sites are directly exposed to the adsorbing material; the adsorption process on internal sites has to be initiated on the ends of the pore, followed by diffusion to the sites at the interior [96, 97].

2.8 Conclusions

With the rapid progress in nanotechnology, better understandings of special physicochemical properties of NMs are being achieved gradually. It is known that the structures and properties of NMs are distinct from their equivalents in bulk at the nanoscale, which has given boosted consideration to miniaturization of devices and agglomerated-assembled materials. However, still some important problems must be addressed, e.g. size-dependent structural evolution from a single atom to bulk matter and their thermal stability. Moreover, few fundamental issues are still unclear, e.g. accurate clustered growth, etc. It is also difficult at present to precisely determine properties of larger clusters containing hundreds of atoms. This challenge can be overcome in the near future through computer simulations and precise experimental techniques.

NMs as a new class of materials have great potential in diverse applications, which have huge promise in both theoretical and practical applications. On the one hand, researchers are very interested in thoroughly understanding some basic scientific issues about the special properties. In order to replace or partially replace the traditional industrial techniques and processes, researchers are doing their best to develop new NMs by using the basis of thermodynamic and kinetic studies over the reaction process. It is believed that, in the following decade, exciting progress would be made to explore in-depth knowledge of special properties of NMs, which will open new doors for NMs to various industries.

References

1 Gupta, C. and Bhardwaj, A. (2020). Summary and future perspectives of nanomaterials and technologies: special emphasis on energy and environment, Chapter 11. In: *Materials Today, Nanomaterials for Sustainable Energy and Environmental Remediation* (eds. M. Naushad, R. Saravanan and K. Raju), 333–353. Elsevier.

2 Durgalakshmi, D., Balakumar, S., Rajendran, S., and Naushad, M. (2020). Functional nanomaterial in energy and environmental science, Chapter 1. In: *Materials Today, Nanomaterials for Sustainable Energy and Environmental Remediation* (eds. M. Naushad, R. Saravanan and K. Raju), 1–23. Elsevier.

3 Llobet, E. (2020). Carbon nanomaterials, Chapter 4. In: *Micro and Nano Technologies, Advanced Nanomaterials for Inexpensive Gas Microsensors* (ed. E. Llobet), 55–84. Elsevier.

4 Kai-Hua Chow, E., Gu, M., and Xu, J. (2020). Carbon nanomaterials: fundamental concepts, biological interactions, and clinical applications, Chapter 14. In: *Micro and Nano Technologies, , Nanoparticles for Biomedical Applications* (eds. E.J. Chung, L. Leon and C. Rinaldi), 223–242. Elsevier.

5 Fan, Y., Yang, B., and Rong, C. (2020). Functionalized carbon nanomaterials for advanced anode catalysts of fuel cells, Chapter 7. In: *Micro and Nano Technologies, Advanced Nanomaterials for Electrochemical-Based Energy Conversion and Storage* (eds. F. Ran and S. Chen), 223–263. Elsevier.

6 Hussain, C.M. and Keçili, R. (2020). *Modern Environmental Analysis Techniques for Pollutants*. Netherlands: Elsevier.

7 Büyüktiryaki, S., Keçili, R., and Hussain, C.M. (2020). Functionalized nanomaterials in dispersive solid phase extraction: advances and prospects. *Trends Anal. Chem.* https://doi.org/10.1016/j.trac.2020.115893.

8 Keçili, R., Büyüktiryaki, S., and Hussain, C.M. (2019). Advancement in bioanalytical science through nanotechnology: past, present and future. *Trends Anal. Chem.* 110: 259–276.

9 Hussain, C.M. and Kharisov, B. (2017). *Advanced Environmental Analysis-Application of Nanomaterials*. The Royal Society of Chemistry.

10 Hussain, C.M. (2017). Magnetic nanomaterials for environmental analysis. In: *Advanced Environmental Analysis-Application of Nanomaterials* (eds. C.M. Hussain and B. Kharisov), 1–13. The Royal Society of Chemistry.

11 Hussain, C.M. (2019). *Handbook of Nanomaterials in Analytical Chemistry: Modern Trends in Analysis*. Elsevier.

12 Büyüktiryaki, S., Sümbelli, Y., Keçili, R., and Hussain, C.M. (2019). Lab-on-chip platforms for environmental analysis. In: *Encyclopedia of Analytical Science*, 3e (eds. P. Worsfold, C. Poole, A. Townshend and M. Miró), 267–273. Academic Press.

13 Keçili, R. and Hussain, C.M. (2018). Recent progress of imprinted nanomaterials in analytical chemistry. *Int. J. Anal. Chem.*: 8503853.

14 Barua, S., Debojeet, S., Shahnaz, N., and Khan, R. (2020). Chemistry of two-dimensional nanomaterials, Chapter 1. In: *Two-Dimensional Nanostructures for Biomedical Technology* (eds. R. Khan and S. Barua), 1–33. Elsevier.

15 Keçili, R., Büyüktiryaki, S., and Hussain, C.M. (2018). Engineered nanosensors based on molecular imprinting technology, Chapter 57. In: *Micro and Nano Technologies, Handbook of Nanomaterials for Industrial Applications*, vol. 2018 (ed. C.M. Hussain), 1031–1046. Elsevier.

16 Felix, L.C., Ede, J.D., Snell, D.A. et al. (2016). Physicochemical properties of functionalized carbon-based nanomaterials and their toxicity to fishes. *Carbon* 104: 78–89.

17 He, Z. and Que, W. (2016). Molybdenum disulfide nanomaterials: structures, properties, synthesis and recent progress on hydrogen evolution reaction. *Appl. Mater. Today* 3: 23–56.

18 Oh, J.Y., Kim, Y.S., Jung, Y. et al. (2016). Preparation and exceptional mechanical properties of bone-mimicking size-tuned graphene oxide@carbon nanotube hybrid paper. *ACS Nano* 10 (2): 2184–2192.

19 Xu, L., Qi, X., Li, X. et al. (2016). Recent advances in applications of nanomaterials for sample preparation. *Talanta* 146: 714–726.

20 Oh, J.Y., Yang, S.J., Park, J.Y. et al. (2015). Easy preparation of self-assembled high-density buckypaper with enhanced mechanical properties. *Nano Lett.* 15 (1): 190–197.

21 Wu, R., Zhou, K., Yue, C.Y. et al. (2015). Recent progress in synthesis, properties and potential applications of SiC nanomaterials. *Prog. Mater Sci.* 72: 1–60.

22 Sadik, O.A., Du, N., Kariuki, V. et al. (2014). Current and emerging technologies for the characterization of nanomaterials. *ACS Sustainable Chem. Eng.* 2 (7): 1707–1716.

23 Tian, J., Xu, J., Zhu, F. et al. (2013). Application of nanomaterials in sample preparation. *J. Chromatogr. A* 1300: 2–16.

24 Iijima, S. (1991). Helical microtubules of graphitic carbon. *Lett. Nat.* 354: 56–58.

25 Mohamed, A.E.-M.A. and Mohamed, M.A. (2020). Carbon nanotubes: synthesis, characterization, and applications, Chapter 2. In: *Micro and Nano Technologies, Carbon Nanomaterials for Agri-Food and Environmental Applications* (ed. K.A. Abd-Elsalam), 21–32. Elsevier.

26 Sharma, A., Patwardhan, A., Dasgupta, K., and Joshi, J.B. (2019). Kinetic study of boron doped carbon nanotubes synthesized using chemical vapour deposition. *Chem. Eng. Sci.* 207: 1341–1352.

27 Mwafy, E.A. and Mostafa, A.M. (2019). Multi walled carbon nanotube decorated cadmium oxide nanoparticles via pulsed laser ablation in liquid media. *Opt. Laser Technol.* 111: 249–254.

28 Mohammad, M.I., Moosa, A.A., Potgieter, H., and Ismael, M.K. (2013). Carbon nanotubes synthesis via arc discharge with a Yttria catalyst. *Int. Sch. Res. Notices*: 785160.

29 Abousalman-Rezvani, Z., Eskandari, P., Roghani-Mamaqani, H., and Salami-Kalajahi, M. (2020). Functionalization of carbon nanotubes by combination of controlled radical polymerization and "grafting to" method. *Adv. Colloid Interface Sci.* 278: 102126.

30 Merum, S., Veluru, J.B., and Seeram, R. (2017). Functionalized carbon nanotubes in bio-world: applications, limitations and future directions. *Mater. Sci. Eng., B* 223: 43–63.

31 Yan, L., Zhao, F., Li, S. et al. (2011). Low-toxic and safe nanomaterials by surface-chemical design, carbon nanotubes, fullerenes, metallofullerenes, and graphenes. *Nanoscale* 3 (2): 362–382.

32 Chen, Y., Zhan, H., Shi, Q.Q. et al. (2019). High-strength macroscopic carbon nanotube fiber and film, Chapter 3. In: *Micro and Nano Technologies, Nanotube Superfiber Materials*, 2e (eds. M.J. Schulz, V. Shanov, Z. Yin and M. Cahay), 65–76. William Andrew Publishing.

33 Mansor, M.R., Fadzullah, S.H.S.M., Masripan, N.A.B. et al. (2019). Comparison between functionalized graphene and carbon nanotubes: effect of morphology and surface group on mechanical, electrical, and thermal properties of nanocomposites, Chapter 9. In: *Micro and Nano Technologies, Functionalized Graphene Nanocomposites and their Derivatives* (eds. M. Jawaid, R. Bouhfid and A. el Kacem Qaiss), 177–204. Elsevier.

34 Reghunadhan, A., Kalarikkal, N., and Thomas, S. (2018). Mechanical property analysis of nanomaterials, Chapter 7. In: *Micro and Nano Technologies, Characterization of Nanomaterials* (eds. S.M. Bhagyaraj, O.S. Oluwafemi, N. Kalarikkal and S. Thomas), 191–212. Woodhead Publishing.

35 Bellucci, S., Gaggiotti, G., Marchetti, M. et al. (2006). Atomic force microscopy characterization of carbon nanotubes. *J. Phys. Conf. Ser.* 61, Proceedings of the International Conference On Nanoscience and Technology (Icn&T 2006) 30 July to 4 August 2006, Basel, Switzerland.

36 Bao, M. and Wang, W. (1996). Future of microelectromechanical systems (MEMS). *Sensor. Actuat. A* 56: 135–141.

37 Lindström, S. and Andersson-Svahn, H. (2011). Miniaturization of biological assays – overview on microwell devices for single-cell analyses. *Biochim. Biophys. Acta* 1810: 308–316.

38 Ríos, Á. and Zougagh, M. (2015). Modern qualitative analysis by miniaturized and microfluidic systems. *Trends Anal. Chem.* 69: 105–113.

39 Mitsudo, K., Kurimoto, Y., Yoshioka, K., and Suga, S. (2018). Miniaturization and combinatorial approach in organic electrochemistry. *Chem. Rev.* 118: 5985–5999.

40 Dixit, C.K., Kadimisetty, K., and Rusling, J. (2018). 3D-printed miniaturized fluidic tools in chemistry and biology. *Trends Anal. Chem.* 106: 37–52.

41 Sharma, V. (2020). Optical properties of tin oxide nanomaterials, Chapter 4. In: *Metal Oxides, Tin Oxide Materials* (ed. M.O. Orlandi), 61–99. Elsevier.

42 Li, J. and Zhang, J.Z. (2009). Optical properties and applications of hybrid semiconductor nanomaterials. *Coord. Chem. Rev.* 253: 3015–3041.

43 Cadena, M.J., Sung, S.H., Boudouris, B.W. et al. (2016). Nanoscale mapping of dielectric properties of nanomaterials from kilohertz to megahertz using ultrasmall cantilevers. *ACS Nano* 10 (4): 4062–4071.

44 Das, S., Banerjee, S., and Sinha, T.P. (2016). Structural and AC conductivity study of CdTe nanomaterials. *Physica E Low Dimens. Syst. Nanostruct.* 78: 73–78.

45 Duan, J. and Zhan, J. (2015). Recent developments on nanomaterials-based optical sensors for Hg^{2+} detection. *Sci. China Mater.* 58 (3): 223–240.

46 Hanifehpour, Y., Hamnabard, N., Veettikkunnu Chandran, A. et al. (2015). $Ln_xPb_{1-x}Te(Ln:Nd^{3+},Yb^{3+})$ nanomaterials: synthesis, characterization, physical properties, and optical properties. *J. Ind. Eng. Chem.* 24: 251–258.

47 Wang, X.H., Huang, L.Q., Niu, L.J. et al. (2015). The impacts of growth temperature on morphologies, compositions and optical properties of Mg-doped ZnO nanomaterials by chemical vapor deposition. *J. Alloys Compd.* 622: 440–445.

48 Yurkov, G.Y., Fionov, A.S., Koksharov, Y.A. et al. (2007). Electrical and magnetic properties of nanomaterials containing iron or cobalt nanoparticles. *Inorg. Mater.* 43: 834–844.

49 Ahmed Barhoum, M. and García-Betancourt, L. (2018). Physicochemical characterization of nanomaterials: size, morphology, optical, magnetic, and electrical properties, Chapter 10. In: *Micro and Nano Technologies, Emerging Applications of Nanoparticles and Architecture Nanostructures* (eds. A. Barhoum and A.S.H. Makhlouf), 279–304. Elsevier.

50 Zhang, B. (2018). Magnetic properties of nanomaterials, Chapter 10. In: *Micro and Nano Technologies, Physical Fundamentals of Nanomaterials* (ed. B. Zhang), 387–450. William Andrew Publishing.

51 Das, P., Fatehbasharzad, P., Colombo, M. et al. (2019). Multifunctional magnetic gold nanomaterials for cancer. *Trends Biotechnol.* 37: 995–1010.

52 Zhang, B. (2018). Thermal properties of nanomaterials, Chapter 7. In: *Micro and Nano Technologies, Physical Fundamentals of Nanomaterials* (ed. B. Zhang), 251–289. William Andrew Publishing.

53 Abu-Thabit, N.Y. (2020). Nanomaterials for flexible transparent conductive films and optoelectronic devices, Chapter 23. In: *Advances in Smart Coatings and Thin Films for Future Industrial and Biomedical Engineering Applications* (eds. A.S.H. Makhlouf and N.Y. Abu-Thabit), 619–643. Elsevier.

54 Shih, P.H., Do, T.-N., Gumbs, G., and Lin, M.-F. (2020). Electronic and optical properties of doped graphene. *Physica E Low Dimens. Syst. Nanostruct.* 118: 113894.

55 Zawadzka, A., Płóciennik, P., Korcala, A., and Szroeder, P. (2019). Optical properties of chiral single-walled carbon nanotubes thin films. *Opt. Mater.* 96: 109295.

56 Kajikawa, K. (2018). Application 42 – Sensing based on localized surface plasmon resonance in metallic nanoparticles. In: *Nanoparticle Technology Handbook*, 3e (eds. M. Naito, T. Yokoyama, K. Hosokawa and K. Nogi), 631–633. Elsevier.

57 Alonso, J., Barandiarán, J.M., Barquín, L.F., and García-Arribas, A. (2018). Magnetic nanoparticles, synthesis, properties, and applications, Chapter 1. In: *Micro and Nano Technologies, Magnetic Nanostructured Materials* (eds. A.A. El-Gendy, J.M. Barandiarán and R.L. Hadimani), 1–40. Elsevier.

58 Proenca, M.P., Sousa, C.T., Ventura, J., and Araújo, J.P. (2020). Cylindrical magnetic nanotubes: synthesis, magnetism and applications, Chapter 6. In: *Woodhead Publishing Series in Electronic and Optical Materials, Magnetic Nano- and Microwires*, 2e (ed. M. Vázquez), 135–184. Woodhead Publishing.

59 Kegl, T., Košak, A., Lobnik, A. et al. (2020). Adsorption of rare earth metals from wastewater by nanomaterials: a review. *J. Hazard. Mater.* 386: 121632.

60 Awad, A.M., Jalab, R., Benamor, A. et al. (2020). Adsorption of organic pollutants by nanomaterial-based adsorbents: an overview. *J. Mol. Liq.* 301: 112335.

61 Keçili, R. and Hussain, C.M. (2018). Mechanism of adsorption on nanomaterials, Chapter 4. In: *Nanomaterials in Chromatography* (ed. C.M. Hussain), 89–115. Elsevier.

62 Gusain, R., Kumar, N., and Ray, S.S. (2020). Recent advances in carbon nanomaterial-based adsorbents for water purification. *Coord. Chem. Rev.* 405: 213111.

63 Kondratyuk, P. and Yates, J.T. Jr., (2007). Molecular views of physical adsorption inside and outside of single-wall carbon manotubes. *Acc. Chem. Res.* 40: 995–1004.

64 Chen, C.L., Hu, J., Shao, D.D. et al. (2009). Adsorption behavior of multiwall carbon nanotube/iron oxide magnetic composites for Ni(II) and Sr(II). *J. Hazard. Mater.* 164: 923–928.

65 Long, R.Q. and Yang, R.T. (2001). Carbon nanotubes as a superior sorbent for nitrogen oxides. *Ind. Eng. Chem. Res.* 40: 4288–4291.

66 Long, R.Q. and Yang, R.T. (2001). Carbon nanotubes as superior sorbent for dioxin removal. *J. Am. Chem. Soc.* 123: 2058–2059.

67 Agnihotri, S., Rood, M.J., and Rostam-Abadi, M. (2005). Adsorption equilibrium of organic vapors on single-walled carbon nanotubes. *Carbon* 43: 2379–2388.

68 Di, Z.C., Li, Y.H., Luan, Z.K., and Liang, J. (2004). Adsorption of chromium(VI) ions from water by carbon nanotubes. *Adsorpt. Sci. Technol.* 22: 467–474.

69 Tan, X.L., Fang, M., Chen, C.L. et al. (2008). Counterion effects of nickel and sodium dodecylbenzene sulfonate adsorption to multiwalled carbon nanotubes in aqueous solution. *Carbon* 46: 1741–1750.

70 Wang, S.G., Gong, W.X., Liu, X.W. et al. (2007). Removal of lead(II) from aqueous solution by adsorption onto manganese oxide-coated carbon nanotubes. *Sep. Purif. Technol.* 58: 17–23.

71 Chen, C.H. and Huang, C.C. (2009). Hydrogen adsorption in defective carbon nanotubes. *Sep. Purif. Technol.* 65: 305–310.

72 Gaur, A. and Shim, M. (2008). Substrate-enhanced O_2 adsorption and complexity in the Raman G-band spectra of individual metallic carbon nanotubes. *Phys. Rev. B* 78: 125422.

73 Masenelli-Varlot, K., McRae, E., and Dupont-Pavlovsky, N. (2002). Comparative adsorption of simple molecules on carbon nanotubes dependence of the adsorption properties on the nanotube morphology. *Appl. Surf. Sci.* 196: 209–215.

74 Li, Y.H., Wang, S.G., Wei, J.Q. et al. (2002). Lead adsorption on carbon nanotubes. *Chem. Phys. Lett.* 357: 263–266.

75 Li, Y.H., Ding, J., Luan, Z.K. et al. (2003). Competitive adsorption of Pb^{2+}, Cu^{2+} and Cd^{2+} ions from aqueous solutions by multiwalled carbon nanotubes. *Carbon* 41: 2787–2792.

76 Chen, C.L. and Wang, X.K. (2006). Adsorption of Ni(II) from aqueous solution using oxidized multiwall carbon nanotubes. *Ind. Eng. Chem. Res.* 45: 9144–9149.

77 Chen, C.L., Wang, X.K., and Nagatsu, M. (2009). Europium adsorption on multiwall carbon nanotube/iron oxide magnetic composite in the presence of polyacrylic acid. *Environ. Sci. Technol.* 43: 2362–2367.

78 Chen, C.L., Hu, J., Xu, D. et al. (2008). Surface complexation modeling of Sr(II) and Eu(III) adsorption onto oxidized multiwall carbon nanotubes. *J. Colloid Interface Sci.* 323: 33–41.

79 Goering, J., Kadossov, E., and Burghaus, U. (2008). Adsorption kinetics of alcohols on singlewall carbon nanotubes: an ultrahigh vacuum surface chemistry study. *J. Phys. Chem. C* 112: 10114–10124.

80 Hyung, H. and Kim, J.H. (2008). Natural organic matter (NOM) adsorption to multi-walled carbon nanotubes: effect of NOM characteristics and water quality parameters. *Environ. Sci. Technol.* 42: 4416–4421.

81 Agnihotri, S., Mota, J.P.B., Rostam-Abadi, M., and Rood, M.J. (2006). Theoretical and experimental investigation of morphology and temperature effects on adsorption of organic vapors in single-walled carbon nanotubes. *J. Phys. Chem. B* 110: 7640–7647.

82 Kondratyuk, P., Wang, Y., Johnson, J.K., and Yates, J.T. Jr., (2005). Observation of a onedimensional adsorption site on carbon nanotubes: adsorption of alkanes of different molecular lengths. *J. Phys. Chem. B* 109: 20999–21005.

83 Kondratyuk, P. and Yates, J.T. Jr., (2005). Desorption kinetic detection of different adsorption sites on opened carbon single walled nanotubes: the adsorption of *n*-nonane and CCl_4. *Chem. Phys. Lett.* 410: 324–329.

84 Talapatra, S., Zambano, A.Z., Weber, S.E., and Migone, A.D. (2000). Gases do not adsorb on the interstitial channels of closed-ended single-walled carbon nanotube bundles. *Phys. Rev. Lett.* 85: 138–141.

85 Muris, M., Dufau, N., Bienfait, M. et al. (2000). Methane and krypton adsorption on single-walled carbon nanotubes. *Langmuir* 16: 7019–7022.

86 Muris, M., Dupont-Pavlovsky, N., Bienfait, M., and Zeppenfeld, P. (2001). Where are the molecules adsorbed on single-walled nanotubes? *Surf. Sci.* 492: 67–74.

87 Fujiwara, A., Ishii, K., Suematsu, H. et al. (2001). Gas adsorption in the inside and outside of single-walled carbon nanotubes. *Chem. Phys. Lett.* 336: 205–211.

88 Byl, O., Kondratyuk, P., Forth, S.T. et al. (2003). Adsorption of CF_4 on the internal and external surfaces of opened singlewalled carbon nanotubes: a vibrational spectroscopy study. *J. Am. Chem. Soc.* 125: 5889–5896.

89 Pearce, J.V., Adams, M.A., Vilches, O.E. et al. (2005). Onedimensional and two-dimensional quantum systems on carbon nanotube bundles. *Phys. Rev. Lett.* 95: 185302.

90 Bienfait, M., Zeppenfeld, P., Dupont-Pavlovsky, N. et al. (2004). Thermodynamics and structure of hydrogen, methane, argon, oxygen, and carbon dioxide adsorbed on single-wall carbon nanotube bundles. *Phys. Rev. B* 70: 035410.

91 Heroux, L., Krungleviciute, V., Calbi, M.M., and Migone, A.D. (2006). CF_4 on carbon nanotubes: physisorption on grooves and external surfaces. *J. Phys. Chem. B* 110: 12597–12602.

92 Jiang, J.W., Sandler, S.I., Schenk, M., and Smit, B. (2005). Adsorption and separation of linear and branched alkanes on carbon nanotube bundles from configurational-bias Monte Carlo simulation. *Phys. Rev. B* 72: 045447.

93 LaBrosse, M.R., Shi, W., and Johnson, J.K. (2008). Adsorption of gases in carbon nanotubes: are defect interstitial sites important? *Langmuir* 24: 9430–9439.

94 Babaa, M.R., Dupont-Pavlovsky, N., McRae, E., and Masenelli-Varlot, K. (2004). Physical adsorption of carbon tetrachloride on as-produced and on mechanically opened single walled carbon nanotubes. *Carbon* 42: 1549–1554.

95 Rols, S., Johnson, M.R., Zeppenfeld, P. et al. (2005). Argon adsorption in open-ended single-wall carbon nanotubes. *Phys. Rev. B* 71: 155411.

96 Burde, J.T. and Calbi, M.M. (2007). Physisorption kinetics in carbon nanotube bundles. *J. Phys. Chem. C* 111: 5057–5063.

97 Rawat, D.S., Calbi, M.M., and Migone, A.D. (2007). Equilibration time: kinetics of gas adsorption on closed- and open-ended single-walled carbon nanotubes. *J. Phys. Chem. C* 111: 12980–12986.

3

Adsorption Mechanism on Nanomaterials (NMs)

3.1 Introduction

Nanomaterials (NMs) have received growing attention from many scientists due to their superior features such as large surface area, small size, high stability, high reactivity, and versatile chemistry for further functionalization [1–11]. The chemical and physical properties of the materials at nanoscale (1–100 nm) are different compared to the same materials at macroscale. For example, bulk gold particles are yellow, while gold nanoparticles exhibit red color. This feature is originated from the interaction between restricted electrons of the gold nanoparticles with the light. The decrease in the size of a material also leads to increase in the surface area. This is very important especially for the adsorption processes. More binding sites are available on the surface of the NMs, which are used for the adsorption processes.

In the adsorption processes for the environmental samples, chemical and biological approaches are used for the adsorption of undesired compounds from contaminated samples. In biological processes, microorganisms such as fungi, bacteria, algae, etc., have been used [12–14]. Although this technique is low cost and environmentally friendly, it is challenging and not efficient. On the other hand, conventional materials such as clays and lichens as economical adsorbents are applied for the removal of undesired compounds such as heavy metals, pharmaceuticals, and dye pigments from environmental samples [15–18]. However, these microscale materials have low surface area and low adsorption capacities. Thus, the development of innovative and efficient adsorbents for the adsorption processes is crucial. The drawbacks of the conventional adsorbents can be overcome by using NMs.

So far, many research groups have reported the use of NMs (e.g. nanoparticles, carbon nanotubes [CNTs], nanorods, and graphenes) as efficient adsorbents for the adsorption of various compounds in complex matrices such as environmental [19], biological [20], and food samples [21].

In this chapter, an overview of the recent developments and applications of NMs as efficient adsorbents is provided. It starts with a summary of the adsorption phenomena, which is followed by the demonstration of different adsorption isotherms. Then, adsorption kinetics and thermodynamics are described. In Sections 3.2.3 and 3.2.4, the applications and adsorption mechanisms of NMs such as nanoparticles

Sample Preparation with Nanomaterials: Next Generation Techniques and Applications,
First Edition. Chaudhery Mustansar Hussain, Rüstem Keçili, and Chaudhery Ghazanfar Hussain.

(Ag, Au, TiO_2, ZnO nanoparticles) and carbon-based NMs (CNTs, fullerenes, and graphene) are presented.

3.2 Adsorption Process

Adsorption phenomena is a type of surface-based exothermic process in which molecules of a target compound in gas or liquid state are accumulated at the surface of an adsorbent [22]. In the adsorption processes, the compound that is adsorbed to the adsorbent is named as adsorbate. On the other hand, desorption process is the release of the adsorbed molecules from the adsorbent surface that is opposite process of adsorption. Adsorption process occurs in two routes that are "physical adsorption" also called "physisorption" and "chemical sorption" also called "chemisorption," which depends on the interactions between the target compound(s) and the surface. In *physical adsorption*, the weak forces such as electrostatic interactions and Van der Waals forces are involved. The strong chemical bonds such as covalent bonds are formed between the surface and the adsorbed molecules in *chemical adsorption*. Chemical adsorption is slower than physical adsorption, and usually monomolecular layer (monolayer) is formed on the adsorbent surface, while physical adsorption usually involves the formation of thick multilayer on the surface [23].

Table 3.1 shows the differences between physical adsorption and chemical adsorption.

3.2.1 Adsorption Isotherms

An adsorption isotherm refers to a curve of the amount of adsorbed molecules to the adsorbent surface as a function of the partial pressure or concentration of the

Table 3.1 Comparison between physical adsorption and chemical adsorption.

Physical adsorption	Chemical adsorption
Electrostatic interactions and Van der Waals forces are involved	Covalent bonds are formed between the surface and the adsorbed molecules
Fast	Slow
Reversible	Irreversible
Not very specific	It is specific
Multilayers are formed	Monolayers are formed
Activation energy is not required	Activation energy is required
It usually occurs at low temperature values and decreases with increasing temperature	High temperature is needed

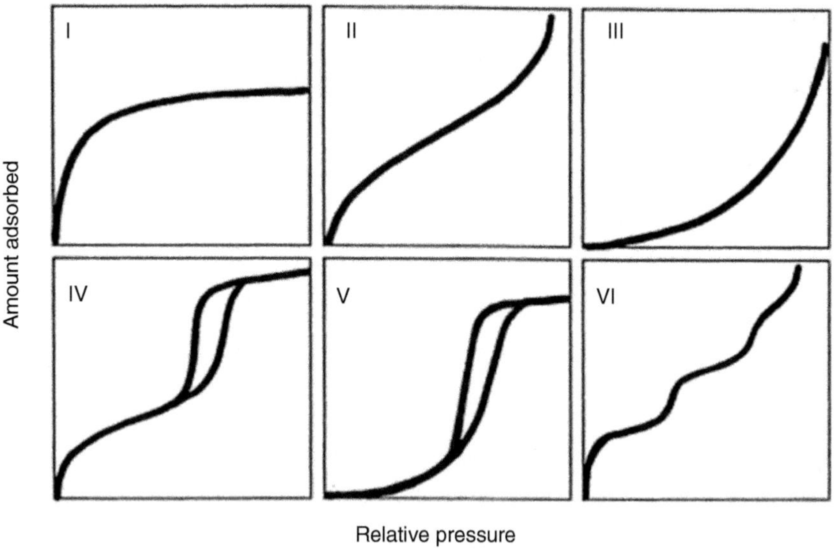

Figure 3.1 Types of adsorption isotherms according to the IUPAC classification. Source: Donohue and Aranovich [25]. © 1998, Elsevier.

adsorbate at a constant temperature [24]. There are six types of adsorption isotherms according to the International Union of Pure and Applied Chemistry (IUPAC) classification for gas adsorption on the surface of solid adsorbents (Figure 3.1) [25]. Type I adsorption isotherms describe the adsorption of gas molecules to the adsorbents having micropores such as activated carbon, and the surface of the adsorbent is covered with monolayer of adsorbed molecules. On the other hand, Type II adsorption isotherm describes the adsorption of gas molecules to the macroporous adsorbents. Type II adsorption isotherm does not display any saturation point as seen in Type I isotherm, and multilayer of adsorbed molecules is formed after the coverage of the adsorbent surface with monolayer adsorbed molecules.

Type III adsorption isotherm refers to multilayer adsorption by weak interactions with low energy between adsorbed molecules and the adsorbent having macropores. Type IV and Type V adsorption models represent the multilayer adsorption together with capillary condensation to the mesoporous adsorbents. Type VI adsorption isotherm describes the stepwise formation of multilayer on the surface of nonporous adsorbent.

Various adsorption isotherm models such as Langmuir, Freundlich, Temkin, Dubinin–Radushkevich, Harkins–Jura and Halsey, Redlich–Peterson, and Brunauer, Emmett, and Teller (BET) are used to get detailed information about the interactions between the surface of the adsorbents and molecules to be adsorbed. These isotherm models are briefly explained in Sections 3.2.1.1–3.2.1.7.

3.2.1.1 Langmuir Isotherm

In 1932, the Langmuir adsorption model was proposed by Irving Langmuir [26], and the main assumptions of this isotherm model are as follows:

- Adsorption occurs at specific binding sites, which are localized on the surface of the adsorbent.
- All adsorption sites on the surface of the adsorbent are identical.
- The surface of the adsorbent is covered with monolayer of adsorbed molecules.
- There is no interaction between the adsorbed molecules on the adsorbent surface.

Langmuir adsorption model can be demonstrated by the following equation.

$$\frac{C_e}{q_m} = \frac{1}{K_L q_{max}} + \frac{C_e}{q_{max}} \tag{3.1}$$

In the aforementioned equation, C_e (mg l^{-1}) and q_m (mg g^{-1}) are the concentration of the molecules at equilibrium and the amount of adsorbed molecules to the surface of the adsorbent at any time, respectively. q_{max} shows the maximum adsorption capacity (mg g^{-1}), and K_L represents the Langmuir constant (l mg^{-1}).

3.2.1.2 Freundlich Isotherm

The Freundlich isotherm model [27] is another approach that is used for the description of multilayer and heterogeneous adsorption of the molecules to the adsorbent surface. This model is demonstrated by Eq. (3.2):

$$\log q_m = \log K_F + \frac{1}{n} \log C_e \tag{3.2}$$

where q_m (mg g^{-1}) represents the amount of adsorbed molecules to the adsorbent surface at any time, C_e (mg l^{-1}) is the equilibrium concentration, n and K_F are the Freundlich constant and Freundlich exponent, respectively. K_F (mg g^{-1}) indicates the adsorption capacity of the adsorbent toward adsorbate, and n is an indicator for the degree of the surface heterogeneity and describes the distribution of the adsorbed molecules to the adsorbent surface. The value of n that is higher than 1 indicates the favorable adsorption of the molecules to the adsorbent surface. The value of higher n reflects the higher intensity of adsorption.

3.2.1.3 Temkin Isotherm

Temkin and Pyzhev proposed another adsorption isotherm in 1940 [28, 29]. In this model, the effect of adsorption heat, which linearly decreases with the coverage of adsorbed molecules layer, is considered. The decrease of the adsorption heat is due to the interactions between the adsorbed molecules on the surface.

The Temkin adsorption isotherm is illustrated by the following equation:

$$q_e = \left(\frac{RT}{b_T}\right) \ln(AC_e) \tag{3.3}$$

where T is the temperature (K), R is the ideal gas constant (8.314 J mol^{-1}K^{-1}), b_T represents the Temkin constant (J mol^{-1}), which depends on the adsorption heat, and A is the equilibrium adsorption constant, which corresponds to the maximum adsorption energy (l mg^{-1}).

3.2.1.4 Dubinin–Radushkevich Model

The adsorption of the molecules into the micropores of the adsorbents is described by the Dubinin–Radushkevich adsorption model [30, 31]. According to this model, molecular adsorption occurs in the micropores (pore filling) instead of adsorption on the surface of the adsorbents that leads to monolayer or multilayer formation. Dubinin–Radushkevich adsorption model is commonly applied to the adsorption processes of the subcritical vapors in the pores of the adsorbents such as zeolites and activated carbons. This model is demonstrated by Eq. (3.4):

$$\ln q_e = \ln q_m - \beta \varepsilon^2 \tag{3.4}$$

where q_e is the amount of adsorbed molecules per mass of adsorbent (mg g^{-1}), q_m represents the adsorption capacity (mg g^{-1}), β is the activity coefficient (mol^2 J^{-2}), which represents the adsorption energy, and ε demonstrates the Polanyi potential as given in the following formula:

$$\varepsilon = RT \ \ln\left(1 + \frac{1}{C_e}\right) \tag{3.5}$$

where R is the ideal gas constant (J mol^{-1} K^{-1}), T is the temperature (K), and C_e is the concentration of the molecules at equilibrium (mg l^{-1}).

3.2.1.5 Harkins–Jura and Halsey Isotherms

These types of isotherms are generally applied for the description of the multilayer adsorption processes to the adsorbent surfaces having heterogeneous pore distribution [32]. Harkins–Jura and Halsey isotherms are described in Eq. (3.6):

$$\left(\frac{1}{qe^2}\right) = \left(\frac{B_{HJ}}{A_{HJ}}\right) - \left(\frac{1}{A_{HJ}}\right) \log C_e \tag{3.6}$$

where B_{HJ} and A_{HJ} represent the Harkins–Jura constants, which can be determined using the linear plot of $\left(\frac{1}{qe^2}\right)$ vs. log C_e.

On the other hand, Halsey isotherm [33] is demonstrated by Eq. (3.7). This model is also applied for the multilayer adsorption processes to the heterogeneous surface of the adsorbents.

$$\ln q_e = \left(\frac{1}{n_H} \ln K_H\right) - \left(\frac{1}{n_H} \ln \frac{1}{C_e}\right) \tag{3.7}$$

where n_H and K_H are the Halsey constants and can be calculated using the slope of the linear plot of ln q_e vs. ln $\frac{1}{C_e}$ for n_H and the intercept of the linear plot of ln q_e vs. ln $\frac{1}{C_e}$ for K_H.

3.2.1.6 Redlich–Peterson Isotherm

Redlich–Peterson isotherm [34] is a combined model that exhibits the features of both Langmuir and Freundlich isotherm models. It is used to the describe the adsorption process to the homogenous and heterogeneous surfaces of adsorbents over a wide range of adsorbate concentrations. This isotherm model shows the features of Freundlich isotherm model at higher adsorbate concentrations, while it obeys the Henry's law at lower adsorbate concentrations. This model is demonstrated by Eq. (3.8):

$$q_e = \frac{K_{RP}C_e}{1 + (\alpha C_e)^{\beta}} \tag{3.8}$$

where K_{RP} ($l\,g^{-1}$) and α ($l\,mg^{-1}$) are Redlich–Peterson constants, β is the exponential value in the range between 0 and 1. If the β value is 0, the isotherm behaves as Langmuir model, while it obeys Henry's law when the β value is 1.

3.2.1.7 BET (Brunauer, Emmett, and Teller) Isotherm

The BET isotherm model is another milestone in the progress of the adsorption science [35–37]. This model represents the multilayer adsorption of the molecules to the adsorbent surface and the assumptions of the BET isotherm are same as for the Langmuir isotherm model. This isotherm model is widely used for the adsorption processes in the gas–solid systems, and the surface area and the porosity of the materials can be determined by using the BET isotherm model. The following equation is used for the description of the BET isotherm model.

$$q_e = \frac{q_s C_{BET} C_e}{(C_s - C_e)\left[1 + (C_{BET} - 1)\frac{C_e}{C_s}\right]} \tag{3.9}$$

where C_e ($mg\,l^{-1}$) is the equilibrium concentration, C_s ($mg\,l^{-1}$) is the saturation concentration, q_e is the amount of adsorbed molecules per mass of adsorbent ($mg\,g^{-1}$) at equilibrium, q_m represents the theoretical saturation capacity ($mg\,g^{-1}$), and C_{BET} ($l\,mg^{-1}$) represents the energy of surface interaction.

3.2.2 Adsorption Kinetics and Thermodynamics

There are three main steps during the adsorption of a compound to the surface of the adsorbent. In the first step, molecular mass transfer from the solution to the adsorbent surface occurs. Then, internal molecular diffusion to the adsorption sites placed on the adsorbent takes place, and the adsorption is then completed in the final step of the process. The investigation of adsorption kinetics and thermodynamics is very important to estimate the adsorption mechanism. For this purpose, the adsorption kinetics models such as the pseudo-first-order, the pseudo-second-order, and intraparticle diffusion models are widely applied to the adsorption systems.

3.2.2.1 Pseudo-first-order Kinetics

The pseudo-first-order model was first proposed by Lagergren [38]. The main assumptions of this kinetic model are as follows:

- Adsorption only occurs at specific binding sites, which are localized on the surface of the adsorbent.
- Adsorption energy does not depend on the formation of layer on the adsorbent surface.
- The first-order equation governs the rate of the adsorption process.
- No interaction occurs between the adsorbed molecules on the surface of the adsorbent.

The equation for the pseudo-first-order kinetic model is given in the following.

$$\ln(q_1 - q_t) = \ln q_1 - k_1 t \tag{3.10}$$

where q_1 represents the amount of adsorbed compound (mg g^{-1}) to the adsorbent at equilibrium, q_t represents the amount of adsorbed compound (mg g^{-1}) to the adsorbent at time t (min), and k_1 (min^{-1}) is the rate constant for the first-order adsorption process.

3.2.2.2 Pseudo-second-order Kinetics

The pseudo-second-order kinetic model has almost the same assumptions as for pseudo-first-order model. The only difference is the rate of the model [39]. It is demonstrated by Eq. (3.11):

$$\frac{t}{q_t} = \frac{1}{k_2 q_2^2} + \frac{1}{q_2} t \tag{3.11}$$

3.2.2.3 Intraparticle Diffusion Model

This model is mainly used for the description of the adsorption process if the adsorbate diffusion to the porous adsorbent is the rate-limiting step. Weber and Morris [40] have proposed this kinetic model and described in the following formula:

$$q_t = k_{id} t^{0.5} + C \tag{3.12}$$

where C is the intercept and represents the boundary layer thickness. The higher C value, the higher effect of boundary layer. q_t represents the amount of adsorbed compound (mg g^{-1}) to the adsorbent at time t (min), and k_{id} is the rate constant of the intraparticle diffusion (mg g^{-1} $min^{0.5}$).

3.2.2.4 Thermodynamic Study

Thermodynamic studies are performed to investigate the temperature effect on the adsorption process. For this purpose, Eqs. (3.13–3.15) are used to calculate various thermodynamic parameters such as enthalpy change (ΔH^0), Gibbs free energy change (ΔG^0), and entropy change (ΔS^0), which are the main characteristics of the adsorption process [41]. The nature of the adsorption process can be predicted by calculating these thermodynamic parameters.

$$\Delta G^0 = -RT \ln K_d \tag{3.13}$$

$$\Delta G^0 = \Delta H^0 - T\Delta S^0 \tag{3.14}$$

$$\ln K_d = \frac{\Delta S^0}{R} - \frac{\Delta H^0}{RT} \tag{3.15}$$

where T is the temperature (K), R represents the ideal gas constant (8.314 J mol^{-1} K^{-1}), and K_d is the equilibrium constant.

ΔG^0 value indicates whether the adsorption process is *spontaneous or not*. If ΔG^0 value is negative, the adsorption takes place *spontaneously. On the other hand,* ΔH^0 gives information about whether the adsorption process is endothermic or not. *The adsorption process is endothermic when the* ΔH^0 is positive value.

3.2.3 Adsorption Process on Nanoparticles

Nanoparticles including metallic nanoparticles (e.g. Ag and Au), metal oxide nanoparticles (e.g. ZnO and TiO_2), and magnetic nanoparticles (MNPs; e.g. Fe_3O_4) are excellent candidates as efficient nanoadsorbents due to their unique properties such as high reactivity, small size, and high surface-area-to-volume ratio, which provide high adsorption capacities toward target compound. The applications of these nanoadsorbents are briefly described in Sections 3.2.3.1–3.2.3.4.

3.2.3.1 Silver Nanoparticles

Silver nanoparticles (Ag NPs) are commonly used for the removal of environmental pollutants such as heavy metals and industrial dyes. Many studies demonstrating the use of Ag NPs for the adsorption processes have been reported. In one of these published studies, efficient removal of Fe(II) ions from aqueous solutions was performed by Asthana et al. [42]. For this purpose, they prepared calcium alginate beads entrapped with Ag NPs and tested their adsorption performance toward Fe(II) ions. Langmuir, Freundlich, Temkin, and Harkins–Jura adsorption isotherm models were applied for the characterization of the adsorption process. The pseudo-first-order, pseudo-second-order, Elovich kinetic models and intraparticle diffusion model were also used for the investigation of the adsorption kinetics. The obtained results showed that adsorption of Fe(II) ions by the prepared Ag NPs-modified adsorbent is well fitted to the Langmuir isotherm model and the pseudo-second-order kinetic model. The adsorption capacity of the adsorbent was calculated as $236.40\,mg\,g^{-1}$.

Gari and Kim prepared graphene oxide (GO) decorated with Ag NPs and used as a nanocomposite adsorbent for the removal of Pb(II) ions from aqueous solutions [43]. The obtained results demonstrated that adsorption process is well described by the Langmuir isotherm model and the pseudo-second-order kinetic model. The adsorption capacity of the prepared nanoadsorbent was found as $312.57\,mg\,g^{-1}$.

In another study [44], Ghaedi et al. prepared activated carbon modified with Ag NPs and used for the adsorption of direct yellow 12 dye from aqueous solutions. The maximum adsorption of dye (c. 98%) was obtained in 10 minutes at pH 1.0, and the prepared adsorbent exhibited high adsorption capacity ($454.54\,mg\,g^{-1}$) toward direct yellow 12. The studies on the adsorption kinetics confirmed that adsorption of direct yellow 12 to activated carbon modified with Ag NPs is well described by pseudo-second-order kinetic model.

Another textile dye methyl orange was efficiently removed by using activated carbon coated with Ag NPs [45]. The adsorption capacity of the prepared adsorbent was found to be $27.48\,mg\,g^{-1}$, and the results obtained from the experiments for the adsorption kinetics showed that adsorption process follows the pseudo-second-order kinetic model.

On the other hand, Ag NPs are efficient antibacterial materials due to their high toxicity toward various microorganisms such as bacteria and viruses. Thus, they are commonly used for the removal of microorganisms from the contaminated water. The adsorbent materials are generally coated with Ag NPs instead of direct use of the Ag NPs in the treatment of water samples due to the agglomeration problems of the nanoparticles. Dankovich and Gray [46] prepared antibacterial filter paper containing Ag NPs and used for the removal of *Escherichia coli* and *Escherichia faecalis* from the contaminated water. The obtained results showed that filter paper containing Ag NPs displayed antibacterial features toward *E. coli* and *E. faecalis*. The authors reported that the prepared Ag NPs-embedded filter paper could be used as an efficient antibacterial material for the treatment of contaminated water.

Ceramic filters modified with Ag NPs were also applied for the treatment of water. For example, Kallman et al. successfully applied porous ceramic filters containing Ag NPs for the efficient removal of *E. coli* [47]. In another study, Vinka and coworkers reported the use of ceramic filters modified with colloidal Ag NPs for the treatment of *E. coli*-contaminated water samples [48]. The obtained results showed that incorporation of Ag NPs increased the removal performance of the ceramic filter and the prepared Ag NPs-embedded ceramic filter exhibits efficient removal performance toward *E. coli* (over 97.8%).

3.2.3.2 Gold Nanoparticles

Gold nanoparticles (Au NPs) are efficient metallic nanoadsorbents similar to Ag NPs due to their easy preparation by using reducing agents such as hydroxyl amine and citrate. Au NPs can also be incorporated into the various adsorbents to increase the adsorption efficiency.

In a study reported by Roosta et al. [49], methylene blue was successfully removed from the aqueous solutions by using activated carbon containing Au NPs. The prepared adsorbent exhibited high affinity toward target dye compound, and the adsorption capacity was found to be 185.18 $mg\,g^{-1}$. Different adsorption isotherm models such as Freundlich, Langmuir, Dubinin-Radushkevich, and Tempkin models and adsorption kinetics models such as pseudo-first-order, pseudo-second-order, Elovich and intraparticle diffusion models were applied for the investigation of adsorption mechanism and kinetics. The obtained results from these investigations showed that the adsorption of methylene blue to the activated carbon modified with gold nanoparticles is well described by the Langmuir adsorption model and pseudo-second-order kinetic model.

Same research group [50] prepared activated carbon modified with Au NPs and used for the removal of alizarin red S dye from aqueous solutions. The prepared adsorbent containing Au NPs exhibited high adsorption capacity (123.4 $mg\,g^{-1}$) toward alizarin red S. It has been reported that c. 15 mg of the adsorbent is sufficient for the efficient removal of dye (over 95%) within five minutes. The obtained results from the adsorption isotherms and kinetics showed that adsorption process is well demonstrated by the Langmuir adsorption model and pseudo-second-order kinetic model.

In another study [51], Solis et al. prepared Au NPs-modified silica particles for the removal of inorganic mercury(II) from aqueous solutions. The obtained results showed that adsorption of mercury(II) ions to the Au NPs-modified silica nanocomposite adsorbent is well fitted to the pseudo-second-order kinetic model, and 96% of the mercury(II) ions was successfully removed.

Say's group has prepared gold–silver nanoclusters having dipicolinic acid (DPA) imprinted nanoshell for selective adsorption of *Bacillus cereus* spores [52]. For this purpose, DPA and methacryloyl iminodiacetic acid–chromium(III) [MAIDA–Cr(III)] were used as the template and metal-chelate functional monomer, respectively. The obtained results from fluorescence measurements showed that the prepared nanoadsorbent exhibits high affinity and selectivity toward DPA. Langmuir and Scatchard adsorption isotherms were applied to describe adsorption behavior of the prepared nanoadsorbent toward target DPA. $K_{affinity}$ values were 2.9×10^6 and $2 \times 10^6\ M^{-1}$, respectively.

3.2.3.3 Zinc Oxide Nanoparticles

Metal oxide nanoparticles such as Zinc oxide (ZnO) nanoparticles have large surface area and exhibit chemical, thermal stability, and antimicrobial feature. Therefore, ZnO nanoparticles are also used as efficient nanoadsorbents. Dehaghi et al. have prepared ZnO-modified chitosan particles and used for the removal of permethrin (a neurotoxin pesticide) from water samples [53]. The obtained results showed that permethrin was successfully removed with a 96% efficiency by using a small amount of ZnO-modified chitosan adsorbent (500 mg).

In another study, activated silica having ZnO nanoparticles was applied for the removal of arsenic(III) ions from aqueous solutions [54]. The results from adsorption isotherms showed that the adsorption of arsenic(III) ions is well fitted to the BET and Langmuir isotherms compared to the Temkin and Freundlich models. The kinetic data indicated that adsorption process follows the pseudo-second-order kinetic model. It was reported that 3.5 g of the prepared adsorbent is sufficient the 99% removal of arsenic(III) ions from aqueous solutions.

Salmani's group has prepared ZnO nanoparticles for the adsorption of cadmium(II) ions from aqueous solutions [55]. The obtained results indicated that adsorption process is well demonstrated by the Langmuir isotherm model and the pseudo-second-order kinetic model.

In a study reported by Sasidharan et al. [56], colloidal ZnO nanoparticles were used for the adsorption of bovine serum albumin (BSA). The results showed that adsorption of BSA to colloidal ZnO nanoadsorbent follows the Freundlich isotherm model and the pseudo-second-order kinetic model.

3.2.3.4 Magnetic Fe_3O_4 Nanoparticles

MNPs are interesting materials that are widely used in many applications such as separation [57], catalysis [58], and drug delivery [59]. MNPs can easily be separated from the complex matrices such as environmental and biological samples by applying an external magnetic field. Thus, no centrifugation or filtration steps are required. Although MNPs exhibit excellent features, there are some drawbacks such as agglomeration of the particles and stability issues. Bare metallic nanoparticles are chemically quite active and easily oxidized that lead to loss of their magnetic features. Thus, these nanoparticles are needed to be coated with a layer (e.g. carbon, silica, or polymers) to protect their stability.

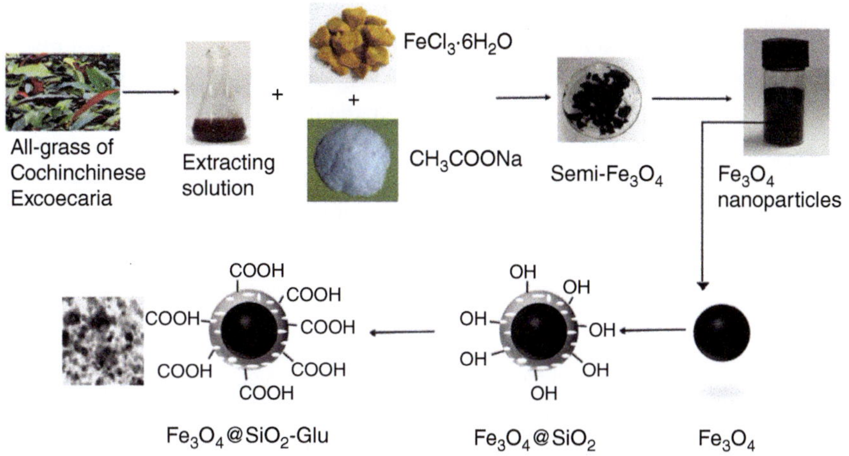

Figure 3.2 The schematic depiction of the preparation of magnetic Fe_3O_4@SiO_2 nanoparticles toward doxorubicin hydrochloride. Source: Reproduced with permission from Cai et al. [62]. © 2019, Elsevier.

Recently, various magnetic Fe_3O_4 nanoparticles have been prepared as efficient adsorbents and used for the adsorption of heavy metals and dyes from aqueous samples. For example, Qi et al. prepared magnetic Fe_3O_4 nanoparticles doped with Ce(III) ions for the removal of antimony(III) ions from aqueous solutions [60]. The prepared magnetic nanoadsorbent exhibit high adsorption behavior toward target antimony(III) ions, and the maximum adsorption capacity was obtained as 224.2 mg g^{-1}.

In another reported study [61], reactive blue 19 dye was successfully removed from wastewater using magnetic Fe_3O_4 nanoparticles modified with L-arginine. The maximum dye removal (96.34%) was achieved at pH 3.0. The obtained data showed that adsorption of reactive blue 19 dye to the magnetic nanoadsorbent follows the pseudo-second-order kinetic model and the Freundlich isotherm model.

Cai et al. reported the development of glutaric anhydride (Glu)-functionalized MNPs composed of Fe_3O_4 and SiO_2 for the efficient adsorption of an anticancer drug doxorubicin hydrochloride [62]. Figure 3.2 shows the schematic depiction of the preparation of magnetic Fe_3O_4@SiO_2 nanoparticles functionalized with Glu toward the target compound doxorubicin hydrochloride. The prepared Glu-functionalized magnetic Fe_3O_4@SiO_2 nanoparticles were efficiently applied for the adsorption of doxorubicin hydrochloride. The results confirmed that the achieved adsorption efficiency of doxorubicin was 92% at the concentration of 20 mg l^{-1} using a very small amount of MNPs (10 mg).

In a study carried out by Fu et al. [63], magnetic Fe_3O_4@chitosan nanoparticles grafted with polyethylenimine were prepared for the effective removal of phosphate from aqueous samples. The obtained results showed that the adsorption of phosphate was well described by the Langmuir adsorption isotherm, and the maximum adsorption was achieved as 48.2 mg g^{-1} at pH 3.0.

Combination of magnetic Fe_3O_4 nanoparticles and molecularly imprinted polymers (MIPs) as composite NMs is also a great choice for the development of magnetic nanocomposite materials. For example, Chen et al. reported the preparation of MIP-based magnetic Fe_3O_4 nanoparticles for the efficient extraction of human serum albumin (HSA) in clinical human serum samples [64]. In their study, the prepared MIP-based MNPs exhibited excellent recognition performance toward the target protein HSA in serum samples. The obtained maximum adsorption capacity was 70.2 mg g^{-1}.

In a study conducted by Yang et al., core–shell magnetic MIPs were prepared for selective removal of indole from fuel oil [65]. In their research, magnetic Fe_3O_4 nanoparticles were synthesized by using coprecipitation technique. Then, surface of the prepared nanoparticles was coated with SiO_2 using 3-(trimethoxysilyl) propyl methacrylate. In the final step, methacrylic acid (MAA) as the functional monomer and ethylene glycol dimethacrylate (EGDMA) as the cross-linker were used for the preparation of selective MIP toward indole on the surface of the modified MNPs. The results obtained from rebinding studies showed that the prepared magnetic MIP composite shows high affinity and selectivity toward indole. The binding capacity of the composite for indole was found to be 50.25 mg g^{-1}.

Zhang et al. prepared MIP-based magnetic nanocomposites for Pb^{2+} extraction from environmental samples [66]. 3-(2-Aminoethylamino)propyltrimethoxysilane (AAPTS) as the functional monomer, tetraethyl orthosilicate (TEOS) as the cross-linker, and template ion Pb^{2+} were used for the preparation of nanocomposites toward Pb^{2+} ions. Different factors that effect the selective extraction of target ion such as pH and sample volume were studied. The results obtained from solid phase extraction (SPE) studies for Pb^{2+} in the presence of other potential interfering ions such as Zn^{2+}, Cd^{2+}, and Hg^{2+} in environmental samples showed that the prepared MIP-based magnetic nanocomposites display high selectivity toward Pb^{2+} ions. The adsorption capacity of the nanocomposites for Pb^{2+} ions was determined as 19.61 mg g^{-1}.

3.2.4 Adsorption Process on Carbon Nanomaterials

Carbon NMs such as activated carbon, CNTs, and GO have been widely applied for the adsorption processes since they exhibit high stability, low density, and large surface area that enhance adsorption capacity. In addition, carbon NMs can easily be modified with different functional groups. These superior features of the carbon NMs make them as excellent nanoadsorbents for the efficient removal of various contaminants from environmental samples.

3.2.4.1 Activated Carbon

Activated carbon as the conventional adsorbent was widely applied for the adsorption processes.

In one of these applications, Tang et al. prepared activated carbon–alginate composite beads for the efficient removal of Bisphenol AF [67]. The obtained results from the kinetic experiments showed that adsorption of Bisphenol AF to activated carbon-based composite adsorbent is well fitted to the pseudo-second-order model. The maximum adsorption of the prepared activated carbon–alginate composite beads was found as 284.6 $mg\,g^{-1}$.

In another reported study [68], L-phenyl alanine was successfully removed from aqueous solutions by using activated carbons. For this purpose, two different activated carbons were prepared. One of them was activated by NaOH and the other one was activated by $ZnCI_2$. Both adsorbents exhibited high adsorption capacity (183.3 and 133.3 $mg\,g^{-1}$, respectively). The data from the thermodynamic studies and adsorption isotherms showed that the adsorption process was exothermic and well demonstrated by the Langmuir model.

Saleh et al. have reported the use of waste tire-based activated carbon having polyethylenimine for the Hg(II) adsorption from aqueous samples [69]. The prepared adsorbent exhibited good adsorption behavior toward target Hg(II) ions. The results from the adsorption thermodynamic and kinetic experiments confirmed that the adsorption of Hg(II) ions to the activated carbon-based adsorbent is an exothermic process and well demonstrated by the pseudo-second-order model. The maximum adsorption capacity was determined as 16.39 $mg\,g^{-1}$.

The removal of carbamazepine, an antiepileptic pharmaceutical compound, from wastewater sampled was successfully achieved by using magnetic activated carbon including Fe_3O_4 [70]. The obtained kinetic data indicated that adsorption process is exothermic and spontaneous. The prepared magnetic nanocomposite adsorbent exhibited high adsorption capacity (182.9 $mg\,g^{-1}$) toward target carbamazepine.

3.2.4.2 Carbon Nanotubes (CNTs)

In 1991, Japanese physicist Sumio Lijima discovered the CNTs [71]. Since then, many scientists worked on the different applications of CNTs. CNTs are composed of graphene sheets, which were rolled into a cylindrical shape and categorized into two groups, which are single-walled carbon nanotubes (SWCNTs) and multiwalled carbon nanotubes (MWCNTs). SWCNTs are composed of one graphene layer, while MWCNTs have two or more layers.

CNTs have been widely applied as efficient nanoadsorbents for the removal of various contaminants (e.g. metal ions, dyes, phenolic compounds, drug substances, etc.) from aqueous solutions. In a study reported by Yang et al. [72], MWCNTs were prepared for the adsorption of Ni(II) ions in aqueous solutions. The adsorption kinetic studies showed that the Ni(II) adsorption to the MWCNTs is well demonstrated by the pseudo-second-order model.

Lu and Chiu [73] compared the adsorption behaviors of the activated carbon, SWCNTs, and MWCNTs toward Zn(II) ions in water samples. The adsorption capacities were found as 13.04, 43.66, and 32.68 $mg\,g^{-1}$, respectively.

The wall of the CNTs exhibits hydrophobic feature due to the π electron density. Thus, hydrophobic interactions could occur between the dye molecules and the wall of the CNTs. Considering this, MWCNTs were also applied for the removal of dye compounds such as direct blue 86 and direct dye 224 dyes [74]. The maximum adsorption capacity values of the prepared MWCNTs toward direct blue 86 and direct dye 224 were 56.2 and 61.3 mg g^{-1}, respectively. The obtained results from the isotherm models and thermodynamic studies indicated that the Freundlich model well demonstrates the adsorption process, and the adsorption type is physical adsorption.

In another crucial report, Sadegh et al. [75] designed and prepared MWCNTs modified with carboxylate groups for the efficient removal of textile dye malachite green from aqueous samples. The results showed that the mechanism of the adsorption process follows the Langmuir isotherm model, and the adsorption capacity of the prepared MWCNTs having carboxylate groups toward the target dye malachite green was found as 49.45 mg g^{-1}.

Magnetic CNTs are also excellent choice for the adsorption processes due to the easy separation of the magnetic adsorbent from the sample solutions. Magnetic MWCNTs were successfully used for the adsorption of Janus green, crystal violet, methylene blue, and thionine dye [76]. The obtained adsorption capacities of the prepared magnetic adsorbent toward target dyes were found as 250, 227.7, 48.1, and 36.4 mg g^{-1}, respectively.

Yan et al. prepared magnetic MWCNTs grafted with guar gum for the efficient removal of dye compounds methylene blue and natural red [77]. The obtained results from the kinetic studies showed that adsorption of dyes is well described by the pseudo-second-order model. The maximum capacity of the prepared magnetic nanoadsorbent was found as 89.85 and 61.92 mg g^{-1} for natural red and methylene blue, respectively.

In another study reported by Zhang et al. [78], magnetic MIP-based MWCNTs composite material was prepared for the efficient removal of bisphenol A (BPA) in aqueous samples. The achieved results from rebinding experiments for BPA in batch mode displayed that the prepared magnetic MIP based-MWCNT composite exhibited great affinity and selectivity toward BPA. The maximum binding capacity for the target for BPA was obtained as 49.26 μmol g^{-1}.

In a research conducted by Yang et al. [79], removal of Cu^{2+} ions from aqueous samples was successfully performed by using the prepared magnetic CNTs (Figure 3.3). The adsorption isotherms of Cu^{2+} ions in aqueous samples are well characterized by Langmuir, Freundlich, and Dubinin–Radushkevich adsorption isotherms models in the temperature range between 303 and 323 K. The achieved maximum adsorption capacity was 23.4 mg g^{-1}.

3.2.4.3 Graphene Oxide (GO)

GO is another carbon-based material that displays excellent features such as large surface area, high stability, and layered structure. GO is prepared by the oxidation of graphene. Graphene shows hydrophobic property that leads to agglomeration problem in aqueous solutions. The oxidation of graphene to GO enhances the dispersion behavior due to the polar groups such as hydroxyl, carboxyl, and epoxide. GO and GO-based composites have already been used as efficient nanoadsorbents for the removal of metal ions [80], dye compounds [81], and pharmaceutical compounds [82] from environmental samples. For example, Raghubanshi et al. prepared GO for the efficient removal of Pb(II) ions from aqueous solutions [83]. The highest

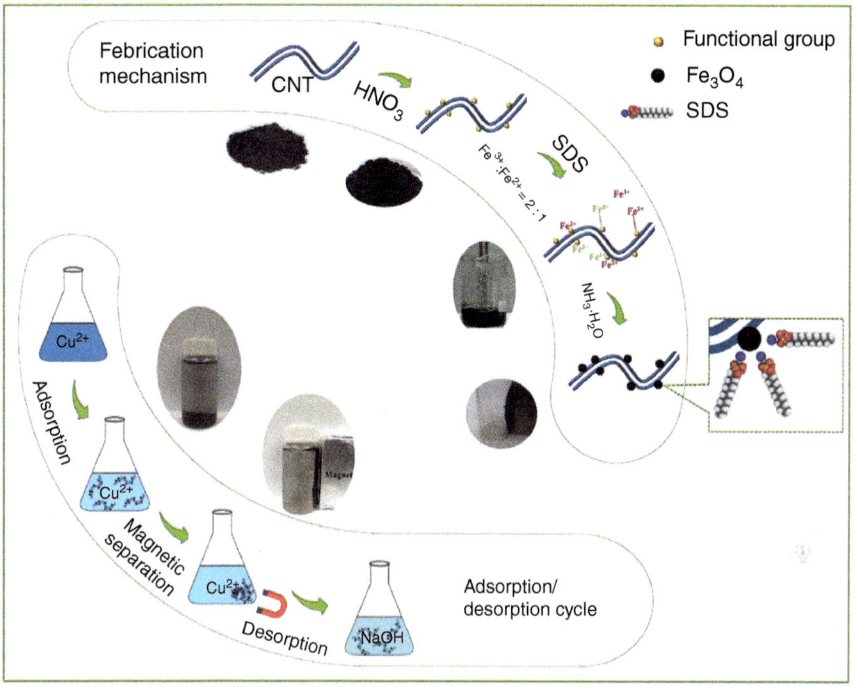

Figure 3.3 The schematic representation of the preparation of magnetic CNTs toward Cu^{2+} ions and adsorption process. Source: Reproduced with permission from Yang et al. [79]. © 2017, Elsevier.

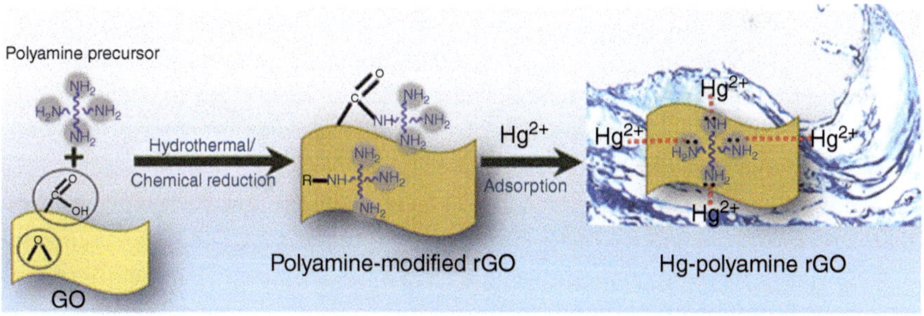

Figure 3.4 The schematic demonstration of the preparation of polyamine-modified rGO toward Hg^{2+} ions. Source: Yap et al. [84]. © 2019, Elsevier.

adsorption capacity of the prepared GO adsorbent was obtained as 120 mg g^{-1}, and the adsorption kinetic was well described by the pseudo-second-order model.

In a research carried out by Yap et al. [84], the removal of Hg^{2+} ions from water samples was successfully performed by using the developed reduced graphene oxide (rGO) modified with polyamine (Figure 3.4). The achieved results indicated that the adsorption of Hg^{2+} ions to rGO having polyamine is well described by the Freundlich isotherm. The maximum adsorption capacity of the prepared polyamine-modified rGO toward the target Hg^{2+} ions was obtained as 63.8 mg g^{-1}.

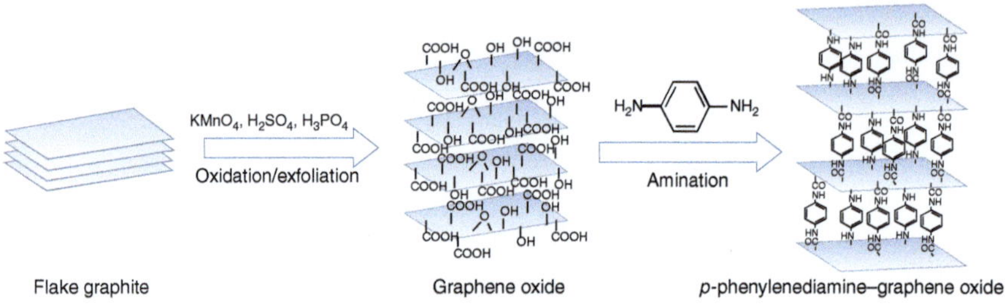

Figure 3.5 The schematic representation of the prepared graphene oxide modified with *p*-phenylenediamine. Source: Wu et al. [86]. © 2016, Elsevier.

In a research performed by Xing et al. [85], Cs(I) ions were efficiently removed from aqueous samples using GO. The achieved results exhibited that the adsorption of Cs(I) ions to the GO was a pH, strength-dependent and exothermic process. In addition, Cs(I) adsorption was well demonstrated by the Langmuir isotherm and pseudo-second-order kinetic model. The obtained adsorption capacity was 95.46 $mg\,g^{-1}$.

Another interesting work was reported by Wu et al. [86]. In their study, a graphene-based nanoadsorbent was developed for the efficient removal of textile dye Congo red from aqueous solutions. For this purpose, *p*-phenylenediamine was incorporated into the nanolayers of graphene as shown in Figure 3.5. The maximum dye removal was achieved at pH 3.0, and the adsorption capacity of the prepared nanoadsorbent was obtained as 892.8 $mg\,g^{-1}$. The adsorption of Congo red dye to the prepared graphene-based nanoadsorbent was well demonstrated by the Langmuir model, and the results obtained from the thermodynamic studies showed that the adsorption process is endothermic and spontaneous.

In another study, zeolite–GO nanocomposite adsorbent was developed for the adsorption of Rhodamine B dye [87]. The maximum adsorption capacity of the nanocomposite was obtained as 64.47 $mg\,g^{-1}$, and the adsorption process followed the Langmuir and Freundlich isotherm models.

Magnetic GO nanocomposites were also successfully applied for the adsorption of dye compounds. Magnetic Fe_3O_4 nanoparticles–sulfonated GO nanocomposite adsorbent was prepared by Zhang et al. and used for the removal of malachite green from aqueous solutions [88]. The results showed that the prepared GO-based magnetic nanoadsorbent exhibits high adsorption capacity toward target dye compound. It has been reported that over 98% of malachite green could efficiently be removed by using very small amount of the magnetic GO nanocomposite (48 mg) within a short time (20 minutes).

Raghua et al. prepared magnetic GO nanocomposites for the removal of eriochrome black T and malachite green from wastewater [89]. The prepared magnetic nanoadsorbent exhibited high adsorption behavior toward eriochrome black T and malachite green with a high adsorption capacity of 160.8 and 179.15 $mg\,g^{-1}$, respectively. The results from the kinetic experiments indicated that the adsorption process follows the pseudo-second-order model. The antibacterial activities of the prepared magnetic GO nanocomposites were also tested toward various bacteria such as *Bacillus subtilis*, *Staphylococcus aureus*, *Salmonella typhimurium*, and *Pseudomonas aeruginosa*. The obtained results showed that the prepared magnetic GO nanocomposites could also be used as a promising antibacterial material for the treatment of contaminated water.

3.3 Conclusions and Future Perspective

The growing number of published research in which nanoadsorbents have been applied for different purposes exhibited that these unique materials are promising adsorbents for the adsorption of many target compounds such as heavy metal, textile dyes, pharmaceutical compounds, organic and inorganic compounds, etc. The demonstrated examples in this chapter highlight the recent progresses in adsorption processes using nanostructured materials over the past years. The designed novel adsorbents in nanoscale as promising materials provide a new approach for efficient adsorption processes in complex matrices such as environmental and biological samples.

In this chapter, we have only focused on the recent reported examples in which NMs were used for the adsorption processes. It is worth to note that these materials, which have superior features such as large surface area, small size, high stability, high reactivity, and versatile chemistry for further functionalization, have recently been applied in many fields (e.g. catalysis, drug delivery, biosensor systems, etc.).

References

1 Büyüktiryaki, S., Keçili, R., and Hussain, C.M. (2020). Functionalized nanomaterials in dispersive solid phase extraction: advances and prospects. *Trends Anal. Chem.* 127: 115893.

2 Hussain, C.M. and Kharisov, B. (2017). *Advanced Environmental Analysis-Application of Nanomaterials*. The Royal Society of Chemistry.

3 Hussain, C.M. (2017). Magnetic nanomaterials for environmental analysis. In: *Advanced Environmental Analysis-Application of Nanomaterials* (eds. C.M. Hussain and B. Kharisov), 1–13. The Royal Society of Chemistry.

4 Hussain, C.M. (2018). *Nanomaterials in Chromatography: Current Trends in Chromatographic Research Technology and Techniques*. Elsevier.

5 Hussain, C.M. (2019). *Handbook of Nanomaterials in Analytical Chemistry: Modern Trends in Analysis*. Elsevier.

6 Büyüktiryaki, S., Sümbelli, Y., Keçili, R., and Hussain, C.M. (2019). Lab-on-chip platforms for environmental analysis. In: *Encyclopedia of Analytical Science*, 3e (eds. P. Worsfold, C. Poole, A. Townshend and M. Miró), 267–273. Academic Press.

7 Keçili, R., Büyüktiryaki, S., and Hussain, C.M. (2019). Advancement in bioanalytical science through nanotechnology: past, present and future. *Trends Anal. Chem.* 110: 259–276.

8 Hussain, C.M. and Keçili, R. (2020). *Modern Environmental Analysis Techniques for Pollutants*. The Netherlands: Elsevier.

9 Keçili, R., Büyüktiryaki, S., and Hussain, C.M. (2020). Micro total analysis systems with nanomaterials, Chapter 8. In: *Handbook of Nanomaterials in Analytical Chemistry* (ed. C.M. Hussain), 185–198. Elsevier.

10 Pyrzynska, K. (2020). Nanomaterials in speciation analysis of metals and metalloids. *Talanta* 212: 120784.

11 Murtada, K. (2020). Trends in nanomaterial-based solid-phase microextraction with a focus on environmental applications – a review. *Trends Environ. Anal. Chem.* 25: e00077.

12 Tan, L., Ning, S., Zhang, X., and Shi, S. (2013). Aerobic decolorization and degradation of azo dyes by growing cells of a newly isolated yeast *Candida tropicalis* TL-F1. *Bioresour. Technol.* 138: 307–313.

13 Gül, U.D. (2013). Treatment of dyeing wastewater including reactive dyes (Reactive Red RB, Reactive Black B, Remazol Blue) and Methylene Blue by fungal biomass. *Water SA* 39 (5): 593–598.

14 Kelewou, H., Merzouki, M., and Lhassani, A. (2014). Biosorption of textile dyes Basic Yellow 2 (BY2) and Basic Green 4 (BG4) by the live yeast *S. cerevisiae*. *J. Mater. Environ. Sci.* 5 (2): 633–640.

15 Hsu, Y.C., Chiang, C.C., and Yu, M.F. (1991). Adsorption behavior of basic dyes on activated clay. *Sep. Sci. Technol.* 32 (15): 2513–2534.

16 Rytwo, G., Nir, S., Crespin, M., and Margulies, L. (2000). Adsorption and interactions of methyl green with montmorillonite and sepiolite. *J. Colloid Interface Sci.* 222 (1): 12–19.

17 Kulkarni, A.N., Kadam, A.A., Kacholec, M.S., and Govindward, S.P. (2014). Lichen *Permelia perlata*: a novel system for biodegradation and detoxification of disperse dye Solvent Red 24. *J. Hazard. Mater.* 276: 461–468.

18 Uluozlu, O.D., Sarı, A., and Tuzen, M. (2010). Biosorption of antimony from aqueous solution by lichen (*Physcia tribacia*) biomass. *Chem. Eng. J.* 163: 382–388.

19 Azodi-Deilami, S., Najafabadi, A.H., Asadi, E. et al. (2014). Magnetic molecularly imprinted polymer nanoparticles for the solid-phase extraction of paracetamol from plasma samples, followed its determination by HPLC. *Microchim. Acta* 181: 1823.

20 Azodi-Deilami, S., Asadi, E., Abdouss, M. et al. (2015). Determination of meloxicam in plasma samples using a highly selective and sensitive voltammetric sensor based on carbon paste electrodes modified by molecularly imprinted polymer nanoparticle-multiwall carbon nanotubes. *Anal. Methods* 7: 1280–1292.

21 Malhotra, B.D., Srivastava, S., Ali, M.A., and Singh, C. (2014). Nanomaterial-based biosensors for food toxin detection. *Appl. Biochem. Biotechnol.* 174 (3): 880–896.

22 Sing, K.S.W., Rouquerol, J., and Rouquerol, F. (1998). *Adsorption by Powders and Porous Solids*. San Diego, CA: *Academic Press*.

23 Dabrowski, A. (2001). Adsorption from theory to practice. *Adv. Colloid Interface Sci.* 93: 135–224.

24 Limousin, G., Gaudet, J.P., Charlet, L. et al. (2007). Sorption isotherms: a review on physical bases, modeling and measurement. *Appl. Geochem.* 22: 249–275.

25 Donohue, M.D. and Aranovich, G.L. (1998). Classification of Gibbs adsorption isotherms. *Adv. Colloid Interface Sci.* 76-77: 137–152.

26 Langmuir, I. (1918). The adsorption of gases on plane surfaces of glass, mica and platinum. *J. Am. Chem. Soc.* 40: 1361–1403.

27 Freundlich, H.M.F. (1906). Über die adsorption in Lösungen. *Z. Phys. Chem.* 57 (A): 385–470.

28 Temkin, M.J. and Pyzhev, V. (1940). Recent modifications to Langmuir isotherms. *Acta Physiochim.* 12: 217–222.

29 Temkin, M.I. and Pyzhev, V. (1940). Kinetics of ammonia synthesis on promoted iron catalyst. *Acta Physicochim. URSS* 12: 327–356.

30 Dubinin, M.M. (1960). The potential theory of adsorption of gases and vapors for adsorbents with energetically non uniform surface. *Chem. Rev.* 60: 235–266.

31 Dubinin, M.M. and Radushkevich, L.V. (1947). The equation of the characteristic curve ofthe activated charcoal. *Proc. Acad. Sci. USSR Phys. Chem. Sect.* 55: 331–337.

32 Harkins, W.D. and Jura, E.J. (1944). The decrease of free surface energy as a basis for the development of equations for adsorption isotherms; and the existence of two condensed phases in films on solids. *J. Chem. Phys.* 12: 112–113.

33 Halsey, G. (1948). Physical adsorption on nonuniform surfaces. *J. Chem. Phys.* 16: 931–937.

34 Redlich, O. and Peterson, D.L. (1959). A useful adsorption isotherm. *J. Phys. Chem.* 63: 1024–1026.

35 Brunauer, S. and Emmett, P.H. (1935). The use of Van der Waals adsorption isotherms in determining the surface area of iron synthetic ammonia catalysts. *J. Am. Chem. Soc.* 57: 1754–1755.

36 Brunauer, S. and Emmett, P.H. (1937). The use of low temperature Van der Waals adsorption isotherms in determining the surface areas of various adsorbents. *J. Am. Chem. Soc.* 59: 2682–2689.

37 Brunauer, S., Emmett, P.H., and Teller, E. (1938). Adsorption of gases in multimolecular layers. *J. Am. Chem. Soc.* 60: 309–319.

38 Lagergren, S. (1898). Zur theorie der sogenannten adsorption geloester stoffe. *Kungl. Svenska Vetenskapsakad. Handl.* 24: 1–39.

39 Ho, Y.S. and McKay, G. (1999). Pseudo second order model for sorption process. *Process Biochem.* 1999 (34): 451–465.

40 Weber, W.J. and Morris, J.C. (1963). Kinetics of adsorption on carbon from solution. *J. Sanit. Eng. Div. Proceed. Am. Soc. Civil Eng.* 89: 31–59.

41 Gueu, S., Yao, B., Adouby, K., and Ado, G. (2007). Kinetics and thermodynamics study of lead adsorption on to activated carbons from coconut and seed hull of the palm tree. *Int. J. Environ. Sci. Technol.* 4 (1): 11–17.

42 Asthana, A., Verma, R., Singh, A.K. et al. (2016). Silver nanoparticle entrapped calcium-alginate beads for Fe(II) removal via adsorption. *Macromol. Symp.* 366: 42–51.

43 Gari, V.R.D.K. and Kim, M. (2015). Removal of Pb(II) using silver nanoparticles deposited graphene oxide: equilibrium and kinetic studies. *Monatsh. Chem.* 146: 1445–1453.

44 Ghaedi, M., Sadeghian, B., Amiri Pebdani, A. et al. (2012). Kinetics, thermodynamics and equilibrium evaluation of direct yellow 12 removal by adsorption onto silver nanoparticles loaded activated carbon. *Chem. Eng. J.* 187: 133–141.

45 Pal, J., Deb, M.K., Deshmukh, D.K., and Verma, D. (2013). Removal of methyl orange by activated carbon modified by silver nanoparticles. *Appl. Water Sci.* 3: 367–374.

46 Dankovich, T.A. and Gray, D.G. (2011). Bactericidal paper impregnated with silver nanoparticles for point-of-use water treatment. *Environ. Sci. Technol.* 45: 1992–1998.

47 Kallman, E.N., Oyanedel-Craver, V.A., and Smith, J.A. (2011). Ceramic filters impregnated with silver nanoparticles for point-of-use water treatment in rural Guatemala. *Environ. Sci. Technol.* 137: 407–415.

48 Vinka, A., Crever, O., and Smith, J.A. (2008). Sustainable colloidal-silver-impregnated ceramic filter for point-of-use water treatment. *Environ. Sci. Technol.* 42: 927–933.

49 Roosta, M., Ghaedi, M., Daneshfar, A. et al. (2014). Optimization of the ultrasonic assisted removal of methylene blue by gold nanoparticles loaded on activated carbon using experimental design methodology. *Ultrason. Sonochem.* 21: 242–252.

50 Roosta, M., Ghaedi, M., and Mohammadi, M. (2014). Removal of Alizarin Red S by gold nanoparticles loaded on activated carbon combined with ultrasound device: optimization by experimental design methodology. *Powder Technol.* 267: 134–144.

51 Solis, K.L., Nam, G.-U., and Hong, Y. (2016). Effectiveness of gold nanoparticle-coated silica in the removal of inorganic mercury in aqueous systems: equilibrium and kinetic studies. *Environ. Eng. Res.* 21 (1): 99–107.

52 Gültekin, A., Ersöz, A., Hür, D. et al. (2009). Gold nanoparticles having dipicolinic acid imprinted nanoshell for *Bacillus cereus* spores recognition. *Appl. Surf. Sci.* 256: 142–148.

53 Dehaghi, S.M., Rahmanifar, B., Moradi, A.M., and Azar, P.A. (2014). Removal of permethrin pesticide from water by chitosan–zinc oxide nanoparticles composite as an adsorbent. *J. Saudi Chem. Soc.* 18: 348–355.

54 Gnanasangeetha, D. and Thambavani, D.S. (2013). ZnO nanoparticle entrenched on activated silica as a proficient adsorbent for removal of As^{3+}. *Int. J. Res. Pharm. Biomed. Sci.* 4 (4): 1295–1304.

55 Salmani, M.H., Zarei, S., Ehrampoush, M.H., and Danaie, S. (2013). Evaluations of pH and high ionic strength solution effect in cadmium removal by zinc oxide nanoparticles. *J. Appl. Sci. Environ. Manage.* 17 (4): 583.

56 Sasidharan, N.P., Chandran, P., and Khan, S.S. (2013). Interaction of colloidal zinc oxide nanoparticles with bovine serum albumin and its adsorption isotherms and kinetics. *Colloids Surf. B: Biointerfaces* 102: 195–201.

57 Bucak, S., Jones, D.A., Laibinis, P.E., and Hatton, T.A. (2003). Protein separations using colloidal magnetic nanoparticles. *Biotechnol. Prog.* 19: 477–484.

58 Kecili, R., Özcan, A., Ersöz, A. et al. (2011). Superparamagnetic nanotraps containing MIP based mimic lipase for biotransformations uses. *J. Nanopart. Res.* 2011 (13): 2073–2079.

59 Prapha, G. and Raj, V. (2016). Preparation and characterization of polymer nanocomposites coated magnetic nanoparticles for drug delivery applications. *J. Magn. Magn. Mater.* 408: 26–34.

60 Qi, Z., Joshi, T.P., Liu, R. et al. (2017). Synthesis of Ce(III)-doped Fe_3O_4 magnetic particles for efficientremoval of antimony from aqueous solution. *J. Hazard. Mater.* 329: 193–204.

61 Dalvand, A., Nabizadeh, R., Ganjali, M.R. et al. (2016). Modeling of reactive blue 19azo dye removal from colored textile wastewater using L-arginine-functionalized Fe_3O_4 nanoparticles: optimization, reusability, kinetic and equilibrium studies. *J. Magn. Magn. Mater.* 404: 179–189.

62 Cai, W., Guo, M., Weng, X. et al. (2019). Adsorption of doxorubicin hydrochloride on glutaric anhydride functionalized $Fe_3O_4@SiO_2$ magnetic nanoparticles. *Mater. Sci. Eng. C* 98: 65–73.

63 Fu, C.-C., Tran, H.-N., Chen, X.-H., and Juang, R.-S. (2020). Preparation of polyaminated Fe_3O_4@chitosan core-shell magnetic nanoparticles for efficient adsorption of phosphate in aqueous solutions. *J. Ind. Eng. Chem.* 83: 235–246.

64 Chen, F., Mao, M., Wang, J. et al. (2020). A dual-step immobilization/imprinting approach to prepare magnetic molecular imprinted polymers for selective removal of human serum albumin. *Talanta* 209: 120509.

65 Yang, W.D., Niu, X., Ni, Z. et al. (2015). Core–shell magnetic molecularly imprinted polymer prepared for selectively removed indole from fuel oil. *Adv. Polym. Technol.* 1: 1–9.

66 Zhang, M., Zhang, Z., Liu, Y. et al. (2011). Preparation of core-shell magnetic ion-imprinted polymer for selective extraction of Pb(II) from environmental samples. *Chem. Eng. J.* 2011 (178): 443–450.

67 Tang, Z., Peng, S., Hu, S., and Hong, S. (2017). Enhanced removal of bisphenol-AF by activated carbon-alginate beads with cetyltrimethyl ammonium bromide. *J. Colloid Interface Sci.* 495: 191–199.

68 Belhamdi, B., Merzougui, Z., Trari, M., and Addoun, A. (2016). A kinetic, equilibrium and thermodynamic study of L-phenylalanine adsorption using activated carbon based on agricultural waste (date stones). *J. Appl. Res. Technol.* 14: 354–366.

69 Saleh, T.A., Sarı, A., and Tuzen, M. (2017). Optimization of parameters with experimental design for the adsorption of mercury using polyethylenimine modified-activated carbon. *J. Environ. Chem. Eng.* 5: 1079–1088.

70 Baghdadi, M., Ghaffari, E., and Aminzadeh, B. (2016). Removal of carbamazepine from municipal wastewater effluent using optimally synthesized magnetic activated carbon: adsorption and sedimentation kinetic studies. *J. Environ. Chem. Eng.* 4: 3309–3321.

71 Iijima, S. (1991). Helical microtubules of graphitic carbon. *Nature* 354: 56–58.

72 Yang, S., Li, J., Shao, D. et al. (2009). Adsorption of Ni(II) on oxidized multi-walled carbon nanotubes: effect of contact time, pH, foreign ions and PAA. *J. Hazard. Mater.* 166: 109–116.

73 Lu, C. and Chiu, H. (2006). Adsorption of zinc(II) from water with purified carbon nanotubes. *Chem. Eng. Sci.* 61: 1138–1145.

74 Kuo, C.-Y., Wu, C.-H., and Wu, J.-Y. (2008). Adsorption of direct dyes from aqueous solutions by carbon nanotubes: determination of equilibrium, kinetics and thermodynamics parameters. *J. Colloid Interface Sci.* 327: 308–315.

75 Sadegh, H., Shahryari-ghoshekandi, R., Agarwal, S. et al. (2015). Microwave-assisted removal of malachite green by carboxylate functionalized multi-walled carbon nanotubes: kinetics and equilibrium study. *J. Mol. Liq.* 206: 151–158.

76 Madrakian, T., Afkhami, A., Ahmadi, M., and Bagheri, H. (2011). Removal of some cationic dyes from aqueous solutions using magnetic-modified multi-walled carbon nanotubes. *J. Hazard. Mater.* 196: 109–114.

77 Yan, L., Chang, P.R., Zheng, P., and Ma, X. (2012). Characterization of magnetic guar gum-grafted carbon nanotubes and the adsorption of the dyes. *Carbohydr. Polym.* 87: 1919–1924.

78 Zhang, Z., Chen, X., Rao, W. et al. (2014). Synthesis and properties of magnetic molecularly imprinted polymers based on multiwalled carbon nanotubes for magnetic extraction of bisphenol A from water. *J. Chromatogr. B* 965: 190–196.

79 Yang, Z.-F., Li, L.-Y., Hsieh, C.-T., and Junag, R.-S. (2018). Co-precipitation of magnetic Fe_3O_4 nanoparticles onto carbon nanotubes for removal of copper ions from aqueous solution. *J. Taiwan Inst. Chem. Eng.* 82: 56–63.

80 Yao, T., Xiao, Y., Wu, X. et al. (2016). Adsorption of Eu(III) on sulfonated graphene oxide: combined macroscopic and modeling techniques. *J. Mol. Liq.* 215: 443–448.

81 Kim, H., Kang, S.-O., Park, S., and Park, H.S. (2016). Adsorption isotherms and kinetics of cationic and anionic dyes on three-dimensional reduced graphene oxide macrostructure. *J. Ind. Eng. Chem.* 21: 1191–1196.

82 Rostamian, R. and Behnejad, H. (2016). A comparative adsorption study of sulfamethoxazole onto graphene and graphene oxide nanosheets through equilibrium, kinetic and thermodynamic modeling. *Process. Saf. Environ. Prot.* 102: 20–29.

83 Raghubanshi, H., Ngobeni, S.M., Osikoya, A.O. et al. (2017). Synthesis of graphene oxide and its application for the adsorption of Pb^{2+} from aqueous solution. *J. Ind. Eng. Chem.* 47: 169–178.

84 Yap, P.L., Tung, T.T., Kabiri, S. et al. (2020). Polyamine-modified reduced graphene oxide: a new and cost-effective adsorbent for efficient removal of mercury in waters. *Sep. Purif. Technol.* 238: 116441.

85 Xing, M., Zhuang, S., and Wang, J. (2020). Efficient removal of Cs(I) from aqueous solution using graphene oxide. *Prog. Nucl. Energy* 119: 103167.

86 Wu, Z.-L., Liu, F., Li, C.-K. et al. (2016). A sandwich-structured graphene-based composite: preparation, characterization, and its adsorption behaviors for Congo red. *Colloids Surf. A Physicochem. Eng. Asp.* 509: 65–72.

87 Cheng, Z.-L., Li, Y.-X., and Liu, Z. (2017). Novel adsorption materials based on graphene oxide/Beta zeolite composite materials and their adsorption performance for rhodamine B. *J. Alloys Compd.* 708: 255–263.

88 Zhang, L., Wu, S., Tai, Y. et al. (2016). Water-soluble magnetic-graphene nanocomposites: use as high-performance adsorbent for removal of dye pollutants. *Fuller Nanotub. Car. N.* 24 (2): 116–122.

89 Raghua, M.S., Kumar, K.Y., Prashanth, M.K. et al. (2017). Adsorption and antimicrobial studies of chemically bonded magnetic graphene oxide-Fe_3O_4 nanocomposite for water purification. *J. Water Process. Eng.* 17: 22–31.

4

Carbon Nanomaterials (CNMs) as Adsorbents for Sample Preparation

4.1 Introduction

Nanotechnology is a rapidly growing and diverse field that impacts many areas of science and engineering and has the potential to revolutionize a vast array of technologies. Engineered nanomaterials (NMs) are defined as materials that are deliberately manufactured to have particles with at least one dimension measuring less than 100 nm [1–5]. Having dimensions in the nanometer range gives NMs new enhanced properties that are not present in the bulk forms of the same material.

The attractiveness of NMs is explained by their unique properties (e.g. mechanical, electrical, magnetic, and optical) at the nanoscale [6–18]. At the nanoscale, electromagnetic forces are dominant and gravitational forces are negligible. Considerably smaller than bulk solids but still larger than individual atoms or molecules, NMs properties do not obey the classical physics either to the absolute quantum mechanical laws but lie in between [19]. Furthermore, the ratio of surface to core atoms is much higher than at the macroscale. Consequently, a slight change in diameter has important impact on the properties. For example, optical properties of gold nanoparticles are considerably different than those of bulk gold since gold colloids solutions appear red, whereas bulk gold is yellow due to the electron confinement influence on the absorption properties.

Actually, the use of NMs is not new. For example, colloidal Au has been extensively studied and applied in medical treatments since the late 1940s [20]. But, examples of its use as a pigment for glass and ceramics date back as far as the Roman era, such as in the Lycurgus Cup [21]. Many of the ideas and possibilities for miniaturization to the atomic and nanoscale were famously outlined by Richard Feynman in his famous lecture "There's Plenty of Room at the Bottom" [22]. However, the most significant surge in NMs research has occurred only in the past decades driven by vast improvements in researchers' ability to image and manipulate matter on the atomic scale and by the realization of the unique, size-dependent properties of materials on length scales of a few nanometers. As a result, the great promise of nanotechnology lies in the possibility of tuning macroscopic materials' properties through the deliberate design of nanoscopic structures.

On the other hand, carbon nanomaterials (CNMs) including single-walled carbon nanotubes (SWCNTs) and multiwalled carbon nanotubes (MWCNTs), graphene,

Sample Preparation with Nanomaterials: Next Generation Techniques and Applications,
First Edition. Chaudhery Mustansar Hussain, Rüstem Keçili, and Chaudhery Ghazanfar Hussain.

and fullerene (FUL) are widely researched for their potential applications in different fields [23–25].

This chapter provides an overview of the latest advancements in design and preparation of CNMs as efficient adsorbents.

4.2 Carbon Nanomaterials (CNMs)

4.2.1 Carbon Nanotubes (CNTs)

Carbon nanotubes (CNTs) were first discovered by the Japanese physicist Sumio Lijimai in 1991 [26]. Since then, many scientists have worked on the different applications of CNTs [27–31]. CNTs are composed of graphene sheets that were rolled into a cylindrical shape and categorized into two groups, which are SWCNTs, double-walled carbon nanotubes (DWCNTs), and MWCNTs. SWCNTs are composed of one graphene layer while DWCNTs and MWCNTs have two or more graphene layers (Figure 4.1).

On the other hand, there are different approaches for the preparation of CNTs. Iijima utilized arc vaporization to produce the first images of MWCNTs [26]. A similar method is by laser evaporation [33], which is capable of producing single-walled CNTs. However, both these methods require a huge amount of energy with a low product yield. Therefore, the third method, catalytic chemical vapor deposition (CVD), is a much preferred method for industrial-scale CNT production [34].

CNTs are among the stiffest and strongest materials ever made, with a Young's modulus, in theory, reaching up to 1000–1200 GPa [35, 36] and experimental tensile strengths of between 11 and 63 GPa [37]. Although the experimental values reported vary significantly, it has been the consensus that CNTs possess much higher mechanical properties compared with most metals [38]. Furthermore, CNTs also exhibit attractive electrical properties, which have been shown to be metallic or semiconducting depending on their structure and diameter [39]. In addition, the CNTs have also been experimentally shown to possess very high thermal conductivities [40–42].

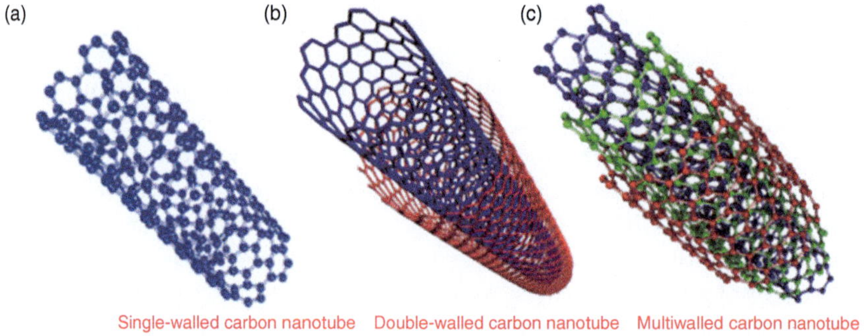

Figure 4.1 The schematic representation of the types of CNTs. (a) single-walled carbon nanotubes (SWCNTs), (b) double-walled carbon nanotubes (DWCNTs) and (c) multiwalled carbon nanotubes (MWCNTs). Source: Kausar and Taherian [32]. © 2019, Elsevier.

4.2.2 Graphene

Graphene is a carbon-based, two-dimensional NM that is likely to be the basic element of groundbreaking new technologies [43]. Graphene was discovered and produced for the first time in 2004, and it has rapidly become one of the most popular and unique NMs [44]. In addition, the Nobel Prize for Physics in 2010 was awarded to Geim and Novoselov in the University of Manchester for their research on graphene [45]. The excellent features of graphene make it the starting point of new technologies in many application areas. Although graphene has a very low thickness, it is more robust than diamond and 300 times stronger than steel. It conducts electricity better than copper, passes light and bends without any deformation, and can be put into any desired shape. It also exhibits great features such as large surface area, high stability, and layered structure. These superior features of graphene make it as an excellent NM in many applications such as separation processes, sensors, and bioimaging.

There are several strategies for the preparation of graphene. Due to the large requirement for research and applications, there has been much investment put into the development of methods to prepare high-quality graphene in bulk. The preparation methods can be classified into two categories, one is "top-down" and the other is "bottom-up." The former one is to break graphite into graphene through approaches such as mechanical cleavage, liquid exfoliation, thermal expansion, and electrochemical exfoliation. The latter one is to synthesize graphene by the techniques such as CVD, arc discharge, epitaxial growth on SiC, and unzipping CNTs [46–49].

On the other hand, the chemical structure of graphene oxide (GO) was proposed by Lerf et al. in which the basal plane of GO is decorated by hydroxyl, epoxy functional groups [50]. Based on a previous model, Dreyer et al. made some modifications to prove that GO also contains carbonyl groups. Carboxylic acid groups probably are along the GO edges, and some organic carbonyl defects are within the sheet. Recent nuclear magnetic resonance demonstrated the presence of five- and six-membered lactol on the edge of the plane and the existence of ester and tertiary alcohols on the surface [51].

The preparation of GO from graphite and then its reduction to graphene are schematically depicted in Figure 4.2.

Since the conductivity of graphene, to a large extent, depends on the long-range π–π conjugation, oxidation in GO largely reduces the mobility and concentration of the charge carriers. Although some of the sp^2 hybridized conjugated regions are preserved, the existence of oxidized regions isolates these conjugated areas. As a result, the oxidized region blocks the long-range conductivity and the percolation pathway cannot be created. The attached functional groups and lattice defects also alter the electronic structure, which make the transport of electrons more difficult [52]. Apart from that, relatively strong interaction between GO sheets and oxidative debris (OD) also alters the electronic structure. Thus, GO is an insulator. To restore some of the electronic conductivity, reduction and/or base washing can be employed.

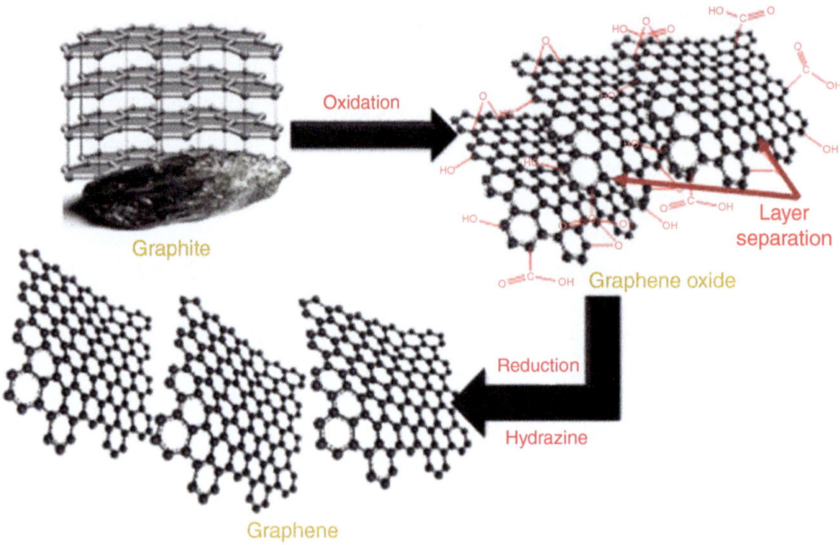

Figure 4.2 The schematic depiction of the preparation of graphene through reduction of the graphene oxide. Source: Kausar and Taherian [32]. © 2019, Elsevier.

GO is widely produced through treating graphite flakes with an oxidizing agent. This oxidization treatment can introduce the polar functional groups onto the surface of each graphene flake that can widen the interlayer distance between the graphene oxide flakes. The interlayer spacing between each GO flake in the bulk graphite oxide is in the range between 6 and 10 Å, depending on the relative humidity [53]. Hummers and Offeman developed a relatively safe and fast approach to produce graphite oxide through oxidization of graphite by sulfuric acid, sodium nitrate, and potassium permanganate in 1958 [54]. This method is still used even today with some modifications [55, 56]. Compared with graphene, the relatively large interlayer space enables graphite oxide to be readily exfoliated by sonication to yield a colloidal suspension of GO through the adoption of a different kind of polar solvent. This preparation method of GO, however, always generates a great amount of poisonous gas and hazardous waste. GO has two important features. It can be produced via cheap raw materials through cost-effective chemical methods on the large scale, and the functional groups on GO enable it to be easily dissolved in water and some polymers such as polyvinyl alcohol (PVA). These two characteristics are essential for large-scale applications of graphene oxide. Rourke et al. proposed that GO actually consists of two distinct parts non-covalently bonded together [57]. One is the highly oxidized graphene-like sheet and the other small component is the highly OD. In acidic and neutral environments, the OD is strongly attached to graphene-like GO sheets through π–π stacking and hydrogen bonding, whereas, in basic environments, OD and graphene-like GO sheets become repulsive due to negative charge on the OD. When they are isolated, they cannot be recombined.

4.2.3 Fullerenes (FULs)

FULs are the zero-dimensional form of graphitic carbon [58–60]. The synthesis of FULs is successfully carried out by using carbon atoms in various shapes including hollow spherical, ellipsoid, and other shapes. Among these, the spherical FUL is known as buckminsterfullerene or buckyball. Structurally FULs are identical to graphite and are composed of irregularly stacked graphene sheets. This type of FULs form pentagonal or hexagonal rings. Various forms of FULs have been found, and their size varies in the range between 30 and 3000 carbon atoms. There are different types of FULs, as shown in Figure 4.3, such as C_{24}, C_{28}, C_{32}, C_{36}, C_{50}, C_{60}, and C_{70}, etc., and the most stable FUL type is C_{60}, which is composed of 12 pentagons and 20 hexagons of sp^2-bonded carbon atoms.

FULs have received great interest in different application areas since their discovery in 1985 [62]. Physical, chemical, and biological features of FULs exhibited their promising applications in the future. It is inferred that size, hydrophobicity, three-dimensionality, and electronic configurations make these NMs an appealing subject in chemistry.

FULs are practically insoluble in water and generally tend to form aggregates in aqueous media [63]. To disperse these in water, prolonged stirring for days is needed. Addition of additives, e.g. surfactants, cyclodextrins (CDs), or sugar polymers, offers a possibility to disperse these in water more efficiently. FULs can also be dispersed in water by first dissolving them in an organic solvent, then adding the solution to water, and evaporating off the organic solvent.

On the other hand, the efficient incorporation of FULs into polymeric networks to increase their processability has become a common practice in the applications. The unique features of FULs, including their spherical shape, conjugated

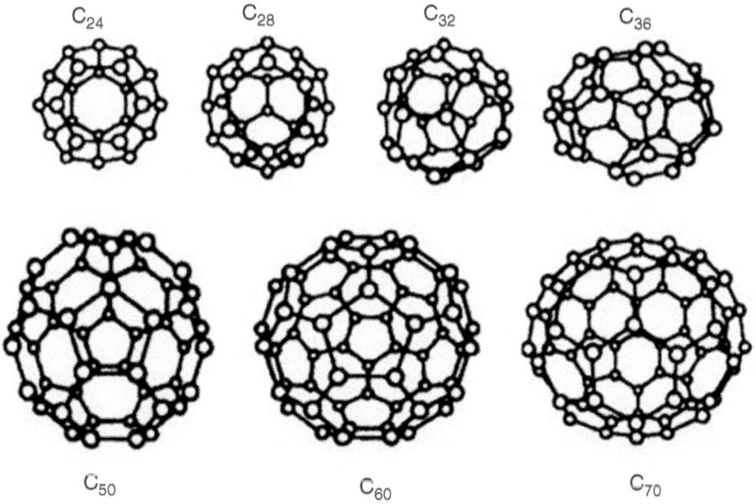

Figure 4.3 The schematic demonstration of the different types of fullerenes. Source: Malhotra and Azahar Ali [61]. © 2017, Elsevier.

three-dimensional π-electronic system, and the ability to display donor–acceptor or π–π interactions, make them attractive candidates for analytical applications [64]. As a result, these unique NMs can be successfully employed as chromatographic stationary nanophases in a wide temperature range varying from 80 to 360 °C, with good thermal stability and high selectivity for aromatic compounds. Chemical functionalization of FULs is often carried out to alter their chemical properties and/or improve miscibility with host polymers.

4.3 Adsorption on CNMs

Adsorption of various compounds (i.e. pollutants) on the CNMs is controlled by hydrophobic, dispersion, and weak dipolar forces. The higher equilibrium rates of CNMs over conventional carbons are attributed to π–π electron donor–acceptor interactions with adsorbates, decreased heterogeneity of adsorption energies, and the absence of pore diffusion as an intermediate mechanism in adsorption. Various carbon-based adsorbents were successfully employed for the pre-concentration of analytes [4]. These include activated carbon, carbon molecular sieves, graphitized carbon black, and porous carbon. These differ in their physicochemical characteristics, such as pore size/shape, surface area, pore volume, and surface functionality. Kinetic and thermodynamic features of carbon-based sorbents including breakthrough volumes, adsorption isotherms, intermolecular interaction mechanisms at the adsorbate/carbon sorbent interface strongly influence pre-concentration [14]. The adsorbent can affect the adsorption by providing high number of available adsorption regions and by facilitating various specific interactions including hydrogen bonding.

The diffusion into the porous structure leads to slow mass transfer, and consequentially the quantitative release of large molecules becomes a limiting factor. On the other hand, in CNTs, the sorption occurs mainly on the outer surface and in the hollow spaces of the open tubes (provided the tubes are uncapped). Therefore, their release by both thermal processes and solvents is relatively simpler and is not limited by diffusion. Therefore, a wide range of compounds from small molecules to large semivolatile compounds can be easily concentrated and desorbed from CNTs [65].

There are other factors that come into play. The exceptionally high aspect ratio (in millions) of CNTs provides a special confinement effect, which leads to a completely different physical behavior when compared to more conventional sorbents. Moreover, outer tube surfaces create a large number of intertubular spaces, which provide specific adsorption sites. As a result, it has been seen that the sorption capacity is higher than what one would expect based on BET surface area measurement. Various approaches have been used to describe the adsorption of on CNTs. The high adsorption capacity of CNTs has been explained through the presence of high-energy adsorption sites, such as CNT defect and interstitial and groove regions between CNT bundles, and the phenomenon of multilayer adsorption during micro-sorption on CNTs. Capillary forces in nanotubes can also be strong, which may draw molecules from vapor or liquid phases by van der

Waals attractive forces and dipole–induced dipole interactions. Hydrophobicity and capillarity can provide ordering and orientation of sorbents on the sorbate. Adsorption studies detail rapid equilibrium rates, high adsorption capacity, low sensitivity to pH range, minimal hysteresis in dispersed CNTs, and consistency with traditional Langmuir, BET, or Freundlich isotherms. Moreover, increase in dispersion energy and the overlapping force of adjacent carbon walls expand sorbent–sorbate and sorbent–sorbent interactions resulting in condensation within the nanotube. The filling of nanotubes may explain why some models describe adsorption capacity above the physical surface area of a CNT.

Adsorption on FULs is a subject of growing experimental and theoretical interest. Few research groups have explored the adsorption properties of FULs and generally focused on the adsorption of simple gases or vapors of organic compounds on C60 solid. Both types of mechanisms, physisorption and chemisorption, have been realized on FULs through theories and experiment. Few studies have reported interfacial interactions of organic molecules with C60 in aqueous media. FULs exhibit strong hydrophobicity, and various methods have been used to disperse FULs in aqueous media [66]. These include sonication, interaction with polymers, redox reactions, and extended mixing [67, 68].

The occurrence of porosity on the FULs' surface has been studied via adsorption. Some research groups by using Kr, N_2, O_2, and CO_2 adsorption, and others using N_2 and O_2 adsorption, proved the presence of micropores in C60. Moreover, research also pointed out that surface area values normally depend on sample preparation and purification, as well as sample age. More recent work indicates that FUL powders with higher degrees of purity consist of nonporous particles. Overall, the porous texture of FULs is very sensitive to the synthesis and purification procedures followed in their preparation.

To understand the adsorption mechanism on FULs, a comparison of desorption behavior of pyrene, phenanthrene, and naphthalene from FULs and CNTs was examined where FULs showed significant desorption in contrast to CNTs. The reason was the difference in their distinct geometries. Spherical FULs resulted in aggregates providing two space structures, i.e. external surface of aggregates, where the adsorption is reversible, and closed interstitial spaces in small aggregates. Interstitial spaces within small aggregates were unavailable for adsorption, but these contributed pore deformations. Rearrangement of small aggregates and penetration of adsorbate molecule into closed interstitial spaces between small aggregates during adsorption led to the molecular entrapment and hysteresis. Whereas CNTs did not form closed interstitial spaces in their aggregates due to their length; hence, no adsorption–desorption hysteresis was observed [69, 70].

4.4 Applications of CNMs

4.4.1 Extraction and Separation Applications

GO and GO-based nanomaterials were successfully applied for the selective recognition and removal of various pollutants from environmental samples. For example,

Ghorbani et al. prepared GO functionalized with ethylene diamine for the removal of Cd(II) and Pb(II) ions from wastewater samples [71]. The effects of various factors such as amount of the adsorbent, shaking rate, pH, and time on the extraction behavior of the prepared ethylene diamine–functionalized GO toward the target ions were investigated. The obtained results confirmed that the prepared ethylene diamine–functionalized GO successfully removed the target Cd(II) and Pb(II) ions from wastewater samples. The extraction percentages for Cd(II) and Pb(II) ions were achieved as 99.4% and 99.6%, respectively.

In another important study conducted by Keshvardoostchokami et al. [72], GO/Ag nanocomposite was developed for the efficient removal of an insecticide pollutant imidacloprid from water samples. The obtained results showed that approximately 63% of the target insecticide imidacloprid was removed from 50 ml water sample containing 10 mg l^{-1} by using 30 mg GO/Ag nanocomposite.

In an interesting study reported by Movahedi et al. [73], toxic Hg(II) ions were successfully removed from aqueous samples by using GO functionalized with benzoylthiourea. The obtained results confirmed that the prepared GO functionalized with benzoylthiourea displayed excellent affinity toward the target Hg(II) ions. The achieved binding capacity was 950 mg g^{-1} under optimum extraction conditions (pH 6.0 and T: 25 °C).

Barati et al. developed molecularly imprinted polymer (MIP)-based magnetic GO/chitosan nanocomposite toward fluoxetine in biological and environmental samples [74]. In their study, MIP-based magnetic GO/chitosan nanocomposite was prepared. The prepared MIP-based magnetic nanocomposite was successfully used for the selective separation of fluoxetine in urine and water samples. The maximum binding capacity was calculated as 66 mg g^{-1}. In another interesting work reported by Liang et al. [75], magnetic Fe_3O_4@SiO_2/GO nanocomposites functionalized with metal-organic framework MIL-101 (Cr) were successfully prepared and applied for the efficient extraction of various triazine herbicides including secbumeton, terbuthylazine, trietazine, atrazine, terbumeton, atraton, and prometon from rice samples. The obtained detection limit values for the target triazine herbicides were in the range between 0.010 and 0.080 μg kg^{-1}.

Membrane technology is widely applied in different separation processes such as treatment of wastewater, gas separation, and desalination, etc. [76–80]. Membrane separation is generally performed based on selective transport of the target compound across the membrane structure [81]. Nanofiltration and reverse osmosis membranes are commonly used for the treatment of wastewater samples. However, thick separating layer restricts their water flux behavior. This drawback can be overcome by incorporating NMs such as graphene, FULs, CNTs, and nanoparticles into the membrane structure. The combination of membranes with the excellent features of these NMs such as graphene provides excellent physical and chemical stability and also high rejection toward the target compound to be separated from the sample.

Prince et al. proposed a facile method to improve the hydrophilic feature of the polyethersulfone hollow-fiber ultrafiltration membrane [82]. For this purpose, graphene-attached polyacrylonitrile-*co*-maleimide having hydroxyl, carboxyl, and amine functional groups was prepared. The characterization of the prepared graphene-based nanocomposite membranes was performed by using scanning electron microscope (SEM), energy-dispersive X-ray (EDX) analyses, pore size, and contact angle measurements. The results showed that the incorporation of graphene-attached polyacrylonitrile-co-maleimide into the polyethersulfone hollow-fiber membrane lead to a decrease of water contact angle (64.5% decrease). On the other hand, oil contact angle increased from 43.6° to 112.5° compared to the control membrane. The water permeability of the prepared graphene-based nanocomposite membrane enhanced by 43%, and the efficient removal of oil was successfully achieved with a rate of 99%.

In another research carried out by Nasseri et al. [83], GO/polysulfone nanocomposite membrane was prepared for the efficient removal of bisphenol A from water samples. For this purpose, three different membranes were prepared by using phase inversion technique. The obtained results indicated that incorporation of GO into the membrane structure considerably increased the permeate flux of the prepared nanocomposite membranes. The maximum bisphenol A (BPA) removal was achieved by using the GO (0.4%)/polysulfone nanocomposite membrane, which has the highest negative zeta potential value (−10.46 mV). The optimum process conditions for BPA removal such as time, initial BPA concentration, pressure, and pH were determined as 10.6 minutes, 7.5 $mg\,l^{-1}$, 1.02 bar, and 5.5, respectively. Under optimum conditions, the removal efficiency of the target pollutant BPA by using the prepared GO (0.4%)/polysulfone membrane was obtained as 93%. Zhang et al. developed a GO-based polyacrylonitrile fiber nanocomposite membrane for the filtration of water–oil emulsion [84]. In their study, polyacrylonitrile nanofibers were produced by electrospinning technique and GO was assembled on the surface of nanofibers. The prepared GO-based polyacrylonitrile nanocomposite membrane exhibited superhydrophilic feature and low oil adhesion behavior. The developed nanocomposite membrane exhibited great rejection value (≥98%) and antifouling performance. Zhu et al. developed a nanocomposite membrane composed of porous reduced graphene oxide (PRGO) and halloysite nanotubes (HNTs) for the removal of salts and dye compounds from water samples [85]. HNTs were used to increase the interlayer spacing of GO (Figure 4.4). After characterization studies by using TEM, XRD, FT-IR, Raman, and energy dispersive spectroscopy (EDS), the prepared sandwich-like nanocomposite membrane was successfully applied for the removal of $MgCl_2$, NaCl, $MgSO_4$, Na_2SO_4, and Reactive Black 5 dye from water samples.

In another interesting study [86], Hu et al. developed a GO-based nanocomposite membrane. The prepared nanocomposite membrane was effectively used for the treatment of water–oil emulsion. The obtained values for water flux of the prepared nanocomposite membrane having GO and control membrane without GO were 677 and 522 $l\,h^{-1}\,m^2\,bar^{-1}$, respectively after 2.5 hours filtration time. On the other hand, the achieved oil rejections of the prepared GO-based nanocomposite membrane and unmodified control membrane were found as 98.7% and 98.1%, respectively after two hours filtration time. It has been reported that incorporation of GO into the membrane structure has significantly increased the performance of the membranes for the separation of water–oil emulsions.

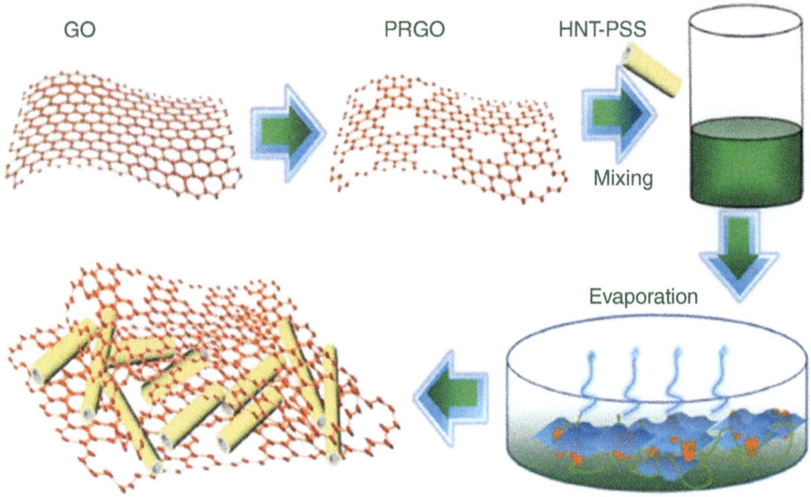

Figure 4.4 The schematic demonstration of the fabrication of GO-based nanocomposite membrane. Source: Zhu et al. [85]. © 2017, Elsevier.

Ho et al. reported the preparation of novel GO-MWCNTs-based polyvinylidene fluoride nanocomposite membrane for the treatment of palm oil mill effluent [87]. The nanocomposite membrane was prepared by using in-situ colloidal precipitation technique. The rejection ability of the prepared GO-MWCNTs-based nanocomposite membrane was investigated using palm oil mill effluent. The obtained results indicated that the developed nanocomposite membrane can be successfully used for the treatment of palm oil mill effluent. The rejection values of phosphorus, chemical oxygen demand, total dissolved solids, hardness, turbidity, total suspended solids, chlorine, and color were enhanced by 6.55%, 75.5%, 1.51%, 21.79%, 81.94%, 100%, 76%, and 86.3%, respectively using GO-MWCNTs-based nanocomposite membrane compared to the control membrane without GO and MWCNTs.

Zhang et al. reported the design and preparation of polydopamine-coated magnetic Fe_3O_4 nanoparticles having MWCNTs for the efficient extraction of various drugs including oxcarbazepine, phenytoin, and carbamazepine in biological samples such as human plasma, urine, and cerebrospinal fluid samples (Figure 4.5) [88]. The developed magnetic nanoadsorbent exhibited excellent extraction behavior toward the target drugs. The maximum extraction was achieved at pH 6 using a small amount of nanoadsorbent (15 mg) within a very short time (1.5 minutes). The obtained detection limit values were in the range between 0.4 and 3.1 ng ml^{-1}.

4.4.2 Chromatographic Applications

CNTs and FULs can be efficiently employed as separation media in liquid chromatography (LC) and gas chromatography (GC). For this purpose, two general approaches were mainly used, which are in packed and in capillary column having CNTs and FULs.

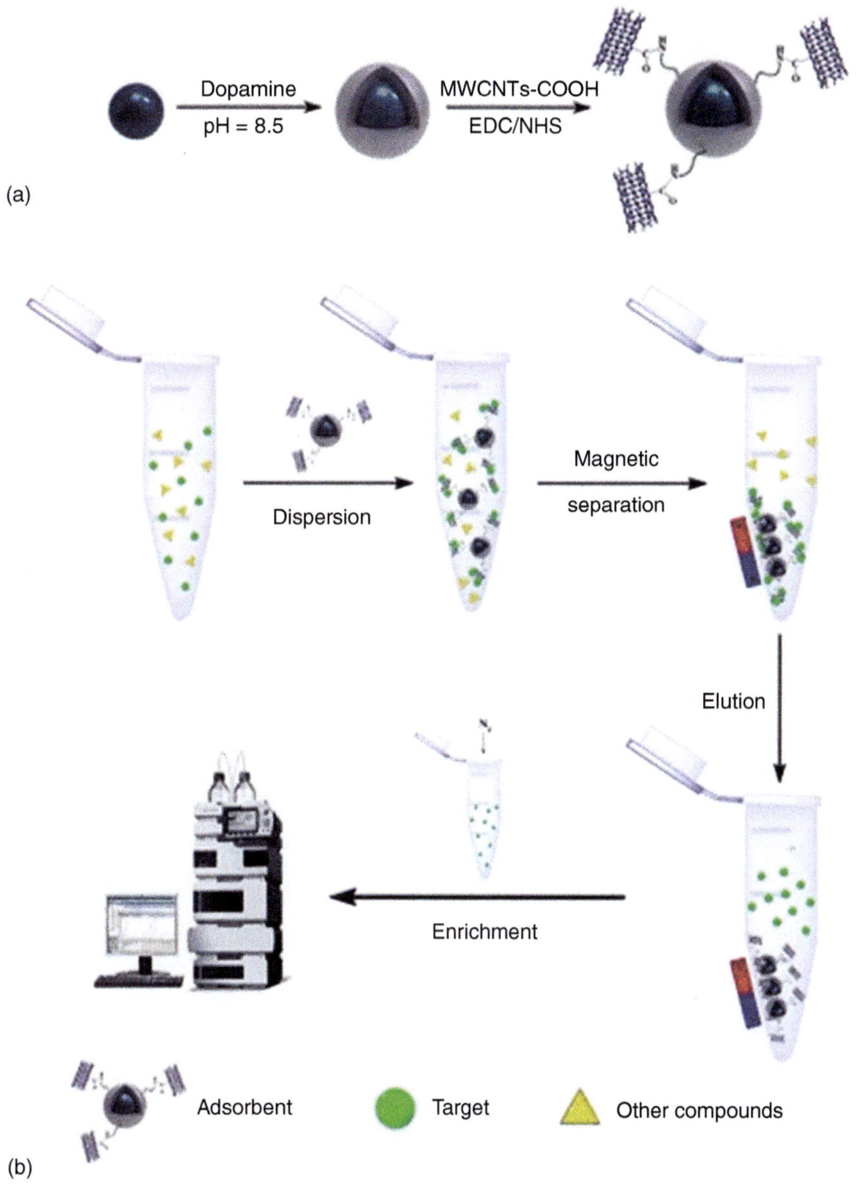

Figure 4.5 The schematic depiction of the (a) preparation of polydopamine-coated magnetic Fe_3O_4 nanoparticles having MWCNTs and (b) dispersive extraction of target drugs. Source: Zhang et al. [88]. © 2018, Springer Nature.

4.4.2.1 Chromatographic Stationary Phases Having CNTs

The use of CNTs as stationary phases in LC was investigated in various studies [89]. The CNTs were mainly employed in combination with another silica-based packing material or merged in a porous polymeric monolith [90]. It was reported that the performances of the both SWCNTs and MWCNTs incorporated into the monolithic columns were improved because of the hydrophobic interactions between CNTs and the target compounds. The incorporation of MWCNTs in reversed-phase LC was also reported [91], and the combination of SWCNT with cellulose trisphenyl-carbamate was reported to improve the enantioselectivity of chiral column [92]. The bonding of SWCNT and functionalized SWCNT with 3-aminopropyl silica gel for the development of stationary phases was investigated by a number of scientists and was applied for the efficient separation of different compounds such as polycyclic aromatic hydrocarbons (PAHs), terpenic compounds, and polychlorinated biphenyl isomers [93, 94]. For example, Aqel et al. reported an interesting research on the effect of MWCNTs' assimilation into benzyl methacrylate monolithic columns in capillary LC [95]. They demonstrated the design and synthesis of polymer-based monolithic materials and their application as stationary phases in the preparation of capillary column. For this purpose, MWCNTs were incorporated into a mixture of comonomers benzyl methacrylate and ethylene dimethacrylate. The efficiencies of the columns and the mobile-phase flow rates were tested by them at room temperature. The incorporation of nonfunctionalized pure CNTs prevents the column bleeding and enables the column to operate at higher temperatures compared to other conventional columns [96]. Such type of modified columns were produced by depositing a thin layer of nanotubes on the surface of stationary phase through CVD technique [97]. It has been claimed that the modification allows a better peak separation and shape and warned to consider the thermal stability carefully during the modification, because it has direct adverse effect on the column performance and reliability [98]. A number of studies in this field that concern mainly the separation of standard mixtures of various compounds can be found in the literature. An example of separation of 11 acidic, neutral, and alkaline organic components (sulfadimidine, resorcinol, aniline, *p*-toluidine, benzyl alcohol, *p*-methoxybenzaldehyde, 2-naphthol, *N*,*N*-dimethylaniline, anisole, 1,3,5-trimethylbenzene, and 2-methoxynaphthalene) on different polystyrene-divinylbenzene (PS-DVB) columns containing 0%, 1%, and 5% MWCNTs is provided.

CNTs easily immobilized onto the walls of silica capillaries through non-covalently or covalently bonding and those modified capillaries used to perform open-tubular CEC (OT-CEC) for the determination of different target compounds [90]. Regarding this, various research papers report the development of CNTs-incorporated porous polymer monolithic capillary columns to achieve the enhanced separation of various small molecules. For example, Chambers et al. demonstrated that the monolithic column without CNTs modifications exhibited very low efficiency of only 1800 plates m^{-1} compared to modified column [99]. The addition of a small quantity of MWCNTs increased the efficiency of the column to greater than 15 000 plates m^{-1} at flow rates of 1 ml min^{-1} and more than 35 000 plates m^{-1} when the flow set at 0.15 ml min^{-1}. The incorporation of small quantity of MWCNTs within porous silica beads helps to increase the affinity of the stationary phase toward aromatic compounds and provides better selectivity and retention to the column.

On the other hand, the CNTs have anticipated the preparation of the stationary phases for GC due to their great adsorbent ability, intrinsic properties, and thermal and chemical stability. In much earlier literature, the application of both MWCNTs and SWCNTs can found in a GC-packed column to separate halogenated and aromatic hydrocarbons, alkanes, ketones, alcohols, esters, ethers, etc. [100]. The GC columns of various dimensions were fabricated by self-assembly of a CNT film onto the capillary surface to prepare stable stationary phase, and it was reported that the MWCNT-modified column showed better selectivity and separation efficiency than SWCNT toward alkanes [101]. The use of SWCNTs as microfabricated stationary phase of GC capillary column has also reported in the literature [102]. In other researches, SWCNTs were embedded to the inner surface of the capillary tubing to enhance the separation of alcohols, alkanes, ketones, aromatic compounds, and racemic mixtures [103, 104].

4.4.2.2 Chromatographic Stationary Phases Having FULs

FULs can also be effectively employed as chromatographic stationary phases for the separation applications. In a study reported by Glausch et al. [105], chemical conjugation of C60 FULs to polysiloxane through amid bond formation was demonstrated to prepare a capillary column for the efficient separation of polychlorinatedbiphenyls (PCBs). Starting from functionalized FULs, the synthesis consists of a single facile reaction step involving the coupling of polyaminopropyl(methyl)siloxane with the reactive ester cyclohexylaminocarbonyl-(*N*)-succinimyloxycarbonyl-1,2-dihydro-[60] FULs. The enhancement in the retention of analytes was observed in moving from tetra-ortho to mono-ortho-substituted PCBs, as the biphenyls with low number of chlorine atoms in the ortho position are liable to assume planar conformations, which provides the charge transfer/dipole interactions between solute and sorbent.

High-resolution separation of other high-boiling-point compounds such as PAHs and phthalic esters was attained on a polysiloxane capillary column modified with FULs [106]. In this case, the polymeric phase was firstly chemically modified by electrophilic addiction to allyl bromide, followed by reaction with sodium azide, and finally by 1,3-dipolar addiction with C60. C12–C32 *n*-alkanes and C1–C10 phthalic esters were effectively separated, as well standard mixtures of PAHs, which eluted in the order of their increasing dispersion force.

Stacking was one of the main interactions governing adsorption, as demonstrated by higher retention times of methyl esters of unsaturated fatty acids in comparison with homologous saturated analytes. Also, alcoholic and aromatic positional isomers were resolved, this evidencing structural selectivity. FULs loading resulted to be an influencing parameter in regard to column performance; specifically, by increasing the amount of CNMs (from 3 to 6 wt%) retention and selectivity were enhanced, although efficiency was not improved. Bonding of C60 imparted a certain polarity to the siloxane phase, resulting in column with higher polarity than commercial SE-30 and similar to that of OV-3.

With the aim to enhance FULs' selectivity for separation of both polar and apolar analytes, different stationary phases based on FULs but with mixed composition have been prepared [107]. Crown ethers and CDs were selected as "model modifiers" with hydrophilic and hydrophobic cavities, respectively. A synergistic effect of FULs and crown ethers was considered to explain the complete separation of structural isomers of butanol on the C60-(dibenzo-24-C8) sorbent, not achieved on the nonmixed phases. In addition, the enhanced retention of heterocyclic compounds bearing nitrogen or sulfur was observed. FULs coupled with CDs were prepared to separate aromatic hydrocarbons, basing on the ability of CD to give inclusion complexes, due to hydrophobic and stacking interactions.

4.5 Conclusions

CNMs, especially in the form of CNTs and FULs, have a tremendous potential in environmental analysis. Their applications as adsorbent materials have received great attention in a wide variety of environmental analytical applications. However, this is only the starting point because there are large numbers of other CNMs with exceptional properties that have not yet been adventured from the environmental analytical point of view. One of the main challenges and research for CNMs is to find cost-effective, scalable production methods that retain the essential properties of these materials. Additionally, the combination of CNMs with other new materials (e.g. quantum dots or ionic liquids) can enhance sorption properties of these, which will help to develop new analytical tools that will simplify the environmental analytical process. Moreover, to bring the CNMs development a step forward, the sorption kinetics on CNMs need to be studied in more detail for understanding the interaction of gases and liquids. Although many cases of success in CNMs were benefited from their unique chemical and physical properties, the applications of CNMs in environmental analysis are still limited in the early stage.

In conclusion, the applications of CNMs in the field of environmental analysis are very interesting and endless; however, the commercial production of CNMs is still some way off, and there is still an urgent need for important breakthroughs. Along with the growth of interest in CNMs, the potential effects on human health and the environment, both adverse and beneficial, need to be considered. Their short- and long-term effects with the human body, immunotoxocity, and phototoxocity will also require detailed exploration. Increasingly, study of their fate and impact in the environment is becoming important due to the discharges already occurring to the environment. The likely further increase in CNMs' discharges along with the dramatic industry growth, and the immense knowledge gaps in risk assessment and management, would necessitate expanding studies in this area.

References

1 Palit, S. and Hussain, C.M. (2020). Green nanomaterials: a sustainable perspective. In: *Green Nanomaterials. Advanced Structured Materials*, vol. 126 (eds. S. Ahmed and W. Ali). Singapore: Springer.

2 Singh, S.B. and Hussain, C.M. (2020). Functionalized nanographene for catalysis, Chapter 5. In: *Micro and Nano Technologies, Handbook of Functionalized Nanomaterials for Industrial Applications* (ed. C.M. Hussain), 111–129. Elsevier.

3 Palit, S. and Hussain, C.M. (2020). Functionalization of nanomaterials for industrial applications: recent and future perspectives, Chapter 1. In: *Micro and Nano Technologies, Handbook of Functionalized Nanomaterials for Industrial Applications* (ed. C.M. Hussain), 3–14. Elsevier.

4 Hussain, C.M. and Keçili, R. (2020). *Modern Environmental Analysis Techniques for Pollutants*. The Netherlands: Elsevier.

5 Büyüktiryaki, S., Keçili, R., and Hussain, C.M. (2020). Functionalized nanomaterials in dispersive solid phase extraction: advances and prospects. *Trends Anal. Chem.* 127: 115893.

6 Keçili, R., Büyüktiryaki, S., and Hussain, C.M. (2019). Advancement in bioanalytical science through nanotechnology: past, present and future. *Trends Anal. Chem.* 110: 259–276.

7 Hussain, C.M. and Kharisov, B. (2017). *Advanced Environmental Analysis-Application of Nanomaterials*. The Royal Society of Chemistry.

8 Hussain, C.M. (2017). Magnetic nanomaterials for environmental analysis. In: *Advanced Environmental Analysis-Application of Nanomaterials* (eds. C.M. Hussain and B. Kharisov). The Royal Society of Chemistry.

9 Keçili, R., Büyüktiryaki, S., Dolak, İ., and Hussain, C.M. (2020). The use of magnetic nanoparticles in sample preparation devices and tools, Chapter 5. In: *Handbook of Nanomaterials in Analytical Chemistry* (ed. C.M. Hussain), 75–95. Elsevier.

10 Keçili, R., Büyüktiryaki, S., and Hussain, C.M. (2020). Membrane applications of nanomaterials, Chapter 7. In: *Handbook of Nanomaterials in Analytical Chemistry* (ed. C.M. Hussain), 159–182. Elsevier.

11 Büyüktiryaki, S., Keçili, R., and Hussain, C.M. (2020). Modern age of analytical chemistry: nanomaterials, Chater 2. In: *Handbook of Nanomaterials in Analytical Chemistry* (ed. C.M. Hussain), 29–40. Elsevier.

12 Keçili, R., Büyüktiryaki, S., and Hussain, C.M. (2020). Micro total analysis systems with nanomaterials, Chapter 8. In: *Handbook of Nanomaterials in Analytical Chemistry* (ed. C.M. Hussain), 185–198. Elsevier.

13 Büyüktiryaki, S., Sümbelli, Y., Keçili, R., and Hussain, C.M. (2019). Lab-on-chip platforms for environmental analysis. In: *Encyclopedia of Analytical Science*, 3e (eds. P. Worsfold, C. Poole, A. Townshend and M. Miró), 267–273. Academic Press.

14 Keçili, R. and Hussain, C.M. (2018). Recent progress of imprinted nanomaterials in analytical chemistry. *Int J. Anal. Chem.* 2018: 1–18. 8503853.

15 Keçili, R. and Hussain, C.M. (2018). Mechanism of adsorption on nanomaterials, Chapter 4. In: *Nanomaterials Chromatography*, 89–115. Elsevier.

16 Keçili, R., Sibel Büyüktiryaki, S., and Hussain, C.M. (2018). Engineered nanosensors based on molecular imprinting technology, Chapter 57. In: *Micro and Nano Technologies, Handbook of Nanomaterials for Industrial Applications* (ed. C.M. Hussain), 1031–1046. Elsevier.

17 Keçili, R., Özcan, A.A., Ersöz, A. et al. (2011). Superparamagnetic nanotraps containing MIP based mimic lipase for biotransformations uses. *J. Nanopart. Res.* 13 (5): 2073–2079.

18 Say, R., Keçili, R., and Ersöz, A. (2016). Molecularly imprinted polymer-based micro- and nanotraps for solid-phase extraction, Chapter 4. In: *Advanced Molecularly Imprinting Materials* (eds. A. Tiwari and L. Uzun). 129–163. Wiley.

19 Kaur, G., Singh, T., and Kumar, A. (2012). Nanotechnology: a review. *Int. J. Educ. Appl. Res.* 2: 50–53.

20 Swanson, J.N. (1949). The value of repeated colloidal gold tests in rheumatoid arthritis. *Ann. Rheum. Dis.* 8: 232–237.

21 Freestone, I., Meeks, N., Sax, M., and Higgitt, C. (2007). The Lycurgus cup – a roman nanotechnology. *Gold Bull.* 270.

22 Feynman, R.P. (1960). There's plenty of room at the bottom. *Eng. Sci.* 23 (5): 22–36.

23 Peng, Z., Liu, X., Zhang, W. et al. (2020). Advances in the application, toxicity and degradation of carbon nanomaterials in environment: a review. *Environ. Int.* 134: 105298.

24 Hajiabadi, S.H., Aghaei, H., Kalateh-Aghamohammadi, M., and Shorgasthi, M. (2020). An overview on the significance of carbon-based nanomaterials in upstream oil and gas industry. *J. Pet. Sci. Eng.* 186: 106783.

25 Xie, F., Yang, M., Jiang, M. et al. (2019). Carbon-based nanomaterials – a promising electrochemical sensor toward persistent toxic substance. *TrAC Trends Anal. Chem.* 119: 115624.

26 Iijima, S. (1991). Helical microtubules of graphitic carbon. *Nature* 354 (6348): 56–58.

27 Mohamed, A.E.-M.A. and Mohamed, M.A. (2020). Carbon nanotubes: synthesis, characterization, and applications, Chapter 2. In: *Micro and Nano Technologies, Carbon Nanomaterials for Agri-Food and Environmental Applications* (ed. K.A. Abd-Elsalam), 21–32. Elsevier.

28 Rather, S.U. (2020). Preparation, characterization and hydrogen storage studies of carbon nanotubes and their composites: a review. *Int. J. Hydrogen Energy* 45 (7): 4653–4672.

29 Selvaraj, M., Hai, A., Banat, F., and Haija, M.A. (2020). Application and prospects of carbon nanostructured materials in water treatment: a review. *J. Water Process. Eng.* 33: 100996.

30 Abousalman-Rezvani, Z., Eskandari, P., Roghani-Mamaqani, H., and Salami-Kalajahi, M. (2020). Functionalization of carbon nanotubes by combination of controlled radical polymerization and "grafting to" method. *Adv. Colloid Interface Sci.* 278: 102126.

31 Patila, M., Chalmpes, N., Dounousi, E. et al. (2020). Use of functionalized carbon nanotubes for the development of robust nanobiocatalysts, Chapter 12. In: *Methods in Enzymology*, vol. 630 (ed. C.V. Kumar), 263–301. Academic Press.

32 Kausar, A. and Taherian, R. (2019). Electrical conductivity behavior of polymer nanocomposite with carbon nanofillers, Chapter 3. In: *Plastics Design Library,*

Electrical Conductivity in Polymer-Based Composites (eds. R. Taherian and A. Kausar), 41–72. William Andrew Publishing.
33 Guo, T., Nikolaev, P., Thess, A. et al. (1995). Catalytic growth of single-walled manotubes by laser vaporization. *Chem. Phys. Lett.* 243 (1–2): 49–54.
34 Harris, P.J.F. (2009). *Carbon Nanotube Science: Synthesis, Properties and Applications.* Cambridge, UK: Cambridge University Press.
35 Wong, E.W., Sheehan, P.E., and Lieber, C.M. (1997). Nanobeam mechanics: elasticity, strength, and toughness of nanorods and nanotubes. *Science* 277 (5334): 1971–1975.
36 Hernandez, E., Goze, C., Bernier, P. et al. (1999). Elastic properties of single-wall nanotubes. *Appl. Phys. A Mater. Sci. Process.* 68 (3): 287–292.
37 Baughman, R.H., Zakhidov, A.A., and de Heer, W.A. (2002). Carbon nanotubes – the route toward applications. *Science* 297 (5582): 787–792.
38 Yu, M.F., Lourie, O., Dyer, M.J. et al. (2000). Strength and breaking mechanism of multiwalled carbon nanotubes under tensile load. *Science* 287 (5453): 637–640.
39 Dresselhaus, M.S., Dresselhaus, G., Saito, R. et al. (2005). Raman spectroscopy of carbon nanotubes. *Phys. Rep. Rev. Sect. Phys. Lett.* 409 (2): 47–99.
40 Kim, Y.A., Kamio, S., Tajiri, T. et al. (2007). Enhanced thermal conductivity of carbon fiber/phenolic resin composites by the introduction of carbon nanotubes. *Appl. Phys. Lett.* 90 (9): 3.
41 Che, J.W., Cagin, T., and Goddard, W.A. (2000). Thermal conductivity of carbon nanotubes. *Nanotechnology* 11 (2): 65–69.
42 Kim, P., Shi, L., Majumdar, A. et al. (2001). Thermal transport measurements of individual multiwalled nanotubes. *Phys. Rev. Lett.* 87 (21): 4.
43 Singh, S.B. and Mustansar, C.M. (2018). Nano-graphene as groundbreaking miracle material: catalytic and commercial perspectives. *ChemistrySelect* 3: 9533–9544.
44 Randviir, E.P., Brownson, D.A.C., and Bank, C.E. (2014). A decade of graphene research: production, applications and outlook. *Mater. Today* 17: 426–432.
45 Gerstner, E.D. (2010). Nobel Prize 2010: Andre Geim & Konstantin Novoselov. *Nat. Phys.* 6: 836–837.
46 Kim, H., Abdala, A.A., and Macosko, C.W. (2010). Graphene/polymer nanocomposites. *Macromolecules* 43 (16): 6515–6530.
47 Young, R.J., Kinloch, I.A., Gong, L., and Novoselov, K.S. (2012). The mechanics of graphene nanocomposites: a review. *Compos. Sci. Technol.* 72 (12): 1459–1476.
48 Sutter, P.W., Flege, J.I., and Sutter, E.A. (2008). Epitaxial graphene on ruthenium. *Nat. Mater.* 7 (5): 406–411.
49 Sutter, P., Hybertsen, M.S., Sadowski, J.T., and Sutter, E. (2009). Electronic structure of few-layer epitaxial graphene on Ru(0001). *Nano Lett.* 9 (7): 2654–2660.
50 Lerf, A., He, H., Forster, M., and Klinowski, J. (1998). Structure of graphite oxide revisited. *J. Phys. Chem. B* 102 (23): 4477–4482.

51 Dreyer, D.R., Park, S., Bielawski, C.W., and Ruoff, R.S. (2010). The chemistry of graphene oxide. *Chem. Soc. Rev.* 39 (1): 228–240.

52 Pei, S.F. and Cheng, H.M. (2012). The reduction of graphene oxide. *Carbon* 50 (9): 3210–3228.

53 Park, S. and Ruoff, R.S. (2009). Chemical methods for the production of graphenes. *Nat. Nanotechnol.* 4 (4): 217–224.

54 Hummers, W.S. and Offeman, R.E. (1958). Preparation of graphitic oxide. *J. Am. Chem. Soc.* 80 (6): 1339.

55 Wilson, N.R., Pandey, P.A., Beanland, R. et al. (2009). Graphene oxide: structural analysis and application as a highly transparent support for electron microscopy. *ACS Nano* 3 (9): 2547–2556.

56 Zhao, J., Pei, S., Ren, W. et al. (2010). Efficient preparation of large-area graphene oxide sheets for transparent conductive films. *ACS Nano* 4 (9): 5245–5252.

57 Rourke, J.P., Pandey, P.A., Moore, J.J. et al. (2011). The real graphene oxide revealed: stripping the oxidative debris from the graphene-like sheets. *Angew. Chem. Int. Ed.* 50 (14): 3173–3177.

58 Dresselhaus, M.S., Dresselhaus, G., and Eklund, P.C. (1996). *Science of Fullerenes and Carbon Nanotubes: Their Properties and Applications.* Academic Press.

59 Deng, S.-L., Tan, Y.-Z., and Xie, S.-Y. (2011). Synthetic chemistry of fullerenes, Chapter 12. In: *Modern Inorganic Synthetic Chemistry* (eds. R. Xu, W. Pang and Q. Huo), 249–267. Elsevier.

60 Moussa, F. (2018). Fullerene and derivatives for biomedical applications, Chapter 5. In: *Nanobiomaterials* (ed. R. Narayan), 113–136. Woodhead Publishing.

61 Malhotra, B.D. and Md, A.A. (2018). Functionalized carbon nanomaterials for biosensors, Chapter 2. In: *Micro and Nano Technologies, Nanomaterials for Biosensors* (eds. B.D. Malhotra and A.A. Md), 75–103. William Andrew Publishing.

62 Kroto, H.W., Heath, J.R., O'brien, S.C. et al. (1985). C60: buckminster fullerene. *Nature* 318: 162–163.

63 Pan, Y., Liu, X., Zhang, W. et al. (2020). Advances in photocatalysis based on fullerene C60 and its derivatives: properties, mechanism, synthesis, and applications. *Appl. Catal., B* 265: 118579.

64 Yan, L., Zhao, F., Li, S. et al. (2011). Low-toxic and safe nanomaterials by surface-chemical design, carbon nanotubes, fullerenes, metallofullerenes, and graphenes. *Nanoscale* 3 (2): 362–382.

65 Hussain, C.M. (2014). Carbon nanomaterials as adsorbents for environmental analysis. In: *Nanomaterials for Environmental Protection* (eds. B.I. Kharisov, O.V. Kharissova and H.V.R. Dias), 217–236. Wiley.

66 Deguchi, S., Alargova, R.G., and Tsujii, K. (2001). Stable dispersions of fullerenes, C60 and C70, in water preparation and characterization. *Langmuir* 17 (19): 6013–6017.

67 Sakurai, T., Wang, X.D., Hashizume, T. et al. (1995). Adsorption of fullerenes on Cu(111) and Ag(111) surfaces. *Appl. Surf. Sci.* 87–88: 405–413.

68 Gritti, F. and Guiochon, G. (2004). Adsorption isotherms of the fullerenes C60 and C70 on a tetraphenylporphyrin-bonded silica. *J. Chromatogr. A* 1053 (1–2): 59–69.

69 Xu, B.-S. (2008). Prospects and research progress in nano onion-like fullerenes. *New Carbon Mater.* 23 (4): 289–301.

70 Samonin, V., Nikonova, V., and Podvyaznikov, M. (2008). Sorption properties of fullerene-modified activated carbon with respect to metal ions. *Prot. Met.* 44 (2): 190–192.

71 Ghorbani, M., Shams, A., Seyedin, O., and Lahoori, N.A. (2018). Magnetic ethylene diamine-functionalized graphene oxide as novel sorbent for removal of lead and cadmium ions from wastewater samples. *Environ. Sci. Pollut. Res.* 25: 5655–5667.

72 Keshvardoostchokami, M., Bigverdi, P., Zamani, P.A., and Piri, F. (2018). Silver@graphene oxide nanocomposite: synthesize and application in removal of imidacloprid from contaminated waters. *Environ. Sci. Pollut. Res.* 25: 6751–6761.

73 Movahedi, F., Masrouri, H., and Tayyebi, H. (2018). Highly efficient adsorption behavior of benzoylthiourea functionalized graphene oxide with respect to the removal of Hg(II) from aqueous solutions: isothermal, kinetic and thermodynamic studies. *Res. Chem. Intermed.* 44: 5419–5438.

74 Barati, A., Azemi, E., Dadfarnia, S. et al. (2017). Synthesis/characterization of molecular imprinted polymer based on magnetic chitosan/graphene oxide for selective separation/preconcentration of fluoxetine from environmental and biological samples. *J. Ind. Eng. Chem.* 46: 212–221.

75 Liang, L., Wang, X., Sun, Y. et al. (2018). Magnetic solid-phase extraction of triazine herbicides from rice using metal-organic framework MIL-101(Cr) functionalized magnetic particles. *Talanta* 179: 512–519.

76 Lan, Y., Groenen-Serrano, K., Coetsier, C., and Causserand, C. (2018). Nanofiltration performances after membrane bioreactor for hospital wastewater treatment: Fouling mechanisms and the quantitative link between stable fluxes and the water matrix. *Water Res.* 146: 77–87.

77 Ben-Mansour, R., Li, H., and Habib, M.A. (2018). Thin film membrane for CO_2 separation with sweeping gas method. *Energy* 144: 619–626.

78 AlQahtani, M.S. and Mezghani, K. (2018). Thermally rearranged polypyrrolone membranes for high-pressure natural gas separation applications. *J. Nat. Gas Sci. Eng.* 51: 262–270.

79 Shokri Doodeji, M., Zerafat, M.M., Yousefi, M.H., and Sabbaghi, S. (2018). Effect of OH-treatment of PDMS on rejection in hybrid nanofiltration membranes for desalination. *Desalination* 426: 60–68.

80 Goh, P.S. and Ismail, A.F. (2018). A review on inorganic membranes for desalination and wastewater treatment. *Desalination* 434: 60–80.

81 Kubaczka, A. (2014). Prediction of Maxwell–Stefan diffusion coefficients in polymer-multicomponent fluid systems. *J. Membr. Sci.* 470: 389–398.

82 Prince, J.A., Bhuvana, S., Anbharasi, V. et al. (2016). Ultra-wetting graphene-based PES ultrafiltration membrane-a novel approach for successful oil-water separation. *Water Res.* 103: 311–318.

83 Nasseri, S., Ebrahimi, S., Abtahi, M., and Saeedi, R. (2018). Synthesis and characterization of polysulfone/graphene oxide nanocomposite membranes for removal of bisphenol A from water. *J. Environ. Manage.* 205: 174–182.

84 Zhang, J., Xue, Q., Pan, X. et al. (2017). Graphene oxide/polyacrylonitrile fiber hierarchical-structured membrane for ultra-fast microfiltration of oil-water emulsion. *Chem. Eng. J.* 307: 643–649.

85 Zhu, L., Wang, H., Bai, J. et al. (2017). A porous graphene composite membrane intercalated by halloysite nanotubes for efficient dye desalination. *Desalination* 420: 145–157.

86 Hu, X., Yu, Y., Zhou, J. et al. (2015). The improved oil/water separation performance of graphene oxide modified Al_2O_3 microfiltration membrane. *J. Membr. Sci.* 476: 200–204.

87 Ho, K.C., Teow, Y.H., Ang, W.L., and Mohammad, A.W. (2017). Novel GO/OMWCNTs mixed-matrix membrane with enhanced antifouling property for palm oil mill effluent treatment. *Sep. Purif. Technol.* 177: 337–349.

88 Zhang, R., Wang, S., Yang, Y. et al. (2018). Modification of polydopamine-coated Fe_3O_4 nanoparticles with multi-walled carbon nanotubes for magnetic-μ-dispersive solid-phase extraction of antiepileptic drugs in biological matrices. *Anal. Bioanal.Chem.* 410: 3779–3788.

89 Duan, A.-H., Xie, S.-M., and Yuan, L.-M. (2011). Nanoparticles as stationary and pseudo-stationary phases in chromatographic and electrochromatographic separations. *TrAC Trends Anal. Chem.* 30: 484–491.

90 Peng, L.-Q., Ye, L.-H., Cao, J. et al. (2017). Separation of metal ions via capillary electrophoresis using a pseudostationary phase microfunctionalized with carbon nanotubes. *Microchim. Acta* 184: 1747–1754.

91 Kwon, S.H. and Park, J.H. (2006). Intermolecular interactions on multiwalled carbon nanotubes in reversed-phase liquid chromatography. *J. Sep. Sci.* 29: 945–952.

92 Chang, Y.X., Ren, C.X., Ruan, Q., and Yuan, L. (2007). Effect of singlewalled carbon nanotubes on cellulose phenylcarbamate chiral stationary phases. *Chem. Res. Chin. Univ.* 23: 646–649.

93 Chang, Y.X., Zhou, L.L., Li, G.X. et al. (2007). Single-wall carbon nanotubes used as stationary phase in HPLC. *J. Liq. Chromatogr. Related Technol.* 30: 2953–2958.

94 Andre, C., Gharbi, T., and Guillaume, Y.C. (2009). A novel stationary phase based on amino derivatized nanotubes for HPLC separations: theoretical and practical aspects. *J. Sep. Sci.* 32: 1757–1764.

95 Aqel, A., Yusuf, K., Al-Othman, Z.A. et al. (2012). Effect of multi-walled carbon nanotubes incorporation into benzyl methacrylate monolithic columns in capillary liquid chromatography. *Analyst* 137: 4309–4317.

96 Valcarcel, M., Simonet, B.M., Cardenas, S., and Suarez, B. (2005). Present and future applications of carbon nanotubes to analytical science. *Anal. Bioanal.Chem.* 382: 1783–1790.

97 Valcarcel, M., Cárdenas, S., Simonet, B.M. et al. (2008). Carbon nanostructures as sorbent materials in analytical processes. *TrAC Trends Anal. Chem.* 27: 34–43.

98 Speltini, A., Merli, D., Quartarone, E., and Profumo, A. (2010). Separation of alkanes and aromatic compounds by packed column gas chromatography using functionalized multi-walled carbon nanotubes as stationary phases. *J. Chromatogr. A* 1217: 2918–2924.

99 Chambers, S.D., Svec, F., and Fréchet, J.M.J. (2011). Incorporation of carbon nanotubes in porous polymer monolithic capillary columns to enhance the chromatographic separation of small molecules. *J. Chromatogr. A* 1218: 2546–2552.

100 Li, Q.L. and Yuan, D.X. (2003). Evaluation of multi-walled carbon nanotubes as gas chromatographic column packing. *J. Chromatogr. A* 1003: 203–209.

101 Karwa, M. and Mitra, S. (2006). Chromatography of self-assembled singe walled carbon nanotubes. *Anal. Chem.* 78: 2064–2070.

102 Stadermann, M., McBrady, A.D., Dick, B. et al. (2006). Ultrafast gas chromatography on single-wall carbon nanotube stationary phases in microfabricated channels. *Anal. Chem.* 78: 5639–5644.

103 Yuan, L.M., Ren, C.X., Li, L. et al. (2006). Single-walled carbon nanotubes used as stationary phase in GC. *Anal. Chem.* 78: 6384–6390.

104 Zhao, L., Ai, P., Duan, A.-H., and Yuan, L.-M. (2011). Single-walled carbon nanotubes for improved enantioseparations on a chiral ionic liquid stationary phase in GC. *Anal. Bioanal.Chem.* 399: 143–147.

105 Glausch, A., Hirsch, A., Lamparth, I., and Schurig, V. (1998). Retention behaviour of polychlorinated biphenyls on polysiloxane-anchored C60 in gas chromatography. *J. Chromatogr. A* 809: 252–257.

106 Fang, P.F., Zeng, Z.R., Fan, J.H., and Chen, Y.Y. (2000). Synthesis and characteristics of [60] fullerene polysiloxane stationary phase for capillary gas chromatography. *J. Chromatogr. A* 867: 177–185.

107 Kartsova, L.A. and Makarov, A.A. (2004). New fullerene-based stationary phases for gas chromatography. *J. Anal. Chem.* 59: 724–729.

5

Membrane Applications of Nanomaterials (NMs)

5.1 Introduction

Membrane technology is extensively employed in various separation processes including wastewater sample treatment, desalination, gas separation, etc. [1–9]. Membrane separation process is usually conducted based on selective transport of the target compound across the membrane structure [10]. Nanofiltration and reverse osmosis membranes are widely applied for the treatment of wastewater samples. However, thick separating layer restricts their water flux behavior. This drawback can be overcome by incorporating nanomaterials such as graphene, fullerenes, carbon nanotubes (CNTs), and nanoparticles (NPs) into the structure of the membrane. The combination of membrane materials with the superior properties of nanomaterials enables great physical and chemical stability and also high rejection toward the target compound(s) to be separated from the sample.

Nanomaterials have gained great attention from many researchers since they display unique properties such as high physical and chemical stability, versatile chemistry for the functionalization, and high surface area [11–22]. The chemical features of nanomaterials in the scale between 1 and 100 nm are completely different compared to the materials in micro- and macroscale. If the size of a material decreases, its surface area increases. This feature is crucial for the binding processes. More binding sites exist on the nanomaterial surface.

This chapter provides an overview of the latest progresses and examples of new nanomaterial-based membranes. It starts with a brief information about the traditional membrane materials. Then, carbon nanomaterial-based membranes including graphene-based, CNT-based, and fullerene-based membranes are described. In the final sections, nanoparticle-based and molecularly imprinted polymer (MIP)-based membranes and their applications are demonstrated.

5.2 Traditional Membranes

The design and development of new generation membrane materials has received great interest by the researchers in the past decades. However, the industrial applications of nanotechnology-based novel membranes are still under development. Most

Sample Preparation with Nanomaterials: Next Generation Techniques and Applications,
First Edition. Chaudhery Mustansar Hussain, Rüstem Keçili, and Chaudhery Ghazanfar Hussain.

of the traditional membranes are still prepared by using polymers and inorganic materials and applied for the separation processes in industry. A type of these traditional membranes is ceramic membranes that are commonly produced by using metal hydroxides (e.g. ZnO and Al_2O_3, etc.) on the surface of porous supporting materials [23, 24]. These kinds of membranes are widely applied for the industrial separation processes. In addition to their increased chemical stability and mechanical resistance, they can easily be recovered after fouling [25]. However, ceramic membranes have some disadvantages such as high operation costs and their brittle structure that limit their applications in large scale [26–28].

Polymer-based membranes are also promising materials in separation processes since they are low-cost materials and show high performance. Polymer-based porous membranes are generally prepared by using mechanical stretching and phase inversion process. Although these types of membranes show great separation performance toward the target compound, they have some drawbacks such as stability under extreme operational conditions such as corrosive environments, organic solvents, high pressure, and temperature values.

Recently, the thin film-based membranes, which are the major breakthrough in the membranes science, have received a great interest for the separation applications. These composite membrane materials having selective thin layers and the supporting layers are highly flexible for the separation medium [29]. However, the thin film-based membranes exhibit some limitations such as compaction effects caused by multilayer structure under pressure. At higher pressure values, the polymeric structure is reorganized to the tighter fitting structure that leads to lower porosity and lower efficiency in separation process [30].

5.3 Carbon Nanomaterial-based Membranes

Since the carbon nanomaterials have unique features such as excellent mechanical, chemical, and thermal stabilities, large surface area, great optical features, and lower density, etc., they are extensively used in the fabrication of novel nanocomposite membranes. The incorporation of carbon nanomaterials such as graphene, CNTs, and fullerenes into the membrane structure is commonly applied in the development of nanocomposite membranes [31]. The combination of nanomaterials with the membranes provides not only high physical and chemical stability and great rejection and flux behaviors of the developed nanocomposite membranes but also introduces various features such as catalytic and antibacterial properties [32, 33].

5.3.1 Graphene-based Membranes

Graphene was discovered by Geim and Novoselov in 2004 [34]. Since then, many efforts were put into the design and development of new graphene-based functional materials. Due to the excellent hydrophobic feature of graphene, it was widely used in the preparation of functional materials with superhydrophobic features. Recently, these superhydrophobic functional materials were efficiently applied for the separation of water–oil mixture [35, 36].

Graphene oxide (GO), which is the oxidation state of graphene, is a promising candidate since it has various functional groups such as carboxylic acid, hydroxyl, and epoxide that provide good compatibility with other materials such as polymeric composite membranes. When the GO is incorporated in the polymeric membrane structure, the mechanical features of the polymer are significantly improved [37]. The horizontal orientation of the GO in the polymeric membrane structure provides a barrier of GO nanolayers since the nanolayers hinder the diffusion of permeating molecules of the target compound.

Prince et al. [38] proposed a facile method to improve the hydrophilic feature of the polyethersulfone hollow fiber ultrafiltration membrane. For this purpose, graphene-attached polyacrylonitrile-*co*-maleimide having hydroxyl, carboxyl, and amine functional groups was prepared. The characterization of the prepared graphene-based nanocomposite membranes was performed by using scanning electron microscope (SEM), energy-dispersive X-ray (EDX) analyses, pore size, and contact angle measurements. The results showed that the incorporation of graphene-attached polyacrylonitrile-*co*-maleimide into the polyethersulfone hollow fiber membrane leads to a decrease of water contact angle (64.5% decrease). On the other hand, oil contact angle increased from 43.6° to 112.5° compared to the control membrane. The water permeability of the prepared graphene-based nanocomposite membrane was enhanced by 43%, and the efficient removal of oil was successfully achieved with a rate of 99%.

In another research carried out by Nasseri et al. [39], GO/polysulfone nanocomposite membrane was prepared for the efficient removal of bisphenol A from water samples. For this purpose, three different membranes were prepared by using phase inversion technique. The obtained results indicated that the incorporation of GO into the membrane structure considerably increased the permeate flux of the prepared nanocomposite membranes. The maximum bisphenol A (BPA) removal was achieved by using the GO (0.4%)/polysulfone nanocomposite membrane that has the highest negative zeta potential value (−10.46 mV). The optimum process conditions for BPA removal such as time, initial BPA concentration, pressure, and pH were determined as 10.6 minutes, 7.5 $mg\,l^{-1}$, 1.02 bar, and 5.5, respectively. Under optimum conditions, the removal efficiency of the target pollutant BPA by using the prepared GO (0.4%)/polysulfone membrane was obtained as 93%.

Zhang et al. developed a GO-based polyacrylonitrile fiber nanocomposite membrane for the filtration of water–oil emulsion [40]. In their study, polyacrylonitrile nanofibers were produced by electrospinning technique and GO was assembled on the surface of nanofibers. The prepared GO-based polyacrylonitrile nanocomposite membrane exhibited superhydrophilic feature and low oil adhesion behavior. The developed nanocomposite membrane exhibited great rejection value (≥98%) and antifouling performance.

In another study [41], Li et al. designed and prepared a GO membrane grafted with polysulfone (PSf) for the efficient separation of dye compounds including Congo red (CR), acid black 1 (AB) and methyl blue (MB) in aqueous samples. For this purpose, the synthesis of PSf-grafted (G) nanosheets (GO-g-PSf) was conducted through a nucleophilic substitution reaction between the chloromethyl groups of chloromethylated polysulfone (CMPSf) and hydroxyl groups of GO. GO-g-PSf dispersion in *N*-methyl pyrrolidone (NMP) was applied for the preparation of GO-g-PSf nanocomposite membranes for the removal of target dye compounds by applying a vacuum filtration self-assembly approach by using polyamide (PA) composite membranes as support. The schematic depiction of the preparation of GO-based nanocomposite membrane toward dye compounds is shown in Figure 5.1. The achieved results confirmed that the rejection rates of the developed GO-g-PSf nanocomposite membrane for the target dyes CR, AB, and MB were 99.5%, 90.3%, and 99.8%, respectively.

In another interesting study, Hu et al. developed a GO-based nanocomposite membrane [42]. The prepared nanocomposite membrane was effectively used for the treatment of water–oil emulsion. The obtained values for water flux of the prepared nanocomposite membrane having GO and control membrane without GO were 677 and 522 $l\,h^{-1}\,m^2\,bar^{-1}$, respectively, after 2.5-hour filtration time. On the other hand, the achieved oil rejections of the prepared GO-based nanocomposite membrane and unmodified control membrane were found as 98.7% and 98.1%, respectively, after two-hour filtration time. It has been reported that the incorporation of GO into the membrane structure has significantly increased the performance of the membranes for the separation of water–oil emulsions.

Ho et al. reported the design and preparation of novel GO-multiwalled carbon nanotubes (MWCNTs)-based polyvinylidene fluoride nanocomposite membrane for the treatment of palm oil mill effluent [43]. The nanocomposite membrane was prepared by using in situ colloidal precipitation technique. The rejection ability of the prepared GO-MWCNT-based nanocomposite membrane was investigated

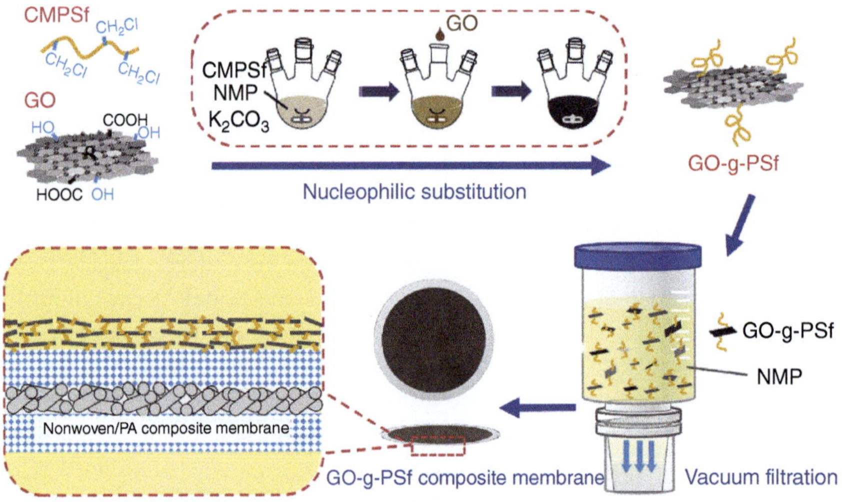

Figure 5.1 The schematic depiction of the preparation of GO-based nanocomposite membrane toward dye compounds. Source: Li et al. [41]. © 2019, Elsevier.

using palm oil mill effluent. The obtained results indicated that the developed nanocomposite membrane can be successfully used for the treatment of palm oil mill effluent. The rejection values of phosphorus, chemical oxygen demand, total dissolved solids, hardness, turbidity, total suspended solids, chlorine, and color were enhanced by 6.55%, 75.5%, 1.51%, 21.79%, 81.94%, 100%, 76%, and 86.3%, respectively, using GO-MWCNT-based nanocomposite membrane compared to the control membrane without GO and MWCNTs.

5.3.2 Carbon Nanotube-based Membranes

Since the discovery of CNTs in 1991 [44], many researchers focused on the use of CNTs in different applications. CNTs are cylindrical structures consisting of rolled graphene sheets and categorized as MWCNTs and single-walled carbon nanotubes (SWCNTs). CNTs have many superiorities such as excellent oleophobicity and hydrophilicity [45] and high mechanical [46], electrical [47], and thermal features [48]. The combination of these unique features with the membrane systems can be successfully carried out for the design and preparation of novel nanocomposite membranes.

In another work published by Zhang et al. [49], a novel MWCNTs-GO-based nanocomposite membrane was designed and prepared for the filtration of Sr^{2+} ions from wastewater. For this purpose, the preparation of GO membranes that are interlinked with MWCNTs on the surface of polyacrylonitrile support was carried out. Then, the prepared nanocomposite membrane was effectively used for the removal of Sr^{2+} ions from wastewater. The results showed that the prepared nanocomposite membrane displayed excellent water flux behavior (four times higher) compared to the conventional membranes. The rejection value for Sr^{2+}-EDTA (ethylenediaminetetraacetic acid) chelate was obtained as 93.4%.

In another research [50], Mahdavi et al. developed MWCNT-based piperazine/polyamide nanocomposite membranes for the removal of salt from water samples. In their study, oxidized and raw MWCNTs were used in the preparation of different nanocomposite membranes. The antifouling features, permeation, and salt rejection behaviors of the prepared raw and oxidized MWCNTs-embedded nanocomposite membranes and unmodified piperazine/polyamide membranes were investigated in details. The results showed that the incorporation of oxidized and raw MWCNTs into the membrane structure significantly increased the salt removal performance of the piperazine/polyamide membranes.

Burç et al. designed and developed a voltammetric sensor composed of polyimide (PI)/MWCNT nanocomposite membrane for the sensitive recognition of paracetamol in commercial tablet samples [51]. For this purpose, PI–MWCNT nanocomposite layer was used for the modification of Pt electrode as the selective membrane (Figure 5.2). In this study, the limit of detection and limit of quantification values were achieved as 0.002 and 0.006 mM, respectively.

Shahzad et al. reported the fabrication of MWCNTs–Al_2O_3-based nanocomposite membranes for the effective removal of Cd^{2+} ions from aqueous solutions [52]. The effects of various parameters such as sintering temperature and initial

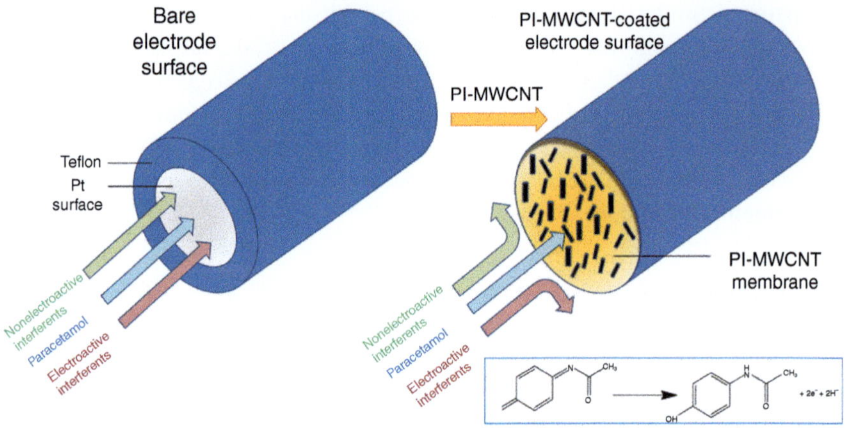

Figure 5.2 The schematic demonstration of bare and PI-MWCNT electrode and the recognition of paracetamol. Source: Burç et al. [51]. © 2019, Elsevier.

compaction load on the water flux behavior, strength, and porosity of the MWCNTs–Al_2O_3-based nanocomposite membranes were investigated. In the final step, the prepared nanocomposite membranes were used for Cd^{2+} removal from aqueous solutions in batch mode. The results confirmed that 93% of Cd^{2+} ions in aqueous solutions were successfully removed by using the prepared nanocomposite membranes.

In an interesting study reported by Saadati and Pakizeh [53], MWCNT-based Pebax–poly(sulfone) nanocomposite membrane was prepared for the separation of water–oil emulsion. In this study, different amounts of MWCNTs in the range from 0.5 to 2.0 wt% were incorporated into the membrane structure. Thermal stability and hydrophilicity of the prepared nanocomposite membrane increased at higher loading of MWCNTs into the membrane structure. The maximum rejection of oil was achieved using the nanocomposite membrane having 2.0 wt% MWCNTs.

Zhang et al. [45] developed CNT-based nanocomposite membranes for the efficient separation of oil–water and catalytic decomposition. For this purpose, polystyrene (PS) microspheres were modified with 3-aminopropyl triethoxysilane (APTES), and Au nanoparticles (Au NPs) were attached to the modified PS microsphere membrane. Then, CNTs layer was deposited on the surface of Au NPs–PS microsphere membrane (Figure 5.3). The obtained CNT-based nanocomposite membrane was successfully used for the catalytic decomposition of nitrophenol (92.6% decomposition) from oil-contaminated water samples and efficient separation of oil–water emulsion.

In another important research [54], CNT-based nanocomposite membranes for the treatment of water–oil emulsions were prepared by Gu et al. In their study, polyacrylic acid was grafted on the CNTs, and Ag nanoparticles were then immobilized. The results confirmed that the obtained CNT-based nanocomposite membranes exhibited excellent superoleophobic features toward different organic solvents such as carbon tetrachloride and dichloromethane. In addition, CNT-based

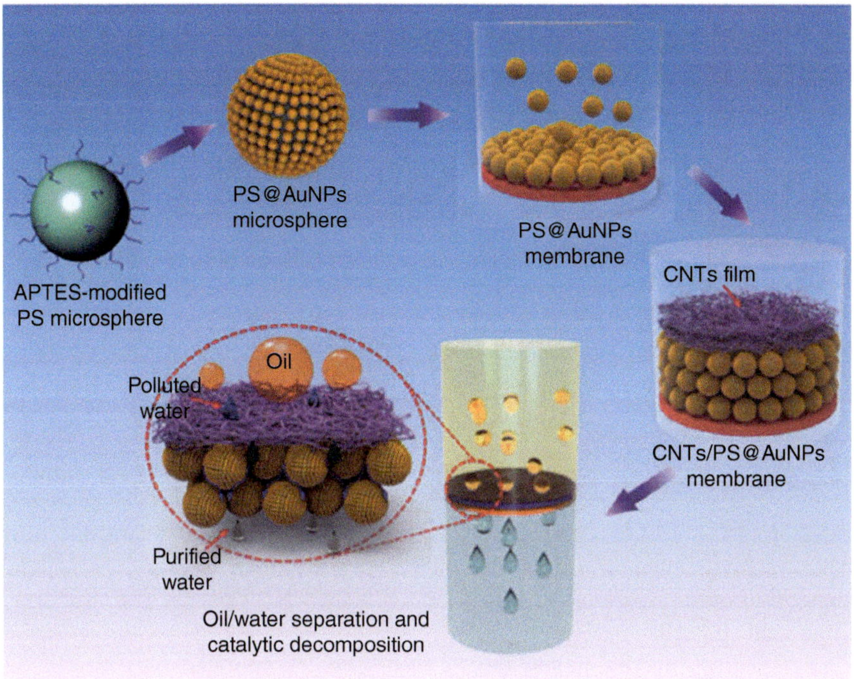

Figure 5.3 The schematic demonstration of the preparation of CNT-based Au NPs–PS nanocomposite membrane. Source: Zhang et al. [45]. © 2020, Royal Society of Chemistry.

nanocomposite membranes showed high superhydrophilicity. The results also indicated that the prepared nanocomposite membranes can be efficiently used for the separation of water–oil emulsions. It was reported that the incorporation of Ag nanoparticles into the nanocomposite membrane structure provides high antibacterial feature that may open a new perspective for the design and development of multifunctional membranes for the separation processes.

Maphutha et al. developed a CNT-based nanocomposite membrane having polyvinyl alcohol layer for the separation of oil from contaminated wastewater [55]. For this purpose, chemical vapor deposition technique was used for the synthesis of CNTs. Then, CNTs were incorporated into the membrane structure. The prepared CNT-based nanocomposite membrane displayed high water flux behavior with excellent rejection of oil from wastewater (over 95%).

Gu et al. reported the preparation of CNT-based nanocomposite membranes with superhydrophobic feature for the efficient removal of oil from water samples [56]. In this study, CNTs were first immobilized on the surface of Al_2O_3 membrane. Then, covalent attachment of PS to the membrane surface was carried out. Figure 5.4 shows the schematic demonstration of the fabrication of CNTs–PS nanocomposite membrane. The prepared nanocomposite membrane was successfully applied for the separation of oil–water emulsion. The efficiency of separation and water flux value were obtained as 99.94% and $5000\,l\,m^{-2}\,h^{-1}\,bar^{-1}$, respectively.

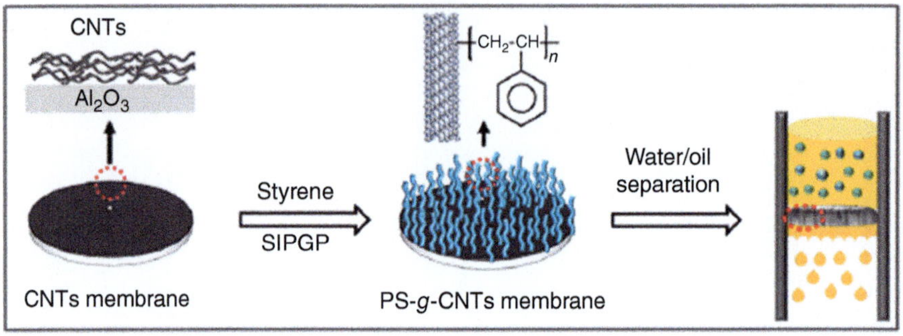

Figure 5.4 The schematic demonstration of the fabrication of CNTs–PS nanocomposite membrane for water–oil separation. Source: Gu et al. [56]. © 2014, Royal Society of Chemistry.

5.3.3 Fullerene-based Membranes

Fullerenes are a class of carbon allotropes that consist of 60 C atoms organized in hexagons and pentagons. The main difference between fullerenes and CNTs is their shapes. Fullerenes are cage-like structures, while the CNTs are tube-like structures in nanoscale [57, 58].

Fullerenes as effective nanomaterials were also incorporated into membrane structures and the developed fullerene-based nanocomposite membranes were successfully used for the treatment of environmental samples such as wastewater. For example, Chen et al. developed a nanocomposite membrane composed of polyvinyl butyral/polyvinylidene fluoride and fullerene coated with F-127 [59]. The characterization studies of the prepared nanocomposite membrane were performed by using FT-IR, SEM, XPS, AFM, and thermal gravimetric analysis (TGA). Then, the prepared nanocomposite membrane was successfully used for the separation of bovine serum albumin, ovalbumin, and bromelain from wastewater samples. The obtained highest rejection value for the target proteins was around 98.6%.

In another interesting study, a fullerene-based sulfonated polyvinyl alcohol nanocomposite membrane for the removal of Cu^{2+} ions from wastewater was developed by Rikame et al. [60]. The developed fullerene-based nanocomposite membrane showed high removal performance toward Cu^{2+} ions in wastewater, and the highest removal degree was found as 73.2%.

Jin et al. reported the preparation of fullerene-based composite membranes for the estrone removal from aqueous samples [61]. In this research, poly(2,6-dimethyl-1,4-phenylene oxide) was modified with fullerene. Two types of nanocomposite membrane were prepared by using different amounts of fullerene (2% and 10%). The results confirmed that the prepared nanocomposite membrane having 10% fullerene exhibited great binding behavior toward the target compound estrone (c. 95% binding).

Another fullerene-based nanocomposite membrane was developed by Saga et al. [62]. In their study, fullerene and sulfonated PS were used for the preparation of nanocomposite membranes. The incorporation of the fullerene into the membrane structure significantly improved the oxidation resistance and decreased the methanol permeability, while the mechanical strength of the membrane was not affected. The results showed that the prepared fullerene-based nanocomposite membranes displayed 30% and 50% lower methanol permeability values than Nafion 117 membrane and PS membrane, respectively.

5.4 Nanoparticle-based Membranes

Nanoparticle-based membranes are also popular materials that are commonly used for separation processes. For example, Yu et al. prepared yttrium nanoparticle-based polysulfone membrane modified with polyvinyl alcohol for the efficient separation of arsenate from water [63]. In this study, yttrium nanoparticles were embedded into the structure of polysulfone membrane having polyvinyl alcohol layer. The developed yttrium nanoparticle-based membrane displayed high removal performance for arsenate in water. The obtained maximum binding capacity of the membrane was 35.56 $mg\,g^{-1}$ at pH 7.0.

In another interesting study [64], gold nanoparticle-based polydopamine–polyethylenimine nanocomposite membranes were developed for the efficient removal of salts from water samples. For this purpose, co-deposition of polydopamine (pDA) and polyethylenimine in the presence of gold nanoparticles was carried out. The prepared nanocomposite membranes successfully removed bivalent cations such as $ZnCl_2$, $BaCl_2$, $NiCl_2$, and $CdCl_2$. The obtained rejection values were 88.3%, 90.6%, 90.4%, and 86.8%, respectively. The results also showed that the prepared gold nanoparticle-based composite membranes exhibited excellent antibacterial activity toward *Escherichia coli* and *Staphylococcus aureus*.

Akbari et al. reported the preparation of polyethylene nanocomposite membranes having silica nanoparticles for humic acid removal from water samples [65]. In their study, the synthesis of poly(ethylene glycol)-grafted silica nanoparticles was carried out by using sol–gel technique and used for the preparation of high-density polyethylene nanocomposite membranes by using thermally induced phase separation technique. The prepared silica nanoparticle-based polyethylene membranes efficiently removed humic acid from water samples. The prepared nanocomposite membranes also exhibited great water flux behavior, which is about 7.45 times higher than the polyethylene membranes without silica nanoparticles.

Karim et al. developed chitosan-based nanocomposite membranes having cellulose nanocrystals for the efficient removal of various dye compounds such as methyl violet 2B, victoria blue 2B, and rhodamine 6G from water samples [66]. For this purpose, cellulose nanocrystals were embedded into the polymeric structure of chitosan. The chitosan-based nanocomposite membranes effectively removed the target dye compounds from water samples (Figure 5.5). The achieved removal efficiencies of rhodamine 6G, methyl violet 2B, and Victoria blue 2B were 70%, 84%, and 98%, respectively.

Figure 5.5 The schematic demonstration of the removal of target dye compounds using chitosan-based nanocomposite membranes having cellulose nanocrystals (a) nanocomposite membrane, (b) methyl violet 2B removal from water samples, (c) rhodamine 6G removal from water samples, and (d) victoria 2B removal from water samples. Source: Reproduced with permission from Karim et al. [66].

5.5 Molecularly Imprinted Polymer (MIP)-based Membranes

MIPs are highly cross-linked materials that possess specific binding sites for the target compound(s) [67–70]. MIPs are prepared by polymerization of suitable functional monomers in the existence of the desired compound also called "template"

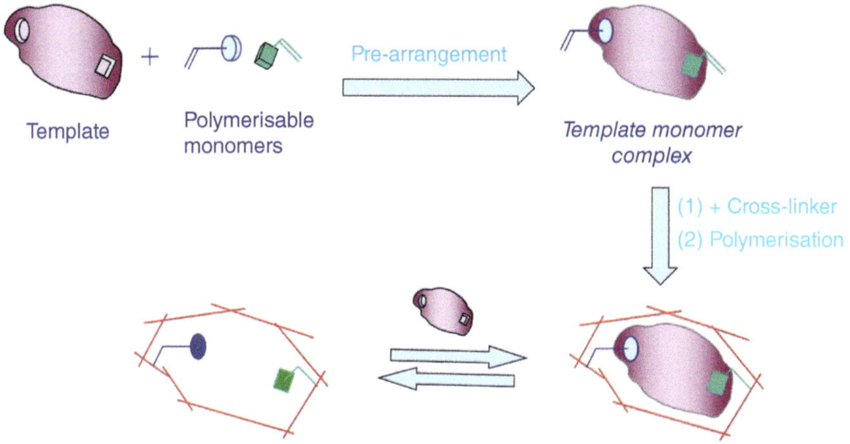

Figure 5.6 The schematic depiction of the molecular imprinting approach. Source: Turiel and Martín-Esteban [71]. © 2019, Elsevier.

and a cross-linker. Figure 5.6 shows the schematic depiction of the molecular imprinting approach.

The design and preparation of novel membranes that exhibit high selectivity and permeation behavior toward the target compound(s) are still a crucial and challenging issue. The recent studies showed that the combination of membrane technology and molecular imprinting technology is an excellent approach for the development of composite membranes. These composite membranes can efficiently be applied for the separation of target compounds [72, 73].

In an important research reported by Roy et al. [74], molecularly imprinted composite membranes bearing ZnS nanoparticles modified with cysteine functional groups were prepared for the removal of As^{3+} and As^{5+} ions from water samples. The obtained results confirmed that the prepared MIP-based nanocomposite membranes exhibited excellent removal behavior toward the target ions. The highest removal capacity values of the composite membrane for the target ions As^{3+} and As^{5+} were found to be 151 and 130 mg g^{-1}, respectively.

Wu et al. designed and prepared MIP-based polydopamine/GO nanocomposite membrane having dendrites-like silver microbeads for the efficient and selective removal of ibuprofen from aqueous samples [75]. The schematic demonstration of the preparation of nanocomposite membrane toward ibuprofen is shown in Figure 5.7. In their study, the maximum adsorption capacity for the target ibuprofen was achieved as 61.55 mg g^{-1}.

In another study [76], Gao et al. developed MWCNTs/MIP-based nanocomposite membranes for the efficient separation of enoxacin from wastewater samples. In their study, MWCNTs were first modified with pDA, and then the target compound enoxacin was imprinted into the pDA/MWCNTs structure. In the final step, pDA/MWCNTs/polyvinylidene fluoride membrane was prepared by using immersion phase inversion technique. Figure 5.8 shows the schematic demonstration of the preparation of MWCNT-based enoxacin-imprinted nanocomposite membrane.

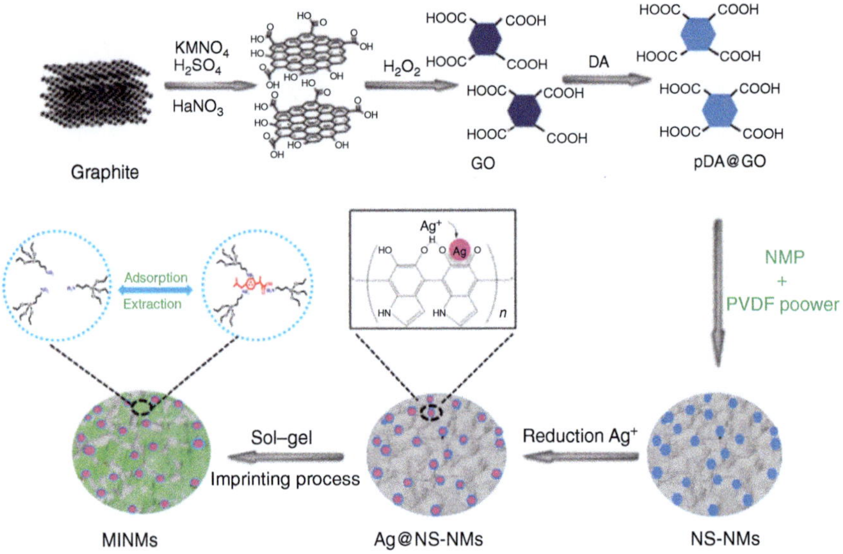

Figure 5.7 The preparation of MIP-based polydopamine/GO nanocomposite membrane having dendrites-like silver microbeads toward ibuprofen. Source: Wu et al. [75]. © 2018, Elsevier.

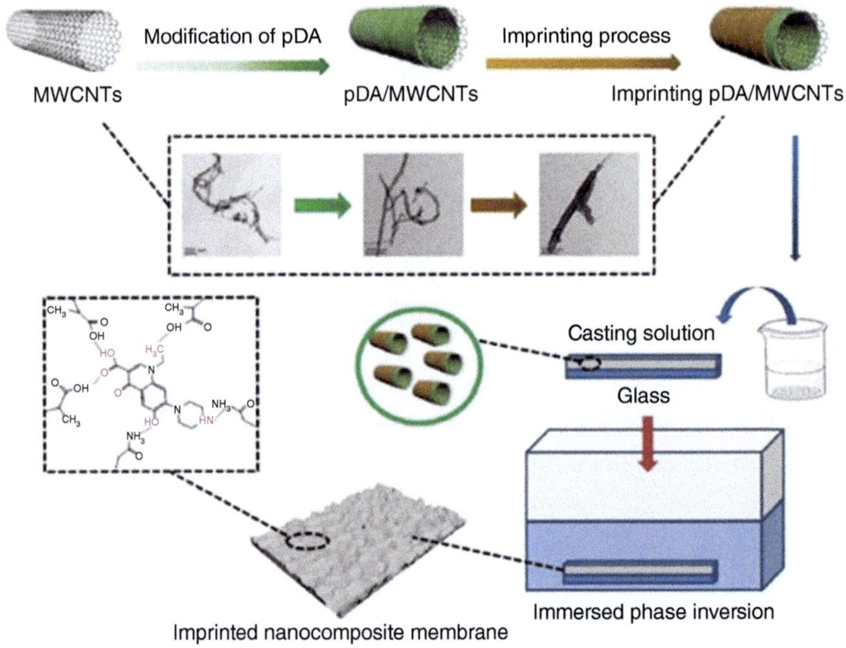

Figure 5.8 The schematic demonstration of the preparation of MWCNT-based enoxacin-imprinted nanocomposite membrane. Source: Gao et al. [76]. © 2018, Elsevier.

The results showed that the prepared MWCNTs/MIP-based nanocomposite membrane displayed great removal performance toward the target compound enoxacin from wastewater samples. The binding capacity of the imprinted membranes for enoxacin was achieved as 31.56 $mg\,g^{-1}$.

Lu et al. developed thermosensitive GO-based ion-imprinted nanocomposite membranes for the selective separation of Eu^{3+} ions [77]. The grafting of silver nanoparticles on the surface of nanocomposite membranes to enhance the antifouling feature was also carried out. Thermosensitive recognition and binding sites of the imprinted membranes were created by using functional monomers acrylamide and *N*-isopropylacrylamide. The prepared GO-based ion-imprinted nanocomposite membranes were successfully applied for the selective separation of Eu^{3+} ions in the existence of Gd^{3+}, La^{3+}, and Sm^{3+}. The binding capacity of the prepared GO-based ion-imprinted nanocomposite membranes toward Eu^{3+} ions was obtained as 101.14 $mg\,g^{-1}$.

In another interesting study carried out by Wang et al. [78], an electrochemical sensor system having boron-embedded MIP and MWCNT-modified carbon paste electrode (CPE) was developed for the sensitive detection of tinidazole (TNZ) in pharmaceutical, serum, and urine samples. The fabrication of MIP/MWCNT nanocomposite membrane-based electrochemical sensor toward TNZ and sensing process is schematically depicted in Figure 5.9. The obtained results indicated that the prepared MIP/MWCNT nanocomposite membrane-based electrochemical sensor displayed excellent affinity and selectivity for the target compound TNZ. The detection limit was achieved as 1.25×10^{-12} M.

Some latest researches on the application of composite membranes based on nanomaterials are shown in Table 5.1.

5.6 Conclusions

The increasing number of research in which membrane technology based on nanomaterials was applied for the separation processes showed that these nanocomposite materials can be efficiently used for the selective separation of target compound(s) from environmental samples. The demonstrated examples in this chapter highlight the recent progresses in the design and fabrication of composite membranes based on nanomaterials such as graphenes, fullerenes, CNTs, and nanoparticles. The combination of membranes with the unique features of nanomaterials provides great physical and chemical stability and also high rejection toward the target compound to be separated from the sample. In addition, the combination of nanotechnology with the membrane technology provides new ways to design and fabrication of novel nanocomposite membrane platforms in many application areas.

In spite of the many successful examples of nanomaterial-based composite membranes in the literature, the technologies applied for the fabrication of nanomaterial-incorporated membrane platforms are still under development and more efforts are needed to produce low-cost, efficient, selective, and reusable

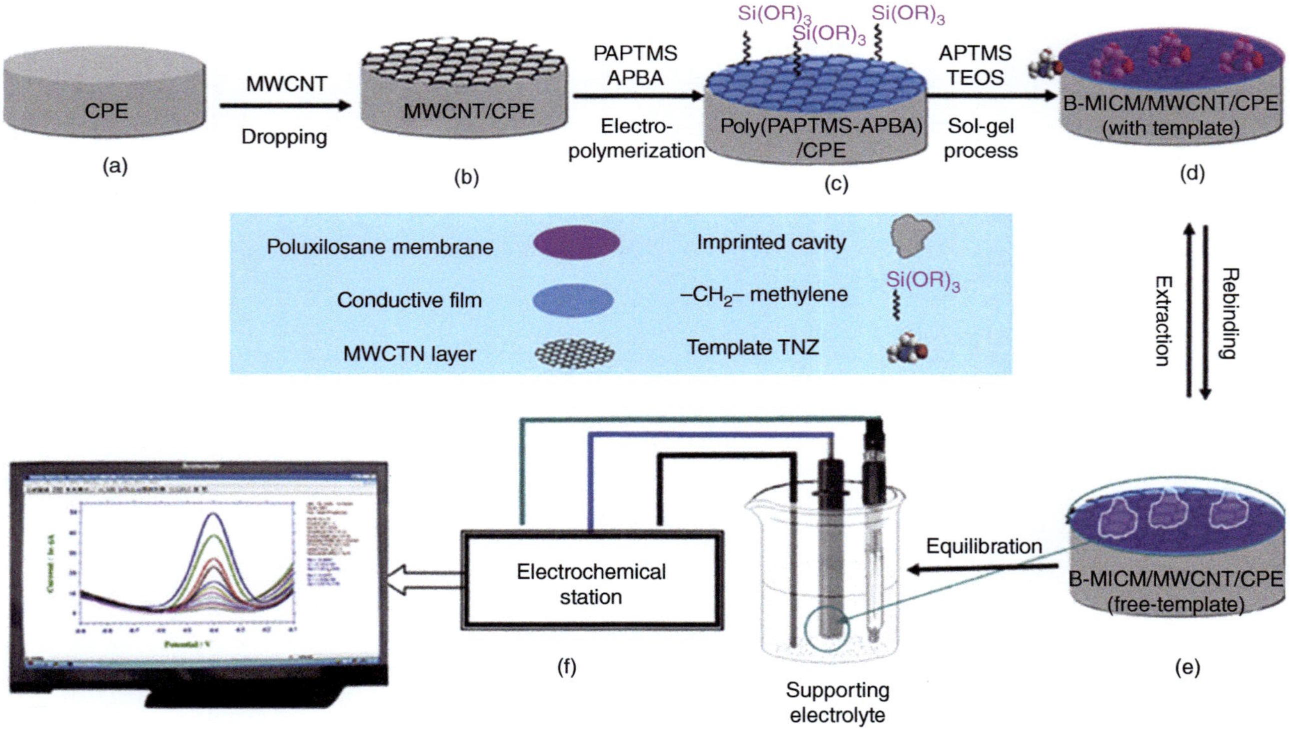

Figure 5.9 The fabrication of MIP/MWCNT nanocomposite membrane-based electrochemical sensor toward TNZ and sensing process. Source: Wang et al. [78]. © 2018, Elsevier.

Table 5.1 Some latest researches on the application of composite membranes based on nanomaterials.

Membrane composition	Analyte	Sample	Reference
MWCNT modified with Fe and Ag nanoparticles/polyethersulfone nanocomposite membrane	Cr(VI)	Water	[79]
MWCNT/graphene oxide-based nanocomposite membrane	Acid red B, methyl blue, and light blue A	Water	[80]
MWCNT modified with tetraethylene pentamine/polyurethane nanocomposite membrane	CO_2/CH_4	Gas	[81]
SiO_2 nanoparticle/MIP-based nanocomposite membrane	Ferulic acid	Water	[82]
MIP-based nanomembrane using nylon 6	Acesulfame	Beverage samples	[83]
MIP/CNT-based nanocomposite membrane	Tartrazine	Beverage samples	[84]
Magnetic Fe_3O_4 nanoparticles/polyvinylidene fluoride nanocomposite membrane	Sodium chloride	Water	[85]
Polyaniline/graphene oxide nanoparticle-based nanocomposite membrane	Pb(II)	Water	[86]
Zinc oxide/graphene oxide nanoparticle-based nanocomposite membrane	Sodium chloride and sodium sulfate	Water	[87]
MIP/graphene oxide nanoparticle-based nanocomposite membrane	Polycyclic aromatic hydrocarbons	Water	[88]
Magnetic Fe_3O_4 nanoparticles/chitosan/polyvinyl alcohol nanofiber-based polyethersulfone nanocomposite membrane	Pb(II) and Cr(VI)	Water	[89]
Molecularly imprinted nanofiber membrane	Bisphenol A	Wastewater	[90]
MIP-based polysulfone nanocomposite membrane	Paclitaxel	Plant extract	[91]
MIP/TiO_2 nanoparticle-based polysulfone nanocomposite membrane	Methyl orange and methylene blue	Water	[92]
MIP-based polyvinylidene fluoride nanocomposite membrane	Artemisinin	Water	[93]
Zinc oxide nanoparticles/ 1-butyl-3-methylimidazolium tetrafluoroborate nanocomposite membrane	CO_2	Gas	[94]

(Continued)

Table 5.1 (Continued)

Membrane composition	Analyte	Sample	Reference
MIP-based alumina nanocomposite membrane	Phenol	Salicylic acid effluent	[95]
MWCNT-based polysulfone hollow fiber nanocomposite membrane	Humic acid and procion red	Water	[96]
Graphene oxide-based polyamide nanocomposite membrane	Sodium chloride and sodium sulfate	Water	[97]
Graphene oxide/polyethylenimine nanocomposite membrane	Phenanthrene, bovine serum albumin, and lysozyme	Water	[98]
TiO_2 nanoparticle-based polysulfone nanocomposite membrane	Cr(VI)	Water	[99]
TiO_2 nanoparticle-based polysulfone nanocomposite membrane doped with Fe	Bisphenol A	Water	[100]
Fluorinated TiO_2 nanoparticle-based polyvinylidene fluoride nanocomposite hollow fiber membrane	CO_2	Gas–liquid	[101]
Nanosilica-based ceramic membrane	Methylene blue	Aqueous solutions	[102]
TiO_2 nanoparticle-based polyvinylidene difluoride and polyacrylonitrile nanocomposite membranes	Phage F2 as the model virus	Drinking water	[103]

membrane platforms before these nanocomposite membrane platforms are commercially available in the market.

References

1 Lan, Y., Groenen-Serrano, K., Coetsier, C., and Causserand, C. (2018). Nanofiltration performances after membrane bioreactor for hospital wastewater treatment: fouling mechanisms and the quantitative link between stable fluxes and the water matrix. *Water Res.* 146: 77–87.

2 Shi, B.-J., Wang, Y., Geng, Y.-K. et al. (2018). Application of membrane bioreactor for sulfamethazine-contained wastewater treatment. *Chemosphere* 193: 840–846.

3 Ben-Mansour, R., Li, H., and Habib, M.A. (2018). Thin film membrane for CO_2 separation with sweeping gas method. *Energy* 144: 619–626.

4 Karimi, M.B., Khanbabaei, G., and Sadeghi, G.M.M. (2017). Vegetable oil-based polyurethane membrane for gas separation. *J. Membr. Sci.* 527: 198–206.

5 AlQahtani, M.S. and Mezghani, K. (2018). Thermally rearranged polypyrrolone membranes for high-pressure natural gas separation applications. *J. Nat. Gas Sci. Eng.* 51: 262–270.

6 James, J.B. and Lin, Y.S. (2017). Thermal stability of ZIF-8 membranes for gas separations. *J. Membr. Sci.* 532: 9–19.

7 Goh, P.S. and Ismail, A.F. (2018). A review on inorganic membranes for desalination and wastewater treatment. *Desalination* 434: 60–80.

8 Shokri Doodeji, M., Zerafat, M.M., Yousefi, M.H., and Sabbaghi, S. (2018). Effect of OH-treatment of PDMS on rejection in hybrid nanofiltration membranes for desalination. *Desalination* 426: 60–68.

9 Chen, P., Ma, X., Zhong, Z. et al. (2017). Performance of ceramic nanofiltration membrane for desalination of dye solutions containing NaCl and Na_2SO_4. *Desalination* 404: 102–111.

10 Kubaczka, A. (2014). Prediction of Maxwell–Stefan diffusion coefficients in polymer–multicomponent fluid systems. *J. Membr. Sci.* 470: 389–398.

11 Ahmadi, M., Elmongy, H., Madrakian, T., and Abdel-Rehim, M. (2017). Nanomaterials as sorbents for sample preparation in bioanalysis: a review. *Anal. Chim. Acta* 958: 1–21.

12 Hussain, C.M. (2017). Magnetic nanomaterials for environmental analysis. In: *Advanced Environmental Analysis-Application of Nanomaterials* (eds. C.M. Hussain and B. Kharisov). The Royal Society of Chemistry.

13 Keçili, R. and Hussain, C.M. (2018). Recent progress of ımprinted nanomaterials in analytical chemistry. *Int. J. Anal. Chem.*: 8503853.

14 Büyüktiryaki, S., Sümbelli, Y., Keçili, R., and Hussain, C.M. (2019). Lab-on-chip platforms for environmental analysis. In: *Encyclopedia of Analytical Science*, 3e (eds. P. Worsfold, C. Poole, A. Townshend and M. Miró), 267–273. Academic Press.

15 Keçili, R., Büyüktiryaki, S., and Hussain, C.M. (2019). Advancement in bioanalytical science through nanotechnology: past, present and future. *TrAC Trends Anal. Chem.* 110: 259–276.

16 Hussain, C.M. and Kharisov, B. (2017). *Advanced Environmental Analysis-Application of Nanomaterials*. The Royal Society of Chemistry.

17 Hussain, C.M. (2018). *Nanomaterials in Chromatography: Current Trends in Chromatographic Research Technology and Techniques*. Elsevier.

18 Hussain, C.M. (2019). *Handbook of Nanomaterials in Analytical Chemistry: Modern Trends in Analysis*. Elsevier.

19 Hussain, C.M. and Keçili, R. (2020). *Modern Environmental Analysis Techniques for Pollutants*. The Netherlands: Elsevier.

20 Keçili, R., Büyüktiryaki, S., and Hussain, C.M. (2020). Micro total analysis systems with nanomaterials, Chapter 8. In: *Handbook of Nanomaterials in Analytical Chemistry* (ed. C.M. Hussain), 185–198. Elsevier.

21 Pyrzynska, K. (2020). Nanomaterials in speciation analysis of metals and metalloids. *Talanta* 212: 120784.

22 Murtada, K. (2020). Trends in nanomaterial-based solid-phase microextraction with a focus on environmental applications – a review. *Trends Environ. Anal. Chem.* 25: e00077.
23 Vercauteren, S., Keizer, K., Vansant, E. et al. (1988). Porous ceramic membranes: preparation, transport properties and applications. *J. Porous. Mater.* 5: 241–258.
24 Sondhi, R., Bhave, R., and Jung, G. (2003). Applications and benefits of ceramic membranes. *Membr. Technol.* 2003 (11): 5–8.
25 Faibish, R.S. and Cohen, Y. (2001). Fouling-resistant ceramic-supported polymer membranes for ultrafiltration of oil-in-water microemulsions. *J. Membr. Sci.* 185: 129–143.
26 Tsuru, T. (2008). Nano/subnano-tuning of porous ceramic membranes for molecular separation. *J. Sol-Gel Sci. Technol.* 46: 349–361.
27 Wu, P., Xu, Y.Z., Huang, Z.X., and Zhang, J.C. (2015). A review of preparation techniques of porous ceramic membranes. *J. Ceram. Process. Res.* 16: 102–106.
28 Pendergast, M.M. and Hoek, E.M.V. (2011). A review of water treatment membrane nanotechnologies. *Energy Environ. Sci.* 4: 1946–1971.
29 Rao, A.P., Desai, N., and Rangarajan, R. (1997). Interfacially synthesized thin film composite RO membranes for sea water desalination. *J. Membr. Sci.* 124: 263–272.
30 Oak, M.S., Kobayashi, T., Wang, H.Y. et al. (1997). pH effect on molecular size exclusion of polyacrylonitrile ultrafiltration membranes having carboxylic acid groups. *J. Membr. Sci.* 123: 185–195.
31 Goh, K., Karahan, H.E., Wei, L. et al. (2016). Carbon nanomaterials for advancing separation membranes: a strategic perspective. *Carbon* 109: 694–710.
32 Liu, S., Zeng, T.H., Hofmann, M. et al. (2011). Antibacterial activity of graphite, graphite oxide, graphene oxide, and reduced graphene oxide: membrane and oxidative stress. *ACS Nano* 5 (9): 6971–6980.
33 Zhang, Q., Fan, X., Wang, H. et al. (2016). Fabrication of Au/CNT hollow fiber membrane for 4-nitrophenol reduction. *RSC Adv.* 6: 41114–41121.
34 Geim, A.K. and Novoselov, K.S. (2007). The rise of graphene. *Nat. Mater.* 6 (3): 183–191.
35 Jayaramulu, K., Datta, K.K.R., Rösler, C. et al. (2016). Biomimetic superhydrophobic/superoleophilic highly fluorinated graphene oxide and ZIF-8 composites for oil-water separation. *Angew. Chem. Int. Ed.* 55 (3): 1178–1182.
36 Li, B., Liu, X., Zhang, X. et al. (2015). Oil-absorbent polyurethane sponge coated with KH-570-modified graphene. *J. Appl. Polym. Sci.* 132 (16): 41821–41828.
37 Le, N.L., Duong, P.H.H., and Nunes, S.P. (2017). Advanced polymeric and organic_inorganic membranes for pressuredriven processes. In: *Comprehensive Membrane Science and Engineering*, 2e (eds. E. Drioli, L. Giorno and E. Fontananova), 120–136. Amsterdam, The Netherlands: Elsevier.
38 Prince, J.A., Bhuvana, S., Anbharasi, V. et al. (2016). Ultra-wetting graphene-based PES ultrafiltration membrane-a novel approach for successful oil-water separation. *Water Res.* 103: 311–318.

39 Nasseri, S., Ebrahimi, S., Abtahi, M., and Saeedi, R. (2018). Synthesis and characterization of polysulfone/graphene oxide nanocomposite membranes for removal of bisphenol A from water. *J. Environ. Manag.* 205: 174–182.

40 Zhang, J., Xue, Q., Pan, X. et al. (2017). Graphene oxide/polyacrylonitrile fiber hierarchical-structured membrane for ultra-fast microfiltration of oil-water emulsion. *Chem. Eng. J.* 307: 643–649.

41 Li, J., Hu, M., Pei, H. et al. (2020). Improved water permeability and structural stability in a polysulfone-grafted graphene oxide composite membrane used for dye separation. *J. Membr. Sci.* 595: 117547.

42 Hu, X., Yu, Y., Zhou, J. et al. (2015). The improved oil/water separation performance of graphene oxide modified Al_2O_3 microfiltration membrane. *J. Membr. Sci.* 476: 200–204.

43 Ho, K.C., Teow, Y.H., Ang, W.L., and Mohammad, A.W. (2017). Novel GO/OMWCNTs mixed-matrix membrane with enhanced antifouling property for palm oil mill effluent treatment. *Sep. Purif. Technol.* 177: 337–349.

44 Iijima, S. (1991). Helical microtubules of graphitic carbon. *Nature* 354: 56–58.

45 Zhang, L., Gu, J., Song, L. et al. (2016). Underwater superoleophobic carbon nanotubes/core-shell polystyrene@Au nanoparticles composite membrane for flow-through catalytic decomposition and oil/water separation. *J. Mater. Chem. A* 4 (28): 10810–10815.

46 Bai, L., Bossa, N., Qu, F. et al. (2017). Comparison of hydrophilicity and mechanical properties of nanocomposite membranes with cellulose nanocrystals and carbon nanotubes. *Environ. Sci. Technol.* 51 (1): 253–262.

47 Sarno, M., Tamburrano, A., Arurault, L. et al. (2013). Electrical conductivity of carbon nanotubes grown inside a mesoporous anodic aluminium oxide membrane. *Carbon* 55: 1022.

48 Namasivayam, M. and Shapter, J. (2017). Factors affecting carbon nanotube fillers towards enhancement of thermal conductivity in polymer nanocomposites: a review. *J. Compos. Mater.* 51 (26): 3657–3668.

49 Zhang, L., Lu, Y., Liu, Y.-L. et al. (2016). High flux MWCNTs-interlinked GO hybrid membranes survived incross-flow filtration for the treatment of strontium-containing wastewater. *J. Hazard. Mater.* 320: 187–193.

50 Mahdavi, M.R., Delnavaz, M., and Vatanpour, V. (2017). Fabrication and water desalination performance of piperazine–polyamide nanocomposite nanofiltration membranes embedded with raw and oxidized MWCNTs. *J. Taiwan Inst. Chem. Eng.* 75: 189–198.

51 Burç, M., Köytepe, S., Titretir Duran, S. et al. (2020). Development of voltammetric sensor based on polyimide-MWCNT composite membrane for rapid and highly sensitive detection of paracetamol. *Measurement* 151: 107103.

52 Shahzad, H.K., Hussein, M.A., Patel, F. et al. (2018). Synthesis and characterization of alumina-CNT membrane for cadmium removal from aqueous solution. *Ceram. Int.* 44 (14): 17189–17198.

53 Saadati, J. and Pakizeh, M. (2017). Separation of oil/water emulsion using a new PSf/pebax/F-MWCNT nanocomposite membrane. *J. Taiwan Inst. Chem. Eng.* 71: 265–276.

54 Gu, J., Xiao, P., Zhang, L. et al. (2016). Construction of superhydrophilic and underwater superoleophobic carbon-based membranes for water purification. *RSC Adv.* 6 (77): 73399–73403.

55 Maphutha, S., Moothi, K., Meyyappan, M., and Iyuke, S.E. (2013). A carbon nanotube-infused polysulfone membrane with polyvinyl alcohol layer for treating oil-containing waste water. *Sci. Rep.* 3: 1509.

56 Gu, J., Xiao, P., Chen, J. et al. (2014). Robust preparation of superhydrophobic polymer/carbon nanotube hybrid membranes for highly effective removal of oils and separation of water-in-oil emulsions. *J. Mater. Chem. A* 2 (37): 15268–15272.

57 Thines, R.K., Mubarak, N.M., Nizamuddin, S. et al. (2017). Application potential of carbon nanomaterials in water and wastewater treatment: A review, *J. J. Taiwan Inst. Chem. E* 72: 116–133.

58 Wang, H., DeSousa, R., Gasa, J. et al. (2007). Fabrication of new fullerene composite membranes and their application in proton exchange membrane fuel cells. *J. Membr. Sci.* 289: 277–283.

59 Chen, G.-E., Zhu, W.-W., Xu, S.-J. et al. (2016). A PVDF/PVB composite UF membrane improved by F-127-wrapped fullerene for protein waste-water separation. *RSC Adv.* 6: 83510–83519.

60 Rikame, S.S., Mungray, A.A., and Mungray, A.K. (2017). Synthesis, characterization and application of phosphorylated fullerene/sulfonated polyvinyl alcohol (PFSP) composite cation exchange membrane for copper removal. *Sep. Purif. Technol.* 177: 29–39.

61 Jin, X., Hu, J.Y., Tint, M.L. et al. (2007). Estrogenic compounds removal by fullerene-containing membranes. *Desalination* 214: 83–90.

62 Saga, S., Matsumoto, H., Saito, K. et al. (2008). Polyelectrolyte membranes based on hydrocarbon polymer containing fullerene. *J. Power Sources* 176: 16–22.

63 Yu, Y., Yu, L., Wang, C., and Chen, J.P. (2018). An innovative yttrium nanoparticles/PVA modified PSF membrane aiming at decontamination of arsenate. *J. Colloid Interface Sci.* 530: 658–666.

64 Lv, Y., Du, Y., Chen, Z.-X. et al. (2018). Nanocomposite membranes of polydopamine/electropositive nanoparticles/polyethyleneimine for nanofiltration. *J. Membr. Sci.* 545: 99–106.

65 Akbari, A., Yegani, R., Pourabbas, B., and Behboudi, A. (2016). Fabrication and study of fouling characteristics of HDPE/PEG grafted silica nanoparticles composite membrane for filtration of humic acid. *Chem. Eng. Res. Des.* 109: 282–296.

66 Karim, Z., Mathewa, A.P., Grahn, M. et al. (2014). Nanoporous membranes with cellulose nanocrystals as functional entity in chitosan: removal of dyes from water. *Carbohydr. Polym.* 112: 668–676.

67 Keçili, R. (2018). Selective recognition of myoglobin in biological samples using molecularly imprinted polymer-based affinity traps. *Int. J. Anal. Chem.* 2018: 4359892.

68 Emir Diltemiz, S., Keçili, R., Ersöz, A., and Say, R. (2017). Molecular imprinting technology in quartz crystal microbalance (QCM) sensors. *Sensors* 17 (3): 454.

69 Keçili, R., Atilir Özcan, A., Ersöz, A. et al. (2011). Superparamagnetic nanotraps containing MIP based mimic lipase for biotransformations uses. *J. Nanopart. Res.* 13 (5): 2073–2079.

70 Kubo, T. and Otsuka, K. (2016). Recent progress for the selective pharmaceutical analyses using molecularly imprinted adsorbents and their related techniques: a review. *J. Pharm. Biomed. Anal.* 130: 68–80.

71 Turiel, E. and Martín-Esteban, A. (2019). Molecularly imprinted polymers-based microextraction techniques. *Trends Anal. Chem.* 118: 574–586.

72 Wu, Y.L., Yan, M., Yan, Y.S. et al. (2014). Fabrication and evaluation of artemisinin-ɪmprinted composite membranes by developing a surface functional monomer-directing prepolymerization system. *Langmuir* 30: 14789–14796.

73 Lu, J., Qin, Y., Meng, M. et al. (2020). Bidirectional molecularly imprinted membranes for selective recognition and separation of pyrimethamine: a double-faced loading strategy. *J. Membr. Sci.* 601: 117917.

74 Roy, E., Patra, S., Madhuri, R., and Sharma, P.K. (2016). A single solution for arsenite and arsenate removal from drinking water using cysteine@ZnS:TiO_2 nanoparticle modified molecularly imprinted biofouling-resistant filtration membrane. *Chem. Eng. J.* 304: 259–270.

75 Wu, X., Wu, Y., Chen, L. et al. (2018). Bioinspired synthesis of pDA@GO-based molecularly imprinted nanocomposite membranes assembled with dendrites-like Ag microspheres for high-selective adsorption and separation of ibuprofen. *J. Membr. Sci.* 553: 151–162.

76 Gao, J., Wu, Y., Cui, J. et al. (2018). Bioinspired synthesis of multi-walled carbon nanotubes based enoxacin-imprinted nanocomposite membranes with excellent antifouling and selective separation properties, *J. J. Taiwan Inst. Chem. E* 91: 468–480.

77 Lu, J., Wu, Y., Lin, X. et al. (2018). Anti-fouling and thermosensitive ion-imprinted nanocomposite membranes based on grapheme oxide and silicon dioxide for selectively separating europium ions. *J. Hazard. Mater.* 353: 244–253.

78 Wang, H., Qian, D., Xiao, X. et al. (2018). Preparation and application of a carbon paste electrode modified with multi-walled carbon nanotubes and boron-embedded molecularly imprinted composite membranes. *Bioelectrochemistry* 121: 115–124.

79 Masheane, M.L., Nthunya, L.N., Malinga, S.P. et al. (2017). Synthesis of Fe-Ag/f-MWCNT/PES nanostructured-hybrid membranes for removal of Cr(VI) from water. *Sep. Purif. Technol.* 184: 79–87.

80 Kang, H., Shi, J., Liu, L. et al. (2018). Sandwich morphology and superior dye-removal performances for nanofiltration membranes self-assembled via graphene oxide and carbon nanotubes. *Appl. Surf. Sci.* 428: 990–999.

81 Gheimasi, K.M., Bakhtiaria, O., and Ahmadi, M. (2018). Preparation and characterization of MWCNT-TEPA/polyurethane nanocomposite membranes for

CO_2/CH_4 separation: experimentaland modeling. *Chem. Eng. Res. Des.* 133: 222–234.

82 Wei, M.-H., Chen, H.-Y., Wang, S. et al. (2017). Synthesis and characterization of hybrid molecularly imprinted membrane with blending SiO_2 nanoparticles for ferulic acid. *J. Inorg. Organomet. Polym.* 27: 586–597.

83 Moein, M.M., Javanbakht, M.K., and Akbari-adergani, B. (2015). Fabrication of a novel electrospun molecularly imprinted nanomembrane coupled with high-performance liquid chromatography for the selective separation and determination of acesulfame. *J. Sep. Sci.* 38: 1372–1379.

84 Yaripour, S., Mohammadi, A., and Nojavan, S. (2016). Electromembrane extraction of tartrazine from food samples: effects of nano-sorbents on membrane performance. *J. Sep. Sci.* 39: 2642–2651.

85 Agbaje, T.A., Al-Gharabli, S., Mavukkandy, M.O. et al. (2018). PVDF/magnetite blend membranes for enhanced flux and salt rejection in membrane distillation. *Desalination* 436: 69–80.

86 Ghaemi, N., Zereshki, S., and Heidari, S. (2017). Removal of lead ions from water using PES based nanocomposite membrane incorporated with polyaniline modified GO nanoparticles: performance optimization by central composite design. *Process. Saf. Environ. Prot.* 111: 475–490.

87 Al Mayyahi, A. (2018). Thin-film composite (TFC) membrane modified by hybrid ZnO-graphene nanoparticles (ZnO-Gr NPs) for water desalination. *J. Environ. Chem. Eng.* 6: 1109–1117.

88 Kibechu, R.W., Ndinteh, D.T., Msagati, T.A.M. et al. (2017). Effect of incorporating graphene oxide and surface imprinting on polysulfone membranes on flux, hydrophilicity and rejection of salt and polycyclic aromatic hydrocarbons from water, *Phys. Chem. Earth* 100: 126–134.

89 Koushkbaghi, S., Zakialamdari, A., Pishnamazi, M. et al. (2018). Aminated-Fe_3O_4 nanoparticles filled chitosan/PVA/PES dual layers nanofibrous membrane for the removal of Cr(VI) and Pb(II) ions from aqueous solutions in adsorption and membrane processes. *Chem. Eng. J.* 337: 169–182.

90 Liu, F., Liu, Q., Zhang, Y. et al. (2015). Molecularly imprinted nanofiber membranes enhanced biodegradation of trace bisphenol A by Pseudomonas aeruginosa. *Chem. Eng. J.* 262: 989–998.

91 Ghasemi, S. and Nematollahzadeh, A. (2018). Molecularly imprinted ultrafiltration polysulfone membrane with specific nano-cavities for selective separation and enrichment of paclitaxel from plant extract. *React. Funct. Polym.* 126: 9–19.

92 Melvin, H.K., Leo, C.P., and Abdullah, A.Z. (2017). Selective removal of dyes by molecular imprinted TiO_2 nanoparticles in polysulfone ultrafiltration membrane. *J. Environ. Chem. Eng.* 5: 3991–3998.

93 Cui, J., Wu, Y., Meng, M. et al. (2016). Bio-inspired synthesis of molecularly imprinted nanocomposite membrane for selective recognition and separation of artemisinin. *J. Appl. Polym. Sci.* 103: 43405.

94 Yoon, K.W., Kim, H., Kang, Y.S., and Kang, S.W. (2017). 1-Butyl-3-methylimidazolium tetrafluoroborate/zinc oxide composite membrane for high CO_2 separation performance. *Chem. Eng. J.* 320: 50–54.

95 Liu, Y., Meng, M., Yao, J. et al. (2016). Selective separation of phenol from salicylic acid effluent over molecularly imprinted polystyrene nanospheres composite alumina membranes. *Chem. Eng. J.* 286: 622–631.

96 Yin, J., Zhu, G., and Deng, B. (2013). Multi-walled carbon nanotubes (MWNTs)/polysulfone (PSU)mixed matrix hollow fiber membranes for enhanced water treatment. *J. Membr. Sci.* 437: 237–248.

97 Yin, J., Zhu, G., and Deng, B. (2016). Graphene oxide (GO) enhanced polyamide (PA) thin-film nanocomposite (TFN) membrane for water purification. *Desalination* 379: 93–101.

98 Zhang, L., Chen, B., Ghaffar, A., and Zhu, X. (2018). Nanocomposite membrane with polyethylenimine-grafted graphene oxide as a novel additive to enhance pollutant filtration performance. *Environ. Sci. Technol.* 52: 5920–5930.

99 Jyothi, M.S., Nayak, V., Padaki, M. et al. (2016). Aminated polysulfone/TiO_2 composite membranes for an effective removal of Cr(VI). *Chem. Eng. J.* 283: 1494–1505.

100 Wang, Q., Yang, C., Zhang, G. et al. (2017). Photocatalytic Fe-doped TiO_2/PSF composite UF membranes: characterization and performance on BPA removal under visible-light irradiation. *Chem. Eng. J.* 319: 39–47.

101 Lin, Y., Xu, Y., Loh, C.H., and Wang, R. (2018). Development of robust fluorinated TiO_2/PVDF composite hollow fiber membrane for CO_2 capture in gas-liquid membrane contactor. *Appl. Surf. Sci.* 436: 670–681.

102 Tolba, G.M.K., Bastaweesy, A.M., Ashour, E.A. et al. (2016). Effective and highly recyclable ceramic membrane based on amorphous nanosilica for dye removal from the aqueous solutions. *Arab. J. Chem.* 9: 287–296.

103 Xiang, Z., Di, C., Zhiwei, W. et al. (2013). Nano-TiO_2 membrane adsorption reactor (MAR) for virus removal in drinking water. *Chem. Eng. J.* 230: 180–187.

6

Surface-Enhanced Raman Spectroscopy (SERS) with Nanomaterials (NMs)

6.1 Introduction

Surface-enhanced Raman Spectroscopy (SERS) is an efficient analytical tool that is extensively used in ultrasensitive sensing applications, and it is capable of significantly enhancing the Raman signals for trace-level detections of the target compounds on the designed substrates [1–10]. Because of this fascinating ability, SERS has received a great interest in the areas of chemistry, physics, biology, and medical sciences to understand the enhancement mechanism and to investigate its potential applications. SERS was first discovered in 1974 by Fleischman et al. [11], who could successfully detect Raman signals of pyridine adsorbed on the surface of a silver electrode. Afterward, a number of studies on this subject such as those of Albrech and Creighton [12], Jeanmaire and Van Duyne investigated the crucial factors of the exaltation impact that are related to the surface state and the nature of the metal, as well as the wavelength of the excitation source [13]. In an interesting research carried out by Moskovits in 1978 [14], it was successfully enabled to distinguish two types of contributions that can be observed in the SERS exaltation, which are chemical and electromagnetic mechanisms.

The research on SERS technique recently focused on various aspects including the design and development of novel SERS-active substrates, investigation of the enhancement mechanism, application of computational approaches, optimization of the parameters for efficient SERS detection, improvement of the sensitivity and reproducibility of SERS and applications of SERS in broad fields.

On the other hand, nanomaterials (NMs) have received great interest from many scientists due to their superior features such as large surface area, small size, high stability, high reactivity, and versatile chemistry for the further functionalization [15–27].

Generated by the surface plasmon resonances of metal nanostructures, strong localized electromagnetic field enhancement can be achieved at locations that are also called "hot spots." The probe molecules located in these locations will provide much strong SERS signals. In the literature, various studies on describing beyond the optical diffraction limit were carried out [28–30].

Various metallic nanostructured SERS substrates were designed and fabricated in order to create hot spots with superior SERS enhancement. For example, Tao

Sample Preparation with Nanomaterials: Next Generation Techniques and Applications,
First Edition. Chaudhery Mustansar Hussain, Rüstem Keçili, and Chaudhery Ghazanfar Hussain.

et al. used cucurbit[*n*]urils (CB[*n*]) to assemble gold nanoparticle–nanoparticle aggregates [29]. In their work, the distance between two connected nanoparticles can be controlled by the size of organic bridged ligands, CB[*n*] molecules, and the molecule can be used as a useful SERS hot-spot nanocontainer to study the SERS enhancement.

Netzer et al. reported that the nanoparticles–nanowire junctions can be formed by a buckling process by studying the SERS activities of an individual junction using a confocal Raman mapping method. In their study, firstyl, aligning of nanowires onto the prestrained polydimethylsiloxane (PDMS) substrates was carried out and then a compressive stress was applied to the substrate to form nanobuckles on the surface of initially "smooth" silver nanowire. Then, the hot spots can be successfully formed in the location of the junction between nanobuckles and nanowire. Finally, the enhanced SERS process can be conducted at these locations [31]. The exploration of SERS theories and fabrication of functional NMs have clearly demonstrated the great potential for SERS technique, ranging from single molecular experiments, biomaterials sensing, to chemical identification. The active interesting topics in SERS research include the design of novel SERS substrates, fabrication of highly uniform and sensitive nanostructured substrates, and expansion of the SERS applications in biomedicine [32–34], sensor platforms [8, 35–39], and bioimaging [40–44].

In this chapter, an overview of SERS technique by using NMs is provided. The chapter starts with the theory and a brief description of SERS. Then, the mechanism of SERS is demonstrated. The determination of SERS enhancement factor (EF) is also described. In Section 6.6, fabrication approaches of SERS substrates are presented. Finally, various applications of SERS including separation, environmental analysis, forensic analysis, and biological applications are discussed.

6.2 Theory of SERS

Electromagnetic and charge transfer enhancements have been commonly accepted as the two main contributions to the giant enhancement. The electromagnetic enhancement can increase the scattering intensity by $\sim 10^4$. The chemical enhancement mechanism contributes to an EF of 10^2. The largest electromagnetic enhancement mechanism occurs for surfaces that are rough on the nanoscale (10–100 nm), while the roughness from atomic to 10 nm scale creates the largest charge transfer enhancement. These two techniques operate multiplicatively. Therefore, there is still disagreement over the relative and absolute EFs for each. This is mainly because electromagnetic and charge transfer models are critical roughness-based models. The schematic demonstration of the SERS technique is shown in Figure 6.1.

6.3 SERS Mechanisms

There are two primary theories, which are electromagnetic enhancement [45] and chemical enhancement [46] for the explanation of the SERS mechanisms. As implied by their names, the electromagnetic enhancement emphasizes on the

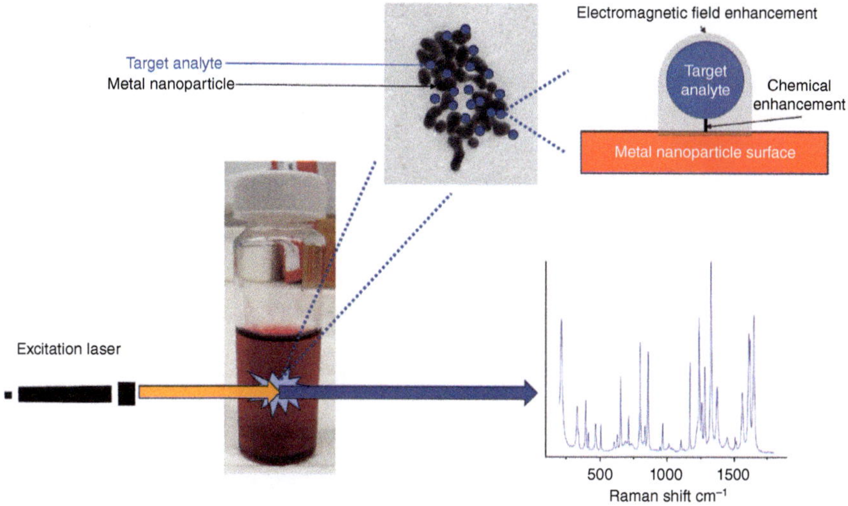

Figure 6.1 The schematic demonstration of SERS technique. Source: Ong TTX, Blanch EW, Jones OAH (2020) Surface Enhanced Raman Spectroscopy in environmental analysis, monitoring and assessment, Science of the Total Environment 720, 137601. © 2020, Elsevier.

amplified electromagnetic fields by the interaction between metals and incident electromagnetic waves, and the chemical enhancement focuses on the changes of electronic states or charge transfers within molecules induced by the interaction between metals and molecules. These two mechanisms differ substantially from each other in theory, leading to different functions and influences on SERS.

Electromagnetic enhancement is believed to

- be universal to and present in all SERS cases and
- contribute the majority of the total enhancement.

On the other hand, chemical enhancement is expected to

- be present only in cases where chemical bonding is formed between molecule and metal,
- produce varying enhancements on molecules depending on their structures, and
- have limited contribution to the total enhancement.

6.3.1 Electromagnetic Enhancement

The electromagnetic enhancement mechanism arises from the excitation of localized surface plasmons, which is collective excitations of the electron gas within the metal that are confined to the surface regions. SERS requires localized surface plasmons on a nanostructured metallic (ideally with a negative real and small positive imaginary dielectric constant) surface, which are excited at the known localized surface plasmon resonance (LSPR) frequency [47]. This frequency is sensitive to the shape, size, material, and morphology of the nanoparticle, as well as the local dielectric environment [48]. Both Au and Ag nanoparticles (Ag NPs) are extensively used for SERS studies while gold is more easily functionalized and biocompatible, Ag NPs have a stronger LSPR absorption, surface plasmons with higher energy, and a resulting greater local electric field enhancement [49]. Sharp corners in metallic nanostructures can dramatically increase the observed SERS intensity. Electromagnetic EFs of up to 1011 have been estimated for jagged nanostructures including nanocubes, nanotriangles, and nanostars [50–53]. Additionally, close association or clustering of nanoparticles has been shown to significantly increase the electromagnetic EF [53].

Nanoscale gaps between closely associated nanoparticles are termed "hot spots" areas in which the SERS enhancement is drastically increased. In the 2 nm gap size between two coupled Ag NPs, which only comprises <1% of the total surface area, adsorbed analytes are dramatically enhanced such that they contribute to >50% of the total achieved signal of SERS [54]. Additionally, when NPs closely associate, the surface plasmons are able to couple, generating broad plasmon resonances that enable excitation from the visible frequencies to the near infrared [55]. Aggregates of Ag NPs will be shown in this thesis to form fractal patterns, resulting in extended networks of SERS-active "hot spots."

Theoretically, SERS effect may be observed by any metal, but Au NPs, Ag NPs, and alkali metals usually provide the strongest enhancement. This could simply be explained as these metals satisfy the resonance conditions at the visible frequencies commonly used in Raman spectroscopy. In conclusion, electromagnetic enhancement process accounts for the majority of the total Raman signal enhancement because of the high resonance of electric field. Moreover, because this mechanism has nothing to do with the analyte, it is expected to provide same impact on the molecules being studied and be active in any SERS case no matter whether the molecule is chemically or physically adsorbed on the surface. Therefore, most of the research on SERS technique has been targeted to this all-purpose mechanism since the discovery of SERS [56–64].

6.3.2 Chemical Enhancement

A further contribution to the enhancement of the Raman signal is given by mechanism of chemical nature. The molecule must be directly adsorbed to the roughened surface in order to experience the chemical enhancement mechanism [65, 66]. This mechanism is site-specific and analyte-dependent. Chemisorption of the molecules on to the roughened surface enables new resonant intermediate states. Therefore, while the electromagnetic enhancement mechanism enables a greater measured signal by locally increasing both the incident and scattered electric fields, the chemical enhancement mechanism increases the signal by directly increasing the Raman cross section of the surface-interacting analytes.

Regarding the magnitude, chemical enhancement typically enables one order or two to the total Raman signal enhancement [67]. Usually, study on chemical enhancement is challenging since chemical enhancement and electromagnetic enhancement are difficult to be separated. To overcome this drawback, Campion et al. performed a test by using an atomically flat Cu (111) surface to exclude contribution of electromagnetic enhancement. They ended up with a chemical EF of 30 in their study [68]. Another crucial feature of chemical enhancement is quite obvious in that it involves interaction with molecules being researched. As a result, the enhancement varies from case to case with different molecules. Furthermore, the molecule has to bond to metal surface in a chemical manner instead of physical in order to induce the enhancement [69]. Overall, chemical enhancement has limited contribution and role in SERS.

6.4 Determination of SERS Enhancement Factor

The quality of a system for SERS analysis is often assessed through the EF, that is, the ratio of the obtained SERS signal to the normal Raman signal. The SERS EF is usually described by using the following equation:

$$\mathrm{EF} = \frac{I_{\mathrm{SERS}}/N_{\mathrm{SERS}}}{I_{\mathrm{NRS}}/N_{\mathrm{NRS}}} \tag{6.1}$$

In this equation, I and N refer to the intensity and the number of molecules within the probe volume, respectively, and the subscript normal Raman scattering (NRS) refers to the normal Raman scattering condition (i.e. with a non-SERS active system) [70]. Since it is uncommon to be able to precisely quantify the number of molecules within the probe volume, a widely applied experimental assumption is that the number of molecules is proportional to the concentration of the analyte used. This yields Eq. (6.2), for what is termed an analytical enhancement factor (AEF):

$$\mathrm{AEF} = \frac{I_{\mathrm{SERS}}/C_{\mathrm{SERS}}}{I_{\mathrm{NRS}}/C_{\mathrm{NRS}}} \tag{6.2}$$

where C_{SERS} and C_{NRS} are the concentrations used in the SERS and normal Raman experiments, respectively. The AEF, while it depends strongly on analyte distribution and the substrate morphology, is a practically useful characterization of the enhancement yielded by a substrate at specific experimental conditions [71].

6.5 Selection Rules

The surface selection rule is an important part of SERS theory, which determines the activity or silence of normal modes and whether the intensity of the normal mode is enhanced or not in the Raman scattering process. There are two popular ways of stating the surface selection rule for SERS. One is to invoke in terms of the image field model, and another one is based on the enhancement of the local electromagnetic fields for the flat and spherical metal surfaces.

6.5.1 Image Field Model

In this model, a dipole located near the surface induces an image dipole within the metal [72]. A dipole parallel to the metal surface has an image dipole in the opposite direction while the image of a dipole perpendicular to the surface is in the same direction as that of its source, which results in screening of parallel dipoles and reinforcement of perpendicular dipoles. Therefore, the adsorbate–metal system can be regarded as molecular dipole–image dipole system.

The surface selection rule for Raman scattering in terms of image charges is the process of finding the image of each dipole, which is equivalent to the addition of a new symmetry operation that characterizes the system involving the adsorbed molecule and the metal. Modes belonging to irreducible representation spanning the x^2, y^2, z^2, xy, and yx components of the derived polarizability will be active while those transforming as xz and yz will be extinguished on the surface, where z axis is selected to be normal to the surface and x and y axes lie in-plane. By considering the image dipole criterion simultaneously, and insisting on normal electric field at the surface, only vibrations, which transform like the z^2 polarization components, are being surface active.

From this point of view, in the image model, the surface effect is replaced by the image dipoles/charges, while the surface detail is neglected in the surface selection rule. The surface rule becomes progressively more valid in spectral regions for which the metal is a good reflector, for instance, in the IR, and it is less valid in the visible or near UV region of the spectrum in which many transition metals are not good reflectors.

6.5.2 Electromagnetic Field Model

In this approach, the surface selection rule arises from the fact that the incident and reflected electric fields are directed at a surface interface in such a way that the tangential component is diminished while the normal component is augmented [73]. Consequently, the components of transition dipole normal to the surface are excited more strongly than the tangential components.

For the molecule on the flat surface, it should be considered that both the state of polarization of the incident ray and the scattered ray. For Raman spectroscopy, there are three classes of vibrational modes with distinct spectral behavior:

(1) those excited only by the normal components of the field and resulting in an induced dipole with a strong component only in the direction perpendicular to the surface,
(2) those excited only by the tangential components of the field and resulting in an induced dipole with a strong component tangential to the surface, and
(3) the mixed case. If z is the surface normal, vibrational mode possessing polarizability components α_{zz} belongs to the first type (most intense spectral feature), irreducible representations α_{xx}, α_{yy}, and α_{xy} belong to the second type (decrease in intensity) and α_{xz} and α_{yz} belong to the third type (very week and commonly absent), where α is the polarizability tensor.

For spherical colloidal metal surface, the squares of the radial and tangential field components are given by the following equation:

$$\overline{E_t^2} \propto 2E_0^2|1 - g|^2; \overline{E_n^2} \propto E_0^2|1 + 2g|^2 \tag{6.3}$$

where E_0 represents the incident field strength

$$g = \frac{\varepsilon - \varepsilon_0}{\varepsilon + 2\varepsilon_0} \tag{6.4}$$

The electromagnetic SERS enhancement is the product of the field intensity associated with the incident and scattered beams. Hence, the three classes of vibrational modes are expected to have the following SERS enhancement.

$$\begin{aligned}
&\alpha_{zz} : \left(\overline{E_n^2}\right)\left(\overline{E_n'^2}\right) \propto |1 + 2g|^2|1 + 2g'|^2 \\
&\alpha_{xz}, \alpha_{yz} : 1/2\left[\overline{E_n^2 E_t'^2} + \overline{E_t^2 E_n'^2}\right] \propto \left[|1 + 2g|^2|1 - g'|^2 + |1 - g|^2|1 + 2g'|^2\right] \\
&\alpha_{xx}, \alpha_{yy}, \text{and } \alpha_{xy} : \overline{E_t^2}\,\overline{E_t'^2} \propto 4|1 - g|^2|1 - g'|^2
\end{aligned} \tag{6.5}$$

where the prime indicates the properties calculated at the Raman shifted frequency, while unprimed quantities are calculated at the incident frequency. The α_{zz} modes show the most enhanced spectral features relative to the others in the red region of the surface–plasma frequency. The $\overline{E_n^2}$ progressively decreases compared with the $\overline{E_t^2}$, which causes diminishing in intensities of α_{zz} modes in the blue region of the surface–plasmon frequency.

6.6 Fabrications of SERS Substrates

Appropriate metal and correct nanostructure are the two crucial components to activate SERS effects. Although alkali metals (Li, Na, and K) [74] and transition metals (Pd, Pt, and Rh) [75–77] are able to induce SERS effects, Au and Ag NPs generally are the preferred choices for SERS applications [78–81] because they are corrosion resistant and their dielectric constants satisfy the resonance condition at Raman excitation wavelengths, which are typically near or in visible region.

To design and fabricate efficient SERS substrates with high qualities, many aspects need to be taken into account. In addition to EF, other important parameters to evaluate SERS substrates for both research and real-world applications include uniformity, repeatability, scalability, shelf life, and cost [82]. In attempting to achieve reliable and scalable SERS substrates, a number of fabrication techniques and approaches have been investigated in the past years such as electron beam lithography [83–85], nanosphere lithography [86, 87], plating [88, 89], laser ablation [90], porous nanoscaffolds [91–93], colloidal aggregation [94], chemical and physical roughening [95–97], oblique angle deposition [98, 99], electrochemical deposition [100, 101], vapor deposition [102, 103] and de-alloying [104, 105], etc.

In summary, the current techniques for SERS substrates fabrication can be categorized as template-assisted fabrication, hybrid fabrication, fabrication by using colloids and direct deposition on flat surface. In addition, nanoscale roughness could be created by aggregating a bunch of nanoparticles in the form of a cluster.

6.6.1 Template-assisted Fabrication

The most powerful and popular approach for SERS substrate fabrication is template-assisted fabrication, during which Au NPs or Ag NPs are deposited over the templated substrates. Apparently, the crucial parameter of this method lies in the preparation of templates. Electron beam lithography is the perfect example to illustrate this fabrication technique since this approach is the most powerful tool to create templates [83, 84, 106, 107]. With the help of this sophisticated lithography approach, nearly any possible pattern at nanometer scale could be precisely written on wafer substrates. But the high cost and low throughput of electron beam lithography technique have impeded its large application areas. To pursue inexpensive alternatives, Van Duyne and coworkers designed and developed nanosphere lithography technique [86, 87, 108, 109]. Compared to electron beam lithography, nanosphere lithography technique replaces resists with polystyrene nanobeads as the pattern or mask materials. Although this method successfully lowers down the manufacturing cost compared to electron beam lithography, it should be pointed out that the layout and release of nanospheres tend to introduce defects and sacrifice the quality of patterns. Similar rationale, except replacing polystyrene nanobeads with SiO_2 nanoparticles, has also been tried out by Jiang and coworkers [110–113]. The simplest template method is to utilize porous nanosubstrates. In this process, porous silicon [114, 115], porous Al_2O_3 [116–120], or other porous nanoscaffold [121, 122] is prepared by chemical etching technique and the roughened surfaces will then serve as templates.

6.6.2 Hybrid Fabrication

Hybrid fabrication involves foreign or third-party materials aside from Au NPs or Ag NPs and base substrates. Specifically, this technique firstly builds bottom layer nanostructures made of foreign materials via various microfabrication techniques and then deposits the Au NPs or Ag NPs over the existing nanostructures. Obviously, this method differs from template method in that third-party materials are introduced into the systems to provide correct nanostructures. The biggest advantage of hybrid method is its flexibility in selecting desired nanostructures. As long as the nanostructure, no matter what it is made of, is available, hybrid method could make use of it by simply covering the nanostructures with SERS-active metal coatings. Many nanostructures, including nanospheres [123, 124], nanotubes [125], nanowires [126], nanorods [127–129], etc. have been demonstrated for SERS applications with hybrid method.

6.6.3 Fabrication by Using Colloids

Colloidal Au or Ag is the most commonly used SERS substrate in laboratory research. By adjusting the chemical reaction temperature, time, concentration, pH, and capping agents, various shaped and sized Au and Ag NPs suspensions were reported for various SERS applications [130–133]. The merits of colloidal metal NPs include low cost and great sensitivity, but the shortcomings of this method are quite obvious, too.

First, the short shelf life of metal colloid restricts its usage even in laboratory environment, not to mention real-world applications.

Second, the large size dispersion has always been an issue for metal colloid, imposing concerns on the repeatability of SERS measurements.

Third, capping agents are typically required for metal colloids to remain stable. These organic molecules, directly attached with metal nanoparticles, would cause interfering signals for Raman measurement.

All in all, despite the exciting results shown with colloid Au and Ag in research community, colloid method is not considered as a mature solution for large-scale production and real-world applications of SERS.

6.6.4 Direct Deposition

In direct deposition technique, metal NPs such as Ag NPs and Au NPs are deposited onto wafer substrates without pre-patterning. Oblique angle deposition technique is a good example to depict this direct deposition approach [134]. In oblique angle deposition technique, a modified physical vapor deposition technique, the substrate surface normal is positioned at a certain angle to the incoming vapor so that the Ag nanorods could be grown preferentially along a specific direction due to a so-called geometrical shadowing effect. Metal island films serve well as another example of direct deposition. In this method, an ultrathin noble metal film is deposited onto wafer substrates and then the substrates are thermal annealed at mildly high temperature. Due to the high surface free energy, the metal film has a tendency to transform into discontinuous island arrays [135, 136].

On the other hand, de-alloying is also an example of direct deposition method for SERS substrates fabrication. In this method, a film of Au and Ag alloy is firstly deposited onto wafer substrates, and then chemical etching is applied to etch out the Ag from the alloy matrix, leaving a porous Au film on substrates as SERS-active component [137, 138].

6.7 Applications of SERS

Owing to its remarkable capability of identifying chemical species and providing structural information at molecular level, SERS has been widely employed in a variety of fields including materials science, environmental science, biological engineering, electrochemistry, food science, forensic science, biomedical diagnostics, etc. In this part of the chapter, important and representative applications of SERS will be briefly demonstrated.

6.7.1 SERS-Based Separation Applications

SERS-based detection processes were successfully combined with various separation techniques such as high-pressure liquid chromatography (HPLC) and gas chromatography (GC), thin layer chromatography (TLC). Heaps and colleagues effectively eluted *p*-nitrothiophenol and caffeine by using a GC column, at the end of which a SERS substrate, formed by a thin film of Ag on ZnSe, was placed. It was kept at low temperature in order to condensate the target compounds from the gas phase. SERS studies were performed after transferring the substrate from the GC system to the Raman instrument [139].

In other studies, Trachta and coworkers [140, 141] successfully analyzed the existence of various pharmaceutical compounds in biological samples including blood and urine. In their studies, the first step was a solvent extraction followed by phase separation by applying centrifugation. This solution was eluted by using an HPLC column and the sample fractions were collected at different times and deposited on the surface of SERS substrates. The eluent was preferred so that it did not interfere with the SERS substrates. SERS substrates were formed by a gelatin-based silver halide, which was photoreduced by laser exposure to prepare the SERS active structures. Several pharmaceutical compounds were successfully identified by employing this approach such as doxepin, codeine, and methadone.

In another interesting work, Wang and colleagues [142] coupled the output of an HPLC system to a capillary tube, on the internal surface of which metal nanoparticles were adsorbed. A different capillary tube was chosen for the efficient analysis of each eluted fraction. The developed system was efficiently employed for the sensitive detection of various compounds including 4,4-bipyridine, 1,4-benzenedithiol, and the thiram pesticide.

In another crucial study [143], the synthesis of Ag NPs-doped sol–gel was successfully carried out by Farquharson and Maksymiuk; and it was efficiently employed for the HPLC-based separation of the analytes and simultaneously played the role of the SERS substrate.

Lucotti et al. successfully analyzed the existence of apomorphine, which is a drug used for the treatment of the Parkinson's disease, in human blood plasma samples [144]. In this study, a small volume of plasma was dropped on the TLC substrate and eluted with ethanol. A drop of colloidal Ag NPs was put on the spots formed after separation and the SERS spectra obtained. A similar approach was applied in a research reported by Vicario et al. for the detection of an anticancer drug irinotecan [145].

6.7.2 SERS-Based Sensor Applications

6.7.2.1 Environmental Analysis

Environmental pollution is one of the important global concerns. Various compounds in environmental samples including drugs display harmful impacts on human body, animals, and plants. Therefore, the development of novel sensor platforms for the selective and sensitive recognition of these compounds in complex matrices is very important. The monitoring of environmental pollution is usually carried out by employing mass spectrometry (MS)-based techniques. Although these techniques are extremely sensitive, they have various drawbacks. These systems are very expensive and require specialized training to carry out the analysis process. In addition, environmental samples also usually require extensive and time-consuming pretreatment stages prior to analysis that can significantly affect the analysis. The development of new analytical methods that matched the sensitively of MS but that could be deployed in the field and require minimum sample volume would be highly advantageous for environmental monitoring. SERS can meet these criteria and overcome the drawbacks of MS-based approaches.

In the following, recent progress related to SERS detection of various environmental pollutants such as pesticides, polycyclic aromatic hydrocarbons, and heavy metal ions with emphasis on the type of substrates, which are successfully applied in SERS-based sensing, is briefly described.

Pesticides, one of the most important classes of organic pollutants, are continuously being released into the environment, mostly owing to a variety of human activities. One such activity is agriculture based on toxic pesticides that generally remain in soil for shorter or longer times and can eventually rapidly reach food and drinkable water. Trace detection of pesticides in soil, lakes, rivers, air, or food is extremely important in order to mitigate the threat of pesticide residues to human health. As stated earlier, SERS is a well-known and powerful analytical technique that is based on nanostructured materials and that is very suitable for the detection of such pollutants as it has the necessary sensitivity and can potentially detect single molecules [146].

One of the earliest steps toward detection of pesticides was taken in 2001 when the adsorption of the dimethyldithiocarbamate derivative fungicides thiram and ziram on Au and Ag surfaces was studied using SERS [147]. It has been shown that SERS exhibits a high sensitivity toward molecules of thiram and also toward molecules of ziram that were dissociated when adsorbed on the metal film. Since then, metallic surfaces based on Au and Ag colloidal nanoparticles earned their valuable place in SERS detection. Therefore, pesticides such as fonofos [148], dimethoate/omethoate [149], phosmet, azinphos-methyl and carbaryl [150], or thiram [151, 152] can be successfully detected using this technique. Moreover, at that time, detection limit as low as 2 ppm was reported, with high potential for further improvement of sensitivity [149].

Recently, more metallic substrates with rather specific geometries have been developed to enable SERS detection of pesticides such as carbamates [153–155]. For example, in order to tune the surface plasmon resonance of the nanoparticles to the laser excitation wavelength of the Raman spectrometer, Saute et al. used Au nanorods as solution-based SERS substrates and detected ultralow levels of different dithiocarbamate fungicides, such as thiram, ferbam, and ziram [153]. The reported sensitivity limits were 8, 4.2, and 11 nM for ferbam, ziram, and thiram, respectively.

In another interesting study [155], Fodjo et al. have not only shown that such a substrate possessed an excellent SERS activity toward various carbamates compared to Ag NPs but have also demonstrated low detection limits of 2.5, 10, 50, and 75 pM for carbaryl, carbofuran, isoprocarb, and propoxur, respectively, under optimized conditions.

Ag nanocubes prepared by a rapid sulfide-mediated polyol method have been further used as SERS-active substrates to detect, besides thiram, pesticides such as paraoxon, with a reported limit of detection as low as 5×10^{-8} mol l^{-1} [156]. Besides Au nanorods and Ag nanocubes, other metal nanostructures, such as Au nanodogbone, were reported as efficient SERS-active colloidal substrates. For example, dogbone-shaped Au nanoparticles of 65 nm led to a limit of detection for thiram of 11.8 nM [157].

SERS was also successfully applied for the sensitive detection of heavy metals. For example, Kandjani et al. reported the design of a SERS-active ZnO/Ag nanoarray device for selective and sensitive detection of Hg^{2+} ions [158]. The specificity of the fabricated platform toward Hg^{2+} ions was achieved through the excellent binding affinity behavior of Hg^{2+} ions toward the decorated Ag NPs to form Hg/Ag amalgam. In this procedure, the developed SERS substrate was first exposed to a solution containing different concentrations of Hg^{2+} ions followed by immersion into rhodamine B (RB). The intensity variation of a characteristic RB Raman band was used for the determination of the concentration of the Hg^{2+} ions in the solution. This ZnO/Ag nanoarrays substrate provides reproducible detection of Hg^{2+} ions with an limit of detection (LOD) of about 0.45 ppb.

In an important study reported by Wang et al. [159], droplet-based microfluidics combined SERS detection was developed for rapid, reproducible, and sensitive trace analysis of Hg^{2+} ions in water. Rhodamine B–bonded gold nanoparticles (Au NPs) were exploited as sensitive SERS platforms for Hg^{2+} ion recognition. In this system, an initial SERS signal was recorded for rhodamine B adsorbed on Au NPs. The interaction between mercury ions and Au NPs leads to the displacement of rhodamine B molecules. This, in turn, causes a decrease of the SERS signal of rhodamine B. By monitoring the variation of the SERS signal also called "turn off" change, quantitative analysis of Hg^{2+} in aqueous solution could be accomplished with an LOD of about 100 ppt.

Similar to Hg^{2+}, Cu^{2+} has a small scattering cross section, which makes its detection in a direct way by SERS impossible. Therefore, strategies similar to the "turn off" sensing of Hg^{2+} ions are well adapted for its SERS detection. For example, silver nanoparticles co-functionalized with cysteine and 3,5-dimethoxy-4-(60-azobenzotriazolyl)phenol (Ag NP conjugates) were proposed for ultrasensitive SERS detection of Cu^{2+} and Hg^{2+} ions in aqueous solutions [160]. The coordination between the metal ions and the glycine residue, which is part of the L-cysteine on the surface of Ag NP, ensures the specificity of the sensor toward Cu^{2+} and Hg^{2+} ions. The addition of SCN^- to the analyte can successfully mask Hg^{2+} ions by allowing for the selective detection of Cu^{2+} ions. The sensitivity of the system is achieved through the formation of Raman hot spots between interconnected nanoparticles as a result of coordination. This SERS-based assay showed an LOD of 10 pM for Cu^{2+} ions and 1 pM for Hg^{2+} ions, which are a few orders of magnitude more sensitive than the typical colorimetric approach based on the aggregation of noble metal nanoparticles.

Li et al. proposed a highly sensitive and selective SERS sensor based on glutathione (GSH)/4-mercaptopyridine (4-MPY)-modified silver nanoparticles (Ag NPs) for trace analysis of As^{3+} ions in aqueous solution [161]. GSH conjugated on the surface of Ag NPs ensures the specific binding of As^{3+} ions through the As–O

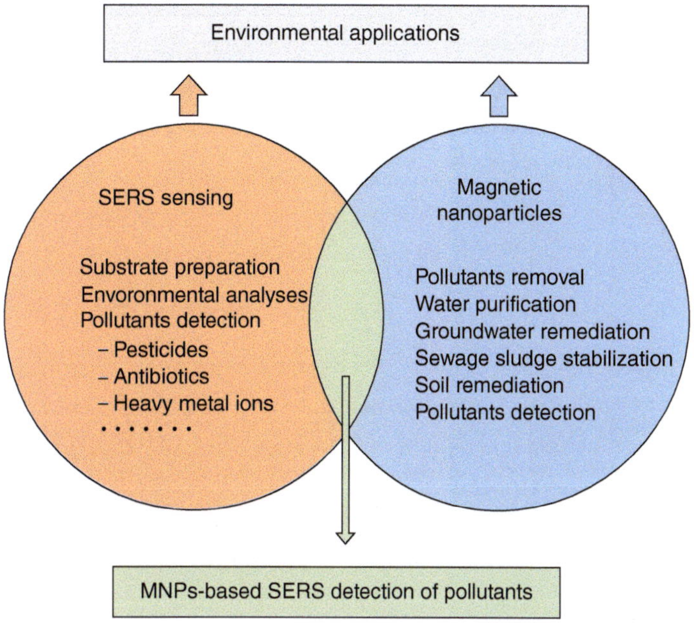

Figure 6.2 The schematic depiction of the environmental applications of SERS and magnetic nanoparticles. Source: Song D et al., (2019) Applications of magnetic nanoparticles in surface-enhanced Raman scattering (SERS) detection of environmental pollutants, Journal of Environmental Sciences, 80, 14–34. © 2018, Elsevier.

linkage, while 4-MPY is used as a Raman reporter molecule. With this procedure, As^{3+} ions could be detected quickly and accurately with an LOD of 0.76 ppb. In addition, the practical applicability of the system was evaluated through the detection of As^{3+} ions in drinking water samples.

On the other hand, magnetic nanoparticles were also efficiently combined with SERS technique for the sensitive recognition of environmental pollutants. Figure 6.2 shows the schematic depiction of the environmental applications of SERS and magnetic nanoparticles.

In an interesting research performed by Du et al. [163], magnetic Fe_3O_4 nanoparticles @Au SERS substrate was developed and efficiently employed in label-free SERS-based detection of the 16 Polycyclic aromatic hydrocarbons (PAHs), which are in United States Environmental Protection Agency (US EPA) priority list.

In another work [164], Tang et al. designed and fabricated Au nanorods superlattice arrays doped with Magnetc naoparticles (MNPs) as SERS substrate, and the developed nanoarray was successfully used for the sensitive detection of 16 different PAHs in standard samples (Figure 6.3).

6.7.2.2 Forensic Analysis

Another interesting application area of SERS-based sensor systems is forensic science. SERS analysis can be efficiently employed in all types of illicit drug sensing, from sensitive detection of these types of drugs in their pure form to effective identification of their metabolites in body fluids. SERS can be especially useful because

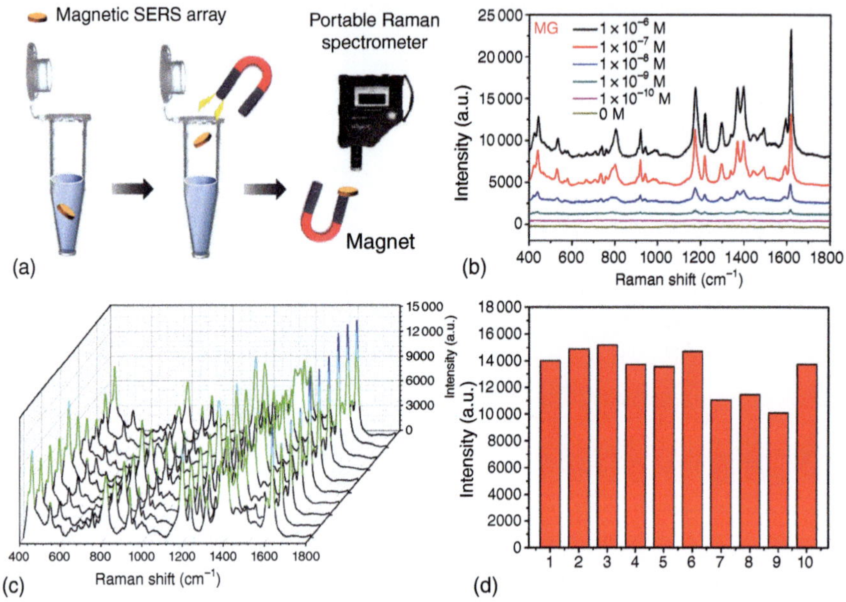

Figure 6.3 (a) The schematic depiction of the magnetic separation and SERS-based detection process. (b) The achieved SERS spectra for different concentrations of malachite green that is a banned fish pesticide. (c, d) The achieved SERS spectra of malachite green acquired from different spots on a single substrate and corresponding histogram for the peak intensity at 1615 cm^{-1}. Source: Tang, S. et al., (2017) Efficient enrichment and self-assembly of hybrid nanoparticles into removable and magnetic SERS substrates for sensitive detection of environmental pollutants. ACS Appl. Mater.Interfaces 9 (8), 7472–7480. © 2017, American Chemical Society.

of the low concentrations of drugs and their metabolites especially exist in toxicology samples. This section of the chapter covers the identification of illicit drugs in solid form as well as in biological samples and controlled substances, which can be relevant in counterfeit pharmaceutical investigations. Some reported examples are briefly demonstrated in the following:

Halouzka et al. designed and prepared a novel substrate, which can be applied with micro-SERS for the sensitive detection of cathinone designer drugs [165]. The researchers applied a coating of copper nanowires to carbon fiber microelectrodes to construct this substrate and found that it ensures the sensitive detection of cathinone drugs down to a range from 10^{-10} to 10^{-12} M.

In another interesting study, Si et al. effectively employed semitransparent plasmon nanosheets to fabricate a new substrate, which can be used to detect benzocaine on banknotes at a level of 0.9×10^{-6} g cm^{-2} [166].

Lee et al. collected trace amounts of new psychoactive substances and legal high drugs from surfaces using wet swabs and utilized polymer-stabilized silver nanoparticles (poly-SERS) for the sensitive detection of those drugs [167].

Yan et al. used the evaporation of monodisperse Au nanospheres on a silicon wafer to create optimal hot spots, which enabled the SERS-based detection of methylenedioxymethamphetamine (MDMA) down to 10 μM and α-methyltryptamine hydrochloride down to 1.01 μM [168].

In 2017, Chen et al. assessed portable SERS-based sensor systems utilizing a silicon chip for the sensitive detection of trinitrotoluene (TNT) [169]. The developed sensor system enabled a very low detection limit for the target compound TNT (~1 pM). The silicon chip was reported to work very well when coupled with a portable SERS system.

In another crucial work reported by Lopez-Lopez et al. [170], SERS technique was successfully utilized for the efficient analysis of gunshot residue. In total, 21 different types of smokeless gunpowders, along with common gunpowder component standards and gunshot residue obtained for the firing of two cartridges with different branding and calibers, were analyzed. The achieved data confirms that for all 21 smokeless gunpowders, the observed peaks largely came from ethyl centralite, diphenylamine derivatives, and diphenylamine, which are all stabilizers existing in the powders that were subjected to analysis. In particular, diphenylamine in gunpowders could be successfully detected.

6.7.2.3 Biological Applications

Sensitive detection and determination of the levels of proteins play a crucial role in biological sciences, especially in early diagnosis and efficient treatment of various diseases. Biosensing through label-free SERS technique is a powerful, facile, and cheap approach to achieve intrinsic structure information of proteins (i.e. myoglobin, hemoglobin and cytochrome *c*) [171].

In a reported study [172], Kahraman et al. provided an overview on label-free techniques for reproducible and sensitive SERS detection of various proteins using colloidal nanoparticles aggregates and metallic NPs.

In an interesting research conducted by Karadan et al. [173], Si nanopillar arrays decorated with Ag NPs were prepared to fabricate SERS-based biosensor platform toward thiophenol. The developed SERS-based biosensor had an EF of 2.4×10^8 for thiophenol. A number of proteins such as human serum albumin, bovine serum albumin (BSA), Alzheimer-disease-related amyloid beta proteins, and lysozyme can be sensitively detected by employing this biosensor.

In another work [174], Szekeres and Kneipp compared the SERS spectra of BSA at different concentrations with Au NPs without aggregating agents. The concentration-dependent agglomerates of the NPs lead to distinct SERS enhancement, indicating that BSA molecules may undergo a change depending on the concentration.

Yang et al. successfully designed and fabricated a SERS-based immunoassay system composed of alpha-fetoprotein antibody-modified core–shell satellite nanostructures as the SERS tags and the nitrocellulose membrane as the solid substrate [175]. The Au@Ag core of the core–shell satellite nanostructures is a prominent SERS platform, and the thin silica interlayer shows long-range plasmon coupling providing great enhancement impact. The achieved results indicated that the developed SERS-based immunoassay system displayed high sensitivity toward the target alpha-fetoprotein. The obtained detection limit was $0.3\ \mathrm{fg\ ml^{-1}}$.

Cheng et al. developed a SERS-based immunoassay platform toward dual prostate-specific antigens (PSAs) for the accurate diagnosis of prostate cancer [176]. For this purpose, magnetic separation SERS immunoassays with two different SERS

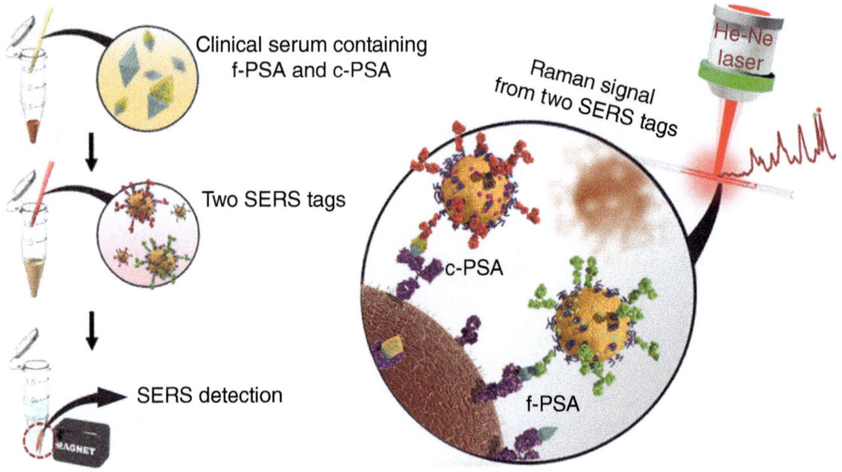

Figure 6.4 The schematic depiction of the simultaneous detection of f-PSA and c-PSA by using SERS-based immunoassay platform. Source: Cheng Z, et al., (2017) Simultaneous detection of dual prostate specific antigens using surface-enhanced Raman scattering-based immunoassay for accurate diagnosis of prostate cancer. ACS Nano, 11, 4926–4933. © 2017, American Chemical Society.

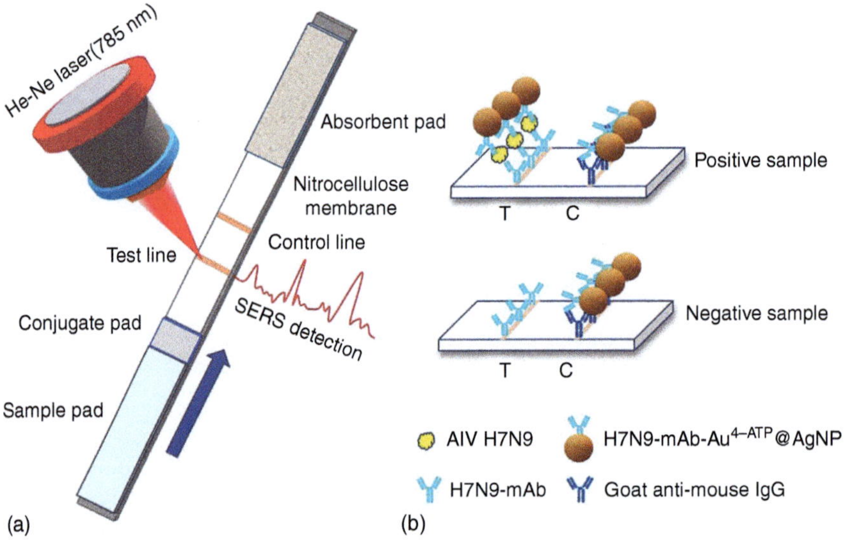

Figure 6.5 (a) The schematic representation of the detection of H7N9 virus by applying SERS-based lateral flow immunoassay. (b) Two brown lines were observed on the test line and control line for positive sample; and only one line appeared on the control line for negative sample. Source: Xiao M et al., (2019) Ultrasensitive detection of avian influenza A (H7N9) virus using surface-enhanced Raman scattering-based lateral flow immunoassay strips, Analytica Chimica Acta 1053, 139–147. © 2019, Elsevier.

tags labeling two kinds of PSA, complexed prostate-specific antigen (c-PSA) and free prostate-specific antigen (f-PSA), were constructed as schematically illustrated in Figure 6.4. In this research, the simultaneous detection enabled the better precision for the free-to-total-PSA ratio in clinical samples. The detection limit values for f-PSA and c-PSA were obtained as 0.012 and 0.15 ng ml^{-1}, respectively.

In another crucial work carried out by Xiao et al. [177], a novel SERS-based lateral flow immunoassay strip was designed and prepared for the ultrasensitive detection of avian influenza A (H7N9) virus. In this work, $AuAg^{4\text{-}ATP}$@AgNPs was chosen as the Raman probe. A specific antibody for H7N9 virus and goat antimouse Immunoglobulin G (IgG) antibody were immobilized on the surface of nitrocellulose membrane as the test and control lines, respectively (Figure 6.5). Accumulation of antibody–virus–antibody–Raman probe complex at the test line could easily be observed by the naked eye and the Raman signal could be achieved by using a portable Raman system. The sensitive detection process by using the developed SERS-based lateral flow immunoassay strip was carried out in 20 minutes. The detection limit for H7N9 virus was achieved as 0.0018 hemagglutination unit (HAU).

6.8 Conclusions

This chapter provides the recent progresses on the different applications of SERS technique by using various NMs including separation, environmental analysis, forensic analysis, and biological applications. SERS technology has numerous superiorities such as excellent sensitivity, rapid process time, minimum sample requirement, and low cost, etc.

Despite great efforts being made to develop SERS technology, there remain critical issues that need to be addressed before the large-scale production and practical applications of SERS. For real-world applications, SERS substrates firstly need to be fabricated in reliable and low-cost ways. Secondly, SERS substrate should produce spectral response consistently. To achieve these aims, micro/nanofabrication techniques need to be further designed and developed. The improved techniques should be capable of:

- controlling nanostructures in terms of size, shape, and geometry,
- growing uniform nanostructures at large scale, and
- integrating with traditional fabrication approaches.

References

1 Ong, T.T.X., Blanch, E.W., and Jones, O.A.H. (2020). Surface enhanced Raman Spectroscopy in environmental analysis, monitoring and assessment. *Sci. Total Environ.* 720: 137601.

2 Cheng, J., Wang, P., and Xiao-Ou, S. (2020). Surface-enhanced Raman spectroscopy for polychlorinated biphenyl detection: recent developments and future prospects. *TrAC Trends Anal. Chem.* 125: 115836.

3 Fan, M., Andrade, G.F.S., and Brolo, A.G. (2020). A review on recent advances in the applications of surface-enhanced Raman scattering in analytical chemistry. *Anal. Chim. Acta* 1097: 1–29.
4 Liu, X., Liu, X., Rong, P., and Liu, D. (2020). Recent advances in background-free Raman scattering for bioanalysis. *TrAC Trends Anal. Chem.* 123: 115765.
5 Huang, C.-C., Cheng, C.-Y., and Lai, Y.-S. (2020). Paper-based flexible surface enhanced Raman scattering platforms and their applications to food safety. *Trends Food Sci. Technol.* https://doi.org/10.1016/j.tifs.2020.04.019.
6 Rajput, P., Sinha, R.K., and Devi, P. (2020). Materials in surface-enhanced Raman spectroscopy-based detection of inorganic water pollutants, Chapter 9. In: *Inorganic Pollutants in Water* (eds. P. Devi, P. Singh and S.K. Kansal), 153–172. Elsevier.
7 Guerrini, L. and Alvarez-Puebla, R.A. (2020). Surface-enhanced Raman scattering (SERS) sensing of nucleic acids, Chapter 2. In: *Frontiers of Nanoscience*, vol. 16 (eds. W.J. Parak and N. Feliu), 9–23. Elsevier.
8 Choi, J.H., El-Said, W.A., and Choi, J.-W. (2020). Highly sensitive surface-enhanced Raman spectroscopy (SERS) platform using core/double shell (Ag/polymer/Ag) nanohorn for proteolytic biosensor. *Appl. Surf. Sci.* 506: 144669.
9 Huang, Q., Fang, C., Muhammad, M., and Yao, G. (2020). Assessment of norfloxacin degradation induced by plasma-produced ozone using surface-enhanced Raman spectroscopy. *Chemosphere* 238: 124618.
10 Yu, X., Li, W., Liang, O. et al. (2020). Molecular orientation and specificity in the identification of biomolecules via surface enhanced Raman spectroscopy. *Anal. Biochem.* 599: 113709.
11 Fleischmann, M., Hendra, P.J., and McQuillan, A.J. (1974). Raman spectra of pyridineadsorbed at a silver electrode. *Chem. Phys. Lett.* 26: 163–166.
12 Albrecht, M.G. and Creighton, J.A. (1977). Anomalously intense Raman spectra of pyridine at a silver electrode. *J. Am. Chem. Soc.* 99: 5215–5217.
13 Jeanmaire, D.L. and Van Duyne, R.P. (1977). Surface Raman spectroelectrochemistry. *J. Electroanal. Chem.* 84: 1–20.
14 Moskovits, M. (1978). Surface roughness and the enhanced intensity of Raman scattering by molecules adsorbed on metals. *J. Chem. Phys.* 69: 4159.
15 Llobet, E. (2020). Carbon nanomaterials, Chapter 4. In: *Micro and Nano Technologies, Advanced Nanomaterials for Inexpensive Gas Microsensors* (ed. E. Llobet), 55–84. Elsevier.
16 Kai-Hua Chow, E., Gu, M., and Xu, J. (2020). Carbon nanomaterials: fundamental concepts, biological interactions, and clinical applications, Chapter 14. In: *Micro and Nano Technologies, Nanoparticles for Biomedical Applications* (eds. E.J. Chung, L. Leon and C. Rinaldi), 223–242. Elsevier.
17 Fan, Y., Yang, B., and Rong, C. (2020). Functionalized carbon nanomaterials for advanced anode catalysts of fuel cells, Chapter 7. In: *Micro and Nano Technologies, Advanced Nanomaterials for Electrochemical-Based Energy Conversion and Storage* (eds. F. Ran and S. Chen), 223–263. Elsevier.

18 Hussain, C.M. and Keçili, R. (2020). *Modern Environmental Analysis Techniques for Pollutants*. The Netherlands: Elsevier.

19 Büyüktiryaki S, Keçili R, Hussain CM (2020) *Functionalized Nanomaterials in Dispersive Solid Phase Extraction: Advances and Prospects, Trends Anal. Chem.*, doi:https://doi.org/10.1016/j.trac.2020.115893

20 Keçili, R., Büyüktiryaki, S., and Hussain, C.M. (2019). Advancement in bioanalytical science through nanotechnology: past, present and future. *Trends Anal. Chem.* 110: 259–276.

21 Hussain, C.M. and Kharisov, B. (2017). *Advanced Environmental Analysis-Application of Nanomaterials*. The Royal Society of Chemistry.

22 Hussain, C.M. (2017). Magnetic nanomaterials for environmental analysis. In: *Advanced Environmental Analysis-Application of Nanomaterials* (eds. C.M. Hussain and B. Kharisov). The Royal Society of Chemistry.

23 Hussain, C.M. (2019). *Handbook of Nanomaterials in Analytical Chemistry: Modern Trends in Analysis*. Elsevier.

24 Büyüktiryaki, S., Sümbelli, Y., Keçili, R., and Hussain, C.M. (2019). Lab-on-chip platforms for environmental analysis. In: *Encyclopedia of Analytical Science*, 3e (eds. P. Worsfold, C. Poole, A. Townshend and M. Miró), 267–273. Academic Press.

25 Keçili, R. and Hussain, C.M. (2018). Recent progress of imprinted nanomaterials in analytical chemistry. *Int. J. Anal. Chem.*: 8503853.

26 Barua, S., Debojeet, S., Shahnaz, N., and Khan, R. (2020). Chemistry of two-dimensional nanomaterials, Chapter 1. In: *Two-Dimensional Nanostructures for Biomedical Technology* (eds. R. Khan and S. Barua), 1–33. Elsevier.

27 Keçili, R., Büyüktiryaki, S., and Hussain, C.M. (2018). Engineered nanosensors based on molecular imprinting technology, Chapter 57. In: *Micro and Nano Technologies, Handbook of Nanomaterials for Industrial Applications* (ed. C.M. Hussain), 1031–1046. Elsevier.

28 Kleinman, S.L., Frontiera, R.R., Henry, A.-I. et al. (2013). Creating, characterizing, and controlling chemistry with SERS hot spots. *Phys. Chem. Chem. Phys.* 15 https://doi.org/10.1039/C1032CP42598J.

29 Tao, C.-A., An, Q., Zhu, W. et al. (2011). Cucurbit[*n*]urils as a SERS hot-spot nanocontainer through bridging gold nanoparticles. *Chem. Commun.* 47 (35): 9867–9869.

30 Farcau, C. and Astilean, S. (2010). Mapping the SERS efficiency and hot-spots localization on gold film over nanospheres substrates. *J. Phys. Chem. C* 114 (27): 11717–11722.

31 Netzer, N.L., Gunawidjaja, R., Hiemstra, M. et al. (2009). Formation and optical properties of compression-induced nanoscale buckles on silver nanowires. *ACS Nano* 3 (7): 1795–1802.

32 Kaewseekhao, B., Nuntawong, N., Eiamchai, P. et al. (2020). Dataset of serum proteomic spectra from tuberculosis patients detected by Raman spectroscopy and surface-enhanced Raman spectroscopy. *Data Brief* 28: 104891.

33 Kaewseekhao, B., Nuntawong, N., Eiamchai, P. et al. (2020). Diagnosis of active tuberculosis and latent tuberculosis infection based on Raman spectroscopy and surface-enhanced Raman spectroscopy. *Talanta* 121: 101916.

34 Carmicheal, J., Hayashi, C., Huang, X. et al. (2020). Label-free characterization of exosome via surface enhanced Raman spectroscopy for the early detection of pancreatic cancer. *Nanomedicine* 16: 88–96.

35 Huang, S., He, Y., Yang, X. et al. (2020). Porous SiO_2@Ni@C and Au nanocages as surface-enhanced Raman spectroscopy platform with use of DNA structure switching for sensitive detection of uracil DNA glycolase. *Sensors Actuators B Chem.* 304: 127273.

36 He, X., Zhou, X., Yu, L., and Wang, X. (2020). Ultrasensitive, recyclable and portable microfluidic surface-enhanced raman scattering (SERS) biosensor for uranyl ions detection. *Sensors Actuators B Chem.* 311: 127676.

37 Ahmed El-Said, W., Alshitari, W., and Choi, J.-w. (2020). Controlled fabrication of gold nanobipyramids/polypyrrole for shell-isolated nanoparticle-enhanced Raman spectroscopy to detect γ-aminobutyric acid. *Spectrochim. Acta A Mol. Biomol. Spectrosc.* 229: 117890.

38 Wu, Y., He, Y., Yang, X. et al. (2018). A novel recyclable surface-enhanced Raman spectroscopy platform with duplex-specific nuclease signal amplification for ultrasensitive analysis of microRNA 155. *Sensors Actuators B Chem.* 275: 260–266.

39 El-Said, W.A. and Choi, J.-w. (2019). High selective spectroelectrochemical biosensor for HCV-RNA detection based on a specific peptide nucleic acid. *Spectrochim. Acta A Mol. Biomol. Spectrosc.* 217: 288–293.

40 Singh, N., Kumar, P., and Riaz, U. (2019). Applications of near infrared and surface enhanced Raman scattering techniques in tumor imaging: a short review. *Spectrochim. Acta A Mol. Biomol. Spectrosc.* 222: 117279.

41 Yamazoe, S., Naya, M., Shiota, M. et al. (2014). Large-area surface-enhanced Raman spectroscopy imaging of brain ischemia by gold nanoparticles grown on random nanoarrays of transparent boehmite. *ACS Nano* 8 (6): 5622–5632.

42 Ardini, M., Huang, J.-A., Sánchez, C.S. et al. (2018). Live ıntracellular biorthogonal imaging by surface enhanced raman spectroscopy using alkyne-silver nanoparticles clusters. *Sci. Rep.* 8: 12652.

43 Malek, K., Jaworska, A., Krala, P. et al. (2013). Imaging of macrophages by surface enhanced Raman spectroscopy (SERS). *Biomed. Spectrosc. Imaging* 2 (4): 349–357.

44 Huang, K.C., Bando, K., Ando, J. et al. (2014). 3D SERS (surface enhanced Raman scattering) imaging of intracellular pathways. *Methods (San Diego, CA)* 68 (2): 348–353.

45 Ding, S.Y., You, E.M., Tian, Z.Q., and Moskovits, M. (2017). Electromagnetic theories of surface-enhanced Raman spectroscopy. *Chem. Soc. Rev.* 46: 4042–4076.

46 Kim, J., Jang, Y., Kim, N.J. et al. (2019). Study of chemical enhancement mechanism in non-plasmonic surface enhanced Raman spectroscopy (SERS). *Front. Chem.* 7: 582.

47 Kosuda, K., Bingham, J., Wustholz, K., and Van Duyne, R. (2010). *Handbook of Nanoscale Optics and Electronics*. Academic Press.

48 Willets, K.A. and Van Duyne, R.P. (2007). Localized surface plasmon resonance spectroscopy and sensing. *Annu. Rev. Phys. Chem.* 58: 267–297.

49 Jiang, R., Chen, H., Shao, L. et al. (2012). Unraveling the evolution and nature of the plasmons in (Au core)–(Ag shell) nanorods. *Adv. Mater.* 24: OP200–OP207.

50 Santos Costa, J.C., Ando, R.A., Sant'Ana, A.C. et al. (2009). High performance gold nanorods and silver nanocubes in surface-enhanced Raman spectroscopy of pesticides. *Phys. Chem. Chem. Phys.* 11: 7491.

51 Hrelescu, C., Sau, T.K., Rogach, A.L. et al. (2009). Single gold nanostars enhance Raman scattering. *Appl. Phys. Lett.* 94: 153113.

52 Fan, Z., Senapati, D., Khan, S.A. et al. (2013). Popcorn-shaped magnetic core-plasmonic shell multifunctional nanoparticles for the targeted magnetic separation and enrichment, label-free SERS imaging, and photothermal destruction of multidrug-resistant bacteria. *Chem. Eur. J.* 2013 (19): 2839–2847.

53 Ding, S.-Y., Yi, J., Li, J.-F. et al. (2016). Nanostructure-based plasmon-enhanced Raman spectroscopy for surface analysis of materials. *Nat. Rev. Mater.* 1: 16021.

54 Fang, J., You, H., Zhu, C. et al. (2007). Thermodynamic and kinetic competition in silver dendrite growth. *Chem. Phys. Lett.* 439: 204–208.

55 Kneipp, K., Kneipp, H., Itzkan, I. et al. (2002). Surface-enhanced Raman scattering and biophysics. *J. Phys. Condens. Matter* 14: R597–R624.

56 Kerker, M. (1984). Electromagnetic model for surface-enhanced Raman scattering (SERS) on metal colloids. *Acc. Chem. Res.* 17: 271–277.

57 Emory, S.R., Haskins, W.E., and Nie, S. (1998). Direct observation of size-dependent optical enhancement in single metal nanoparticles. *J. Am. Chem. Soc.* 120: 8009–8010.

58 Zeman, E.J. and Schatz, G.C. (1987). An accurate electromagnetic theory study of surface enhancement factors for silver, gold, copper, lithium, sodium, aluminum, gallium, indium, zinc, and cadmium. *J. Phys. Chem.* 91: 634–643.

59 Kennedy, B.J., Spaeth, S., Dickey, M., and Carron, K.T. (1999). Determination of the distance dependence and experimental effects for modified SERS substrates based on self-assembled monolayers formed using alkanethiols. *J. Phys. Chem. B* 103: 3640–3646.

60 Haynes, C.L. and Van Duyne, R.P. (2003). Plasmon-sampled surface-enhanced Raman excitation spectroscopy. *J. Phys. Chem. B* 107: 7426–7433.

61 McFarland, A.D., Young, M.A., Dieringer, J.A., and Van Duyne, R.P. (2005). Wavelength-scanned surface-enhanced Raman excitation spectroscopy. *J. Phys. Chem. B* 109: 11279–11285.

62 Dieringer, J.A., McFarland, A.D., Shah, N.C. et al. (2006). Introductory lecture surface enhanced Raman spectroscopy: new materials, concepts, characterization tools, and applications. *Faraday Discuss.* 132: 9–26.

63 Wang, Y., Chen, H., Sun, M. et al. (2017). Ultrafast carrier transfer evidencing graphene electromagnetically enhanced ultrasensitive SERS in graphene/Ag-nanoparticles hybrid. *Carbon* 122: 98–105.

64 Kaneko, S., Watanabe, S., Fujii, S. et al. (2020). The practical electromagnetic effect in surface-enhanced Raman scattering observed by the lithographically fabricated gold nanosquare dimers. *AIP Adv.* 10: 025301.

65 Ricci, M., Becucci, M., and Castellucci, E.M. (2019). Chemical enhancement in the SERS spectra of indigo: DFT calculation of the Raman spectra of indigo-Ag14 complexes. *Vib. Spectrosc.* 100: 159–166.

66 Chong, N.S., Donthula, K., Davies, R.A. et al. (2015). Significance of chemical enhancement effects in surface-enhanced Raman scattering (SERS) signals of aniline and aminobiphenyl isomers. *Vib. Spectrosc.* 81: 22–31.

67 Kambhampati, P., Child, C.M., Foster, M.C., and Campion, A. (1998). On the chemical mechanism of surface enhanced Raman scattering: experiment and theory. *J. Chem. Phys.* 108: 5013–5026.

68 Campion, A., Ivanecky, J.E., Child, C.M., and Foster, M. (1995). On the mechanism of chemical enhancement in surface-enhanced Raman scattering. *J. Am. Chem. Soc.* 117: 11807–11808.

69 Lombardi, J.R. and Birke, R.L. (2008). A unified approach to surface-enhanced Raman spectroscopy. *J. Phys. Chem. C* 112: 5605–5617.

70 Le Ru, E.C., Blackie, E., Meyer, M., and Etchegoin, P.G. (2007). Surface enhanced Raman scattering enhancement factors: a comprehensive study. *J. Phys. Chem. C* 111: 13794–13803.

71 Le Ru, E. and Etchegoin, P. (2009). *Principles of Surface-Enhanced Raman Spectroscopy and Related Plasmonic Effects*. Elsevier.

72 Panicker, C.Y., Anto, P.L., Varghese, H.T., and Philip, D. (2005). Surface selection rules of surface enhanced Raman scattering. *Int. J. Chem. Sci.* 3 (2): 268–270.

73 Ru, E.C.L., Meyer, S.A., Artur, C. et al. (2011). Experimental demonstration of surface selection rules for SERS on flat metallic surfaces. *Chem. Commun.* 47: 3903–3905.

74 Schulze, W., Breithaupt, B., Charlé, K.P., and Kloss, U. (1984). Surface enhanced Raman scattering (SERS) at matrix isolated metal clusters of Ag, Na and K. *Ber. Bunsenges. Phys. Chem.* 88: 308–310.

75 Abdelsalam, M.E., Mahajan, S., Bartlett, P.N. et al. (2007). SERS at structured palladium and platinum surfaces. *J. Am. Chem. Soc.* 129: 7399–7406.

76 Mrozek, M.F., Xie, Y., and Weaver, M.J. (2001). Surface-enhanced Raman scattering on uniform platinum-group overlayers: preparation by redox replacement of underpotential-deposited metals on gold. *Anal. Chem.* 73: 5953–5960.

77 Tian, Z.-Q., Yang, Z.-L., Ren, B. et al. (2006). Surface-enhanced Raman scattering from transition metals with special surface morphology and nanoparticle shape. *Faraday Discuss.* 132: 159–170.

78 Wu, S., Duan, N., He, C. et al. (2020). Surface-enhanced Raman spectroscopic–based aptasensor for Shigella sonnei using a dual-functional metal complex-ligated gold nanoparticles dimer. *Colloids Surf. B: Biointerfaces* 190: 110940.

79 Darienzo, R.E., Chen, O., Sullivan, M. et al. (2020, 2020). Au nanoparticles for SERS: temperature-controlled nanoparticle morphologies and their Raman enhancing properties. *Mater. Chem. Phys.* 240: 122143.

80 Chen, X., Wang, D., Li, J. et al. (2020). A spectroscopic approach to detect and quantify phosmet residues in Oolong tea by surface-enhanced Raman scattering and silver nanoparticle substrate. *Food Chem.* 312: 126016.

81 Kong, L., Dong, N., Tian, G. et al. (2020). Highly enhanced Raman scattering with good reproducibility observed on a flexible PI nanofabric substrate decorated by silver nanoparticles with controlled size. *Appl. Surf. Sci.* 511: 145443.

82 Tripp, R.A., Dluhy, R.A., and Zhao, Y. (2008). Novel nanostructures for SERS biosensing. *Nano Today* 3: 31–37.

83 Sackmann, M., Bom, S., Balster, T., and Materny, A. (2007). Nanostructured gold surfaces as reproducible substrates for surface-enhanced Raman spectroscopy. *J. Raman Spectrosc.* 38: 277–282.

84 Billot, L., Lamy de la Chapelle, M., Grimault, A.S. et al. (2006). Surface enhanced Raman scattering on gold nanowire arrays: evidence of strong multipolar surface plasmon resonance enhancement. *Chem. Phys. Lett.* 422: 303–307.

85 Yu, Q., Guan, P., Qin, D. et al. (2008). Inverted size-dependence of surface-enhanced Raman scattering on gold nanohole and nanodisk arrays. *Nano Lett.* 8: 1923–1928.

86 Jensen, T.R., Malinsky, M.D., Haynes, C.L., and Van Duyne, R.P. (2000). Nanosphere lithography: tunable localized surface plasmon resonance spectra of silver nanoparticles. *J. Phys. Chem. B* 104: 10549–10556.

87 Ormonde, A.D., Hicks, E.C.M., Castillo, J., and Van Duyne, R.P. (2004). Nanosphere lithography: fabrication of large-area Ag nanoparticle arrays by convective self-assembly and their characterization by scanning UV–visible extinction spectroscopy. *Langmuir* 20: 6927–6931.

88 Khan, M.A., Hogan, T.P., and Shanker, B. (2009). Gold-coated zinc oxide nanowirebased substrate for surface-enhanced Raman spectroscopy. *J. Raman Spectrosc.* 40: 1539–1545.

89 Cheng, C., Yan, B., Wong, S.M. et al. (2010). Fabrication and SERS performance of silver-nanoparticle-decorated Si/ZnO nanotrees in ordered arrays. *ACS Appl. Mater. Interfaces* 2: 1824–1828.

90 Neddersen, J., Chumanov, G., and Cotton, T.M. (1993). Laser ablation of metals: a new method for preparing SERS active colloids. *Appl. Spectrosc.* 47: 1959–1964.

91 Lin, H., Mock, J., Smith, D. et al. (2004). Surface-enhanced Raman scattering from silver-plated porous silicon. *J. Phys. Chem. B* 108: 11654–11659.

92 Walsh, R.J. and Chumanov, G. (2001). Silver coated porous alumina as a new substrate for surface-enhanced Raman scattering. *Appl. Spectrosc.* 55: 1695–1700.

93 Myeong-Lok, S., Sung-Jin, C., David, J.B. et al. (2012). A nanoforest structure for practical surfaceenhanced Raman scattering substrates. *Nanotechnology* 23: 095301.

94 Jana, N.R. and Pal, T. (2007). Anisotropic metal nanoparticles for use as surface-enhanced Raman substrates. *Adv. Mater.* 19: 1761–1765.

95 Williamson, T.L., Guo, X., Zukoski, A. et al. (2005). Porous GaN as a template to produce surface-enhanced Raman scattering-active surfaces. *J. Phys. Chem. B* 109: 20186–20191.

96 Christou, K., Knorr, I., Ihlemann, J.R. et al. (2010). Fabrication and characterization of homogeneous surface-enhanced Raman scattering substrates by single pulse UV-laser treatment of gold and silver films. *Langmuir* 26: 18564–18569.

97 Xue, G., Dong, J., and Zhang, M. (1991). Surface-enhanced Raman scattering (SERS) and surface-enhanced resonance Raman scattering (SERRS) on HNO_3^- roughened copper foil. *Appl. Spectrosc.* 45: 756–759.

98 Eiamchai, P., Chananonnawathorn, C., Horprathum, M. et al. (2020). Spatial elemental investigations in nanostructured alloyed Ag/Au SERS substrates by magnetron sputtering oblique-angle co-deposition towards increased performance and shelf life. *Appl. Surf. Sci.* 513: 145748.

99 Henry, W.A., Biris, A.S., Watanabe, F. et al. (2016). Surface-enhanced infrared absorption studies of copper nanostructures formed by oblique-angle deposition. *Chem. Phys. Lett.* 663: 111–114.

100 He, P., Liu, H., Li, Z. et al. (2004). Electrochemical deposition of silver in room-temperature ionic liquids and its surface-enhanced Raman scattering effect. *Langmuir* 20: 10260–10267.

101 Richter, L., Albrycht, P., Księżopolska-Gocalska, M. et al. (2020). Fast and efficient deposition of broad range of analytes on substrates for surface enhanced Raman spectroscopy. *Biosens. Bioelectron.* 156: 112124.

102 Durmanov, N.N., Guliev, R.R., Eremenko, A.V. et al. (2018). Non-labeled selective virus detection with novel SERS-active porous silver nanofilms fabricated by electron beam physical vapor deposition. *Sensors Actuators B Chem.* 257: 37–47.

103 Emamian, S., Eshkeiti, A., Narakathu, B.B. et al. (2015). Gravure printed flexible surface enhanced Raman spectroscopy (SERS) substrate for detection of 2,4-dinitrotoluene (DNT) vapor. *Sensors Actuators B Chem.* 217: 129–135.

104 Erlebacher, J., Aziz, M.J., Karma, A. et al. (2001). Evolution of nanoporosity in dealloying. *Nature* 410: 450–453.

105 Biener, J., Nyce, G.W., Hodge, A.M. et al. (2008). Nanoporous plasmonic metamaterials. *Adv. Mater.* 20: 1211–1217.

106 Reilly, T.H., Corbman, J.D., and Rowlen, K.L. (2007). Vapor deposition method for sensitivity studies on engineered surface-enhanced Raman scattering-active substrates. *Anal. Chem.* 79: 5078–5081.

107 De Jesús, M.A., Giesfeldt, K.S., Oran, J.M. et al. (2005). Nanofabrication of densely packed metal polymer arrays for surface-enhanced Raman spectrometry. *Appl. Spectrosc.* 59: 1501–1508.

108 Haynes, C.L. and Van Duyne, R.P. (2001). Nanosphere lithography: a versatile nanofabrication tool for studies of size-dependent nanoparticle optics. *J. Phys. Chem. B* 105: 5599–5611.

109 Hulteen, J.C., Treichel, D.A., Smith, M.T. et al. (1999). Nanosphere lithography: size-tunable silver nanoparticle and surface cluster arrays. *J. Phys. Chem. B* 103: 3854–3863.

110 Jiang, P. (2006). Large-scale fabrication of periodic nanostructured materials by using hexagonal non-close-packed colloidal crystals as templates. *Langmuir* 22: 955–3958.

111 Jiang, P. and McFarland, M.J. (2004). Large-scale fabrication of wafer-size colloidal crystals, macroporous polymers and nanocomposites by spin-coating. *J. Am. Chem. Soc.* 126: 13778–13786.

112 Lin, T.-H., Linn, N.C., Tarajano, L. et al. (2009). Electrochemical SERS at periodic metallic nanopyramid arrays. *J. Phys. Chem. C* 113: 1367–1372.

113 Liu, X., Sun, C.-H., Linn, N.C. et al. (2009). Wafer-scale surface-enhanced Raman scattering substrates with highly reproducible enhancement. *J. Phys. Chem. C* 113: 14804–14811.

114 Chan, S., Kwon, S., Koo, T.W. et al. (2003). Surface-enhanced Raman scattering of small molecules from silver-coated silicon nanopores. *Adv. Mater.* 15: 1595–1598.

115 Chattopadhyay, S., Lo, H.-C., Hsu, C.-H. et al. (2005). Surface-enhanced Raman spectroscopy using self-assembled silver nanoparticles on silicon nanotips. *Chem. Mater.* 17: 553–559.

116 Broglin, B.L., Andreu, A., Dhussa, N. et al. (2007). Investigation of the effects of the local environment on the surface-enhanced Raman spectra of striped gold/silver nanorod arrays. *Langmuir* 23: 4563–4568.

117 Gu, G.H., Kim, J., Kim, L., and Suh, J.S. (2007). Optimum length of silver nanorods for fabrication of hot spots. *J. Phys. Chem. C* 111: 7906–7909.

118 Lombardi, I., Cavallotti, P.L., Carraro, C., and Maboudian, R. (2007). Template assisted deposition of Ag nanoparticle arrays for surface-enhanced Raman scattering applications. *Sensors Actuators B Chem.* 125: 353–356.

119 Schierhorn, M., Lee, S.J., Boettcher, S.W. et al. (2006). Metal–silica hybrid nanostructures for surface-enhanced Raman spectroscopy. *Adv. Mater.* 18: 2829–2832.

120 Du, Y., Shi, L., He, T. et al. (2008). SERS enhancement dependence on the diameter and aspect ratio of silver-nanowire array fabricated by anodic aluminium oxide template. *Appl. Surf. Sci.* 255: 1901–1905.

121 Henley, S.J., Carey, J.D., and Silva, S.R.P. (2006). Silver-nanoparticle-decorated carbon nanoscaffolds: application as a sensing platform. *Appl. Phys. Lett.* 89: 183120.

122 Sun, Y., Liu, K., Miao, J. et al. (2010). Highly sensitive surface-enhanced Raman scattering substrate made from superaligned carbon nanotubes. *Nano Lett.* 10: 1747–1753.

123 Dick, L.A., McFarland, A.D., Haynes, C.L., and Van Duyne, R.P. (2001). Metal film over nanosphere (MFON) electrodes for surface-enhanced Raman spectroscopy (SERS): improvements in surface nanostructure stability and suppression of irreversible loss. *J. Phys. Chem. B* 106: 853–860.

124 Bao, L.-L., Mahurin, S.M., Liang, C.-D., and Dai, S. (2003). Study of silver films over silica beads as a surface-enhanced Raman scattering (SERS) substrate for detection of benzoic acid. *J. Raman Spectrosc.* 34: 394–398.

125 Li, X., Chen, G., Yang, L. et al. (2010). Multifunctional Au-coated TiO_2 nanotube arrays as recyclable SERS substrates for multifold organic pollutants detection. *Adv. Funct. Mater.* 20: 2815–2824.

126 Deng, S., Fan, H.M., Zhang, X. et al. (2009). An effective surface-enhanced Raman scattering template based on a Ag nanocluster–ZnO nanowire array. *Nanotechnology* 20: 175705.

127 Tatsunori, S., Yuto, T., Ryo, N. et al. (2008). Surface enhanced Raman scattering properties using Au-coated ZnO nanorods grown by two-step, off-axis pulsed laser deposition. *J. Phys. D. Appl. Phys.* 41: 235304.

128 Chen, L., Luo, L., Chen, Z. et al. (2009). ZnO/Au composite nanoarrays as substrates for surface-enhanced Raman scattering detection. *J. Phys. Chem. C* 114: 93–100.

129 Song, W., Han, X., Chen, L. et al. (2010). Site-specific deposition of Ag nanoparticles on ZnO nanorod arrays via galvanic reduction and their SERS applications. *J. Raman Spectrosc.* 41: 907–913.

130 Rivas, L., Sanchez-Cortes, S., García-Ramos, J.V., and Morcillo, G. (2000). Mixed silver/gold colloids: a study of their formation, morphology, and surface-enhanced Raman activity. *Langmuir* 16: 9722–9728.

131 Nickel, U., zu Castell, A., Pöppl, K., and Schneider, S. (2000). A silver colloid produced by reduction with hydrazine as support for highly sensitive surface-enhanced Raman spectroscopy. *Langmuir* 16: 9087–9091.

132 Rycenga, M., Hou, K.K., Cobley, C.M. et al. (2009). Probing the surface-enhanced Raman scattering properties of Au–Ag nanocages at two different excitation wavelengths. *Phys. Chem. Chem. Phys.* 11: 5903–5908.

133 Rycenga, M., Cobley, C.M., Zeng, J. et al. (2011). Controlling the synthesis and assembly of silver nanostructures for plasmonic applications. *Chem. Rev.* 111: 3669–3712.

134 Driskell, J.D., Shanmukh, S., Liu, Y. et al. (2008). The use of aligned silver nanorod arrays prepared by oblique angle deposition as surface enhanced Raman scattering substrates. *J. Phys. Chem. C* 112: 895–901.

135 Xin, S. and Hao, L. (2013). Gold nanoisland arrays by repeated deposition and postdeposition annealing for surface-enhanced Raman spectroscopy. *Nanotechnology* 24: 355706.

136 Chursanova, M., Dzhagan, V., Yukhymchuk, V. et al. (2009). Nanostructured silver substrates with stable and universal SERS properties: application to organic molecules and semiconductor nanoparticles. *Nanoscale Res. Lett.* 5: 403–409.

137 Ding, Y., Kim, Y.J., and Erlebacher, J. (2004). Nanoporous gold leaf: "ancient technology"/advanced material. *Adv. Mater.* 16: 1897–1900.

138 Ciesielski, P.N., Scott, A.M., Faulkner, C.J. et al. (2008). Functionalized nanoporous gold leaf electrode films for the immobilization of photosystem I. *ACS Nano* 2: 2465–2472.

139 Heaps, D.A. and Gri_ths, P.R. (2005). Off-line direct deposition gas chromatography/surface-enhanced Raman scattering and the ramifications for on-line measurements. *Appl. Spectrosc.* 59: 1305–1309.

140 Trachta, G., Schwarze, B., Sägmüller, B. et al. (2004). Combination of high-performance liquid chromatography and SERS detection applied to the analysis of drugs in human blood and urine. *J. Mol. Struct.* 693: 175–185.

141 Sägmüller, B., Schwarze, B., Brehm, G. et al. (2003). Identification of illicit drugs by a combination of liquid chromatography and surface-enhanced Raman scattering spectroscopy. *J. Mol. Struct.* 661–662: 279–290.

142 Wang, W., Xu, M., Guo, Q. et al. (2015). Rapid separation and on-line detection by coupling high performance liquid chromatography with surface-enhanced Raman spectroscopy. *RSC Adv.* 5: 47640–47646.

143 Farquharson, S. and Maksymiuk, P. (2003). Simultaneous chemical separation and surface-enhanced Raman spectral detection using silver-doped sol–gels. *Appl. Spectrosc.* 57: 479–482.

144 Lucotti, A., Tommasini, M., Casella, M. et al. (2012). TLC-surface enhanced Raman scattering of apomorphine in human plasma. *Vib. Spectrosc.* 62: 286–291.

145 Vicario, A., Sergo, V., Toffoli, G., and Bonifacio, A. (2015). Surface-enhanced Raman spectroscopy of the anti-cancer drug irinotecan in presence of human serum albumin. *Colloids Surf. B: Biointerfaces* 127: 41–46.

146 Kneipp, J., Kneipp, H., and Kneipp, K. (2008). SERS-a single-molecule and nanoscale tool for bioanalytics. *Chem. Soc. Rev.* 37: 1052–1060.

147 Sánchez-Cortés, S., Domingo, C., García-Ramos, J.V., and Aznárez, J.A. (2001). Surface-enhanced vibrational study (SEIR and SERS) of dithiocarbamate pesticides on gold films. *Langmuir* 17: 1157–1162.

148 Vongsvivut, J., Robertson, E.G., and McNaughton, D. (2010). Surface-enhanced Raman spectroscopic analysis of fonofos pesticide adsorbed on silver and gold nanoparticles. *J. Raman Spectrosc.* 41: 1137–1148.

149 Guerrini, L., Sanchez-Cortes, S., Cruz, V.L. et al. (2011). Surface-enhanced Raman spectra of dimethoate and omethoate. *J. Raman Spectrosc.* 42: 980–985.

150 Liu, B., Zhou, P., Liu, X. et al. (2013). Detection of pesticides in fruits by surface-enhanced Raman spectroscopy coupled with gold nanostructures. *Food Bioprocess Technol.* 6: 710–718.

151 Guo, K., Xiao, R., Zhang, X. et al. (2015). Silver nanopartical over AuFON substrate for enhanced Raman readout and their application in pesticide monitoring. *Molecules* 20: 6299–6309.

152 Khlebtsov, B.N., Khanadeev, V.A., Panfilova, E.V. et al. (2015). Gold nanoisland films as reproducible SERS substrates for highly sensitive detection of fungicides. *ACS Appl. Mater. Interfaces* 7: 6518–6529.

153 Saute, B., Premasiri, R., Ziegler, L., and Narayanan, R. (2012). Gold nanorods as surface enhanced Raman spectroscopy substrates for sensitive and selective detection of ultra-low levels of dithiocarbamate pesticides. *Analyst* 137: 5082–5087.

154 Zheng, X., Chen, Y., Chen, Y. et al. (2012). High performance Au/Ag core/shell bipyramids for determination of thiram based on surface-enhanced Raman scattering. *J. Raman Spectrosc.* 43: 1374–1380.

155 Fodjo, E.K., Riaz, S., Li, D.-W. et al. (2012). Cu@Ag/β-$AgVO_3$ as a SERS substrate for the trace level detection of carbamate pesticides. *Anal. Methods* 4: 3785–3791.

156 Wang, B., Zhang, L., and Zhou, X. (2014). Synthesis of silver nanocubes as a SERS substrate for the determination of pesticide paraoxon and thiram. *Spectrochim. Acta A* 121: 63–69.

157 Saute, B. and Narayanan, R. (2011). Solution-based direct readout surface enhanced Raman spectroscopic (SERS) detection of ultra-low levels of thiram with dogbone shaped gold nanoparticles. *Analyst* 136: 527–532.

158 Kandjani, A.E., Sabri, Y.M., Mohammad-Taheri, M. et al. (2014). Detect, remove and reuse: a new paradigm in sensing and removal of Hg(II) from wastewater via SERS-active ZnO/Ag nanoarrays. *Environ. Sci. Technol.* 49: 1578–1584.

159 Wang, G., Lim, C., Chen, L. et al. (2009). Surface-enhanced Raman scattering in nanoliter droplets: towards high-sensitivity detection of mercury(II) ions. *Anal. Bioanal. Chem.* 394: 1827–1832.

160 Li, F., Wang, J., Lai, Y. et al. (2013). Ultrasensitive and selective detection of copper(II) and mercury(II) ions by dye-coded silver nanoparticle-based SERS probes. *Biosens. Bioelectron.* 39: 82–87.

161 Li, J., Chen, L., Lou, T., and Wang, Y. (2011). Highly sensitive SERS detection of As^{3+} ions in aqueous media using glutathione functionalized silver nanoparticles. *ACS Appl. Mater. Interfaces* 3: 3936–3941.

162 Song, D., Yang, R., Long, F., and Zh, A. (2019). Applications of magnetic nanoparticles in surface-enhanced Raman scattering (SERS) detection of environmental pollutants. *J. Environ. Sci.* 80: 14–34.

163 Du, J., Xu, J., Sun, Z., and Jing, C. (2016). Au nanoparticles grafted on Fe_3O_4 as effective SERS substrates for label-free detection of the 16 EPA priority polycyclic aromatic hydrocarbons. *Anal. Chim. Acta* 915: 81–89.

164 Tang, S., Li, Y., Huang, H. et al. (2017). Efficient enrichment and self-assembly of hybrid nanoparticles into removable and magnetic SERS substrates for sensitive detection of environmental pollutants. *ACS Appl. Mater. Interfaces* 9 (8): 7472–7480.

165 Halouzka, V., Halouzkova, B., Jirovsky, D. et al. (2017). Copper nanowire coated carbon fibers as efficient substrates for detecting designer drugs using SERS. *Talanta* 165: 384–390.

166 Si, K.J., Guo, P., Shi, Q., and Cheng, W. (2015). Self-assembled nanocube-based plasmene nanosheets as soft surface-enhanced Raman scattering substrates toward direct quantitative drug identification on surfaces. *Anal. Chem.* 87: 5263–5269.

167 Lee, W.W.Y., Silverson, V.A.D., Jones, L.E. et al. (2016). Surface-enhanced Raman spectroscopy of novel psychoactive substances using polymer-stabilized Ag nanoparticle aggregates. *Chem. Commun.* 52: 493–496.

168 Yan, X., Li, P., Zhou, B. et al. (2017). Optimal hotspots of dynamic surfaced-enhanced Raman spectroscopy for drugs quantitative detection. *Anal. Chem.* 89: 4875–4881.

169 Chen, N., Ding, P., Shi, Y. et al. (2017). Portable and reliable surface-enhanced Raman scattering silicon chip for signal-on detection of trace trinitrotoluene explosive in real systems. *Anal. Chem.* 89: 5072–5078.

170 Lopez-Lopez, M., Merk, V., García-Ruiz, C., and Kneipp, J. (2016). Surface-enhanced Raman spectroscopy for the analysis of smokeless gunpowders and macroscopic gunshot residues. *Anal. Bioanal. Chem.* 408: 4965–4973.

171 Zong, C., Xu, M., Xu, L.J. et al. (2018). Surface-enhanced Raman spectroscopy for bioanalysis: reliability and challenges. *Chem. Rev.* 118: 4946–4980.

172 Kahraman, M., Mullen, E.R., Korkmaz, A., and Wachsmann-Hogiu, S. (2017). Fundamentals and applications of SERS-based bioanalytical sensing. *Nano* 6: 831–852.

173 Karadan, P., Aggarwal, S., Anappara, A.A. et al. (2018). Tailored periodic Si nanopillar based architectures as highly sensitive universal SERS biosensing platform. *Sensors Actuators B Chem.* 254: 264–271.

174 Szekeres, G.P. and Kneipp, J. (2019). SERS probing of proteins in gold nanoparticle agglomerates. *Front. Chem.* 7: 30.

175 Yang, Y., Zhu, J., Zhao, J. et al. (2019). Growth of spherical gold satellites on the surface of Au@Ag@SiO_2 core-shell nanostructures used for an ultrasensitive SERS immunoassay of alpha-fetoprotein. *ACS Appl. Mater. Interfaces* 11: 3617–3626.

176 Cheng, Z., Choi, N., Wang, R. et al. (2017). Simultaneous detection of dual prostate specific antigens using surface-enhanced Raman scattering-based immunoassay for accurate diagnosis of prostate cancer. *ACS Nano* 11: 4926–4933.

177 Xiao, M., Xie, K., Dong, X. et al. (2019). Ultrasensitive detection of avian influenza A (H7N9) virus using surface-enhanced Raman scattering-based lateral flow immunoassay strips. *Anal. Chim. Acta* 1053: 139–147.

7

Nanomaterials (NMs) for Biological Sample Preparations

7.1 Introduction

Biomolecules fall into the same size scale as nanomaterials (NMs). Therefore, NMs can be designed to interact with biomolecules. The merge of biomedical sciences and nanotechnology was then logical, creating a new area of research called "bionanotechnology". This area can then be seen as the efficient combination of biological systems including cells, proteins, peptides, and nucleic acids, with organic and/or inorganic materials for the attainment of functional nanostructured constructs [1].

Bionanotechnology also concerns the improvement and application of instruments, initially designed to generate and manipulate nanostructured materials to study fundamental biological processes and structures [1, 2]. Therefore, one of the main areas to benefit from what bionanotechnology has to offer is medicine, where the excellent features of NMs and nanosystems can be exploited to give rise to new methods of biosensing [3, 4], imaging [5–7], therapy [8–11], and tissue engineering [12–15] toward a better healthcare [16, 17].

Since the boom of bionanotechnology, and driven by their intriguing features, a vast collection of NMs were designed, prepared, and employed as building blocks for NMs–biomolecules hybrid systems. In such a short time, the NM landscape grew immensely, and NMs have already revolutionized biomedical studies improving the detection limits of certain relevant analytes, providing better contrast or fluorescence for bioimaging, new approaches and routes of localized therapy, better drug delivery, and many other advances. One popular technique is to harness the intrinsic fluorescent features of NMs or introduce fluorescent dyes to NM systems with other advantageous properties for applications in biomedicine. Fluorescent-based methods are one of the most commonly used methods because of their high sensitivity, versatility, and simplicity. NMs including quantum dots (QDs) [18], silicon and silica nanostructures [19], polymeric nanoparticles [20], noble metal nanoparticles [21], upconversion nanoparticles [22], and carbon-based NMs [23] are examples of NMs applied in fluorescent assays due to their photophysical and photochemical features.

This chapter provides an overview of the latest advancements on the design and development of novel NMs for the biological sample preparation processes. The chapter starts with the use of NMs in diagnostic platforms. Then, lab-on-a-chip

Sample Preparation with Nanomaterials: Next Generation Techniques and Applications,
First Edition. Chaudhery Mustansar Hussain, Rüstem Keçili, and Chaudhery Ghazanfar Hussain.

(LOC) platforms using NMs are described. Finally, biomedical and sensor applications of NMs for biological samples are demonstrated.

7.2 The Use of NMs in Diagnostic Platforms

NMs are an extremely exciting class of materials that have gained great attention from many researchers since they display unique properties such as high physical and chemical stability, versatile chemistry for the functionalization, and high surface area [24–41]. The chemical features of nanomaterials in the scale between 1 and 100 nm are completely different compared to the materials in micro- and macroscale.

NMs-based sensor systems have been successfully employed for the sensitive detection of a wide range of biomolecules, including proteins (e.g. antigens, enzymes, and hormones) [18, 42], nucleic acids (e.g. DNA, mRNA, and miRNA) [43–45], and small molecules (e.g. ATP, various ions) [46, 47].

There are many superiorities of NMs in diagnostic test applications. For example, many alternative signaling mechanisms are available, which can yield various different signals that enable flexibility of assay design. For example, signals can take the form of changes in electrical impedance [48] or color [49]. Strong signal intensities can be obtained by exploiting unique nanoscale phenomena, allowing sensitive detection and low limits of detection. It is now possible, with advances in materials characterization methods, to finely tune the surface chemistries of NMs enabling effective optimization of the performance of nanodiagnostic platforms in a range of environments. Because of their small size, NMs are associated with extremely large ensemble surface areas that provide a high ratio of sensing molecules to NMs, further improving sensitivities and reaction kinetics. By applying NMs, it is possible to use lower amounts of signaling molecules and lower volumes of reagents and samples to carry out assays. As a result of this, diagnostic platforms housed on a small piece of silicon, known as "lab-on-a-chip" devices, are becoming increasingly common [50].

7.2.1 The Optimization of NMs in Diagnostic Platforms

Optimization of the features of NMs is a crucial process for their use in, and the design of, high-performance diagnostics. Many aspects of the NMs such as nanoparticles must be optimized, including their structure and synthesis. It is crucial to note that the prepared nanoparticles should have a high degree of uniformity. In addition, nanoparticles should retain their chemical features and stability in a variety of conditions. Colloidal suspensions of nanoparticles are particularly useful for biosensing because they allow the design of homogenous assays, which do not require any washing steps and so minimize the time required to carry out a diagnostic test.

Each of the structural components of nanoparticles can be optimized, including the core, surface, and sensing interface. This allows precise tuning of their properties and also means that toxicity effect can be minimized. A current challenge for nanoparticle core design is the development of innovative approaches and strategies which limit their toxicity. This is particularly important for QDs, which often contain cores composed of heavy metals such as cadmium and selenide. Recent advances have used less toxic materials, such as silicon [51] and carbon [52], to produce novel classes of QD. However, this has generally come at a cost to their optical properties, with these nanoparticles showing reduced fluorescence intensities compared to traditional QDs.

The nanoparticle core is an important factor in determining its physical and electronic properties. These properties are extremely sensitive to minor variations in size, shape, and crystal structure. This not only allows the selection of various nanoparticle properties under different conditions but also means that nanoparticle synthesis must be strictly controlled and reproducible. The surface of a nanoparticle provides a barrier between the core and the environment, and it therefore ensures that the nanoparticle retains its stability, and other properties, in various conditions. The surface typically consists of capping molecules that directly bind to the nanoparticle shell. These molecules are able to prevent aggregation of nanoparticles and their adsorption to other molecules. This can also aid the stability in various aqueous solutions despite pH changes and can serve as the molecule to which functional molecules are conjugated. The optimization of surface capping is therefore essential for the ideal functioning of the nanoparticle. A thick capping layer would be more likely to provide improved protection against the formation of nonspecific interactions with other molecules that might also be present in solution, such as proteins, lipids, and other biomolecules, and therefore prevent the formation of a "corona" of biomolecules around the nanoparticle [53]. However, most work in this field is focused on the production of capping layers that are as thin as possible, as it is desirable to minimize the distance between the nanoparticle core and the sensing molecules on its surface, so that distance-dependent interactions such as resonance energy transfer (RET) are as efficient as possible.

The most significant advances in surface capping techniques in recent years have used molecules that have been specifically designed to be hydrophilic and have a low molecular weight. In the case of QDs, the use of zwitterionic surface ligands has given rise to exceptionally stable and functional particles [54]. Zwitterions are neutral molecules with both positive and negative charges. When used for the surface capping of nanoparticles, they can form electrostatic interactions and hydrogen bonds with water molecules that prevent the formation of nonspecific interactions with other biomolecules.

7.2.2 Biofunctionalization of NMs in Diagnostic Platforms

Biofunctionalization, also known as bioconjugation, is the process of immobilizing biological molecules to the surface of the nanostructure (such as DNA or peptides) [55]. The conjugation strategy used depends on the application, the nanostructure, and the biological molecule to be attached. Conventional strategies that employ covalent modification often require multiple stages of washing and purification, can cross-link nanoparticles, and can affect their stability. Biofunctionalization strategies should ideally have little or no effect on the properties of the nanomaterial, be of high yield, and allow optimal functioning of the sensing mechanism. More recent strategies include using streptavidin–biotin interactions and polyhistidine tags. However, more precise control over biofunctionalization is desirable and can be achieved by using strain-promoted cycloadditions [56] and enzyme-based additions [57].

On the other hand, in molecular diagnostics, NMs can either act as labels or supports. As labels, the NMs are usually connected to receptors of the analyte and are able to detect a binding event and produce a response. As a support, the nanomaterial will perform the function of a surface, to which immobilized receptors would typically be attached, on which recognition events would occur. The analyte–receptor mechanism will yield either a physical or chemical response that in turn alters the signal from the NM.

Gold nanoparticles (AuNPs) are particularly useful NMs because of their unique features such as stability, relatively straightforward synthesis, and the way in which they interact with light. The free electrons of AuNPs resonate in response to the oscillating electric fields of a ray of light. These resonant oscillations are referred to as surface plasmons. The spatial confinement of these surface plasmons, known as localized surface plasmon resonance (LSPR), causes AuNPs to absorb blue-green light (~450 nm) and reflect red light (~700 nm), resulting in the red color that is characteristic of AuNPs. The reflection of light in the visible spectrum allows AuNP-based sensors to give colorimetric signals. The LSPR peak can shift in response to external factors that, for example, cause AuNPs to aggregate. In addition, AuNPs can act as label-free sensors by studying their surface-enhanced Raman scattering (SERS) in response to the binding of certain analytes.

7.3 NMs-based Lab-on-a-chip (LOC) Platforms

In the recent years, there has been a great effort in the development of novel approaches for the sensitive detection and treatment of diseases and ailments. Currently, the detection of various diseases is carried out by using expensive instruments. There are various issues that should be considered when performing these experiments. First of all, bulky equipment that can be pretty expensive to buy and/or repair is used in the experiments. Second, with these equipment having high dimensions, it can be certain that large amounts of sample and or reagent have to be used. The third issue is the time needed to perform each experiment. Because of these issues, innovative techniques and approaches need to be addressed. Basically, a system for the sensitive detection has to be designed, which is not only light and portable but should also be rapid in detection and must take as little fluid as possible.

On the other hand, the fast advancements of the microelectronic technology since 1960s established semiconductor production techniques in order to enhance the density of transistors in integrated circuits. In 1980s, these techniques led to the first design and fabrication of the microelectromechanical systems (MEMS) [58]. Further refinement of the fabrication technologies of silicon-based MEMS made it straight forward to design and preparation of simple microfluidic platforms.

The first demonstration of microfluidic systems was carried out by Terry et al. [59]. In their study, a microfluidic system was prepared by using a silicon wafer, and the prepared systems were successfully used as a miniaturized gas analysis system in terms of gas chromatography (GC) with a thermal conductivity detector. The microfluidic system was fabricated by applying photolithography technique and chemical etching. The efforts on the design and fabrication of microfluidic platforms did not pick up momentum until the early 1990s, when Manz et al. first introduced the concept of micro total analysis system (μTAS) [60]. According to this approach, the injection, separation, and detection systems were found in the same interface. This research was not only defining the μTAS systems but it was also a realization at the same time for scientists to use the technique in life sciences. Since then, the developed miniaturized systems were successfully applied for different applications such as separation [61–63], detection [64–66], or used as a reactor [67–70]. Then, scientists realized the potential of these unique systems having great advantages in other fields of science besides analytical chemistry.

Microfluidic techniques can be successfully applied for the design and fabrication of LOC platforms that enables to produce a number of microchannels existing on the surface of single chip which fits in your hand. The microchannels enable the handling of fluids in quantities as low as a few picoliters as well as the manipulation of biochemical reactions at very small volumes. In addition to microchannels, LOC platforms also contain various components such as electrodes, integrated pumps, electrical field valves, and electronics. Figure 7.1 shows a schematic deposition of an LOC system having various electrical and fluidic connections. During the fabrication process of the system, electrodes, microfluidic structures, and various back-end processes are needed to obtain a fully functional LOC system [71, 72]. As can be seen from the Figure 7.1, sealing of microfluidic structures can be performed by using a cover layer that is patterned or drilled for the formation of openings for electrical/fluidic interfaces. In addition, the system can be integrated into the peripheral equipment (i.e. pumps and valves) using inlet and outlet ports.

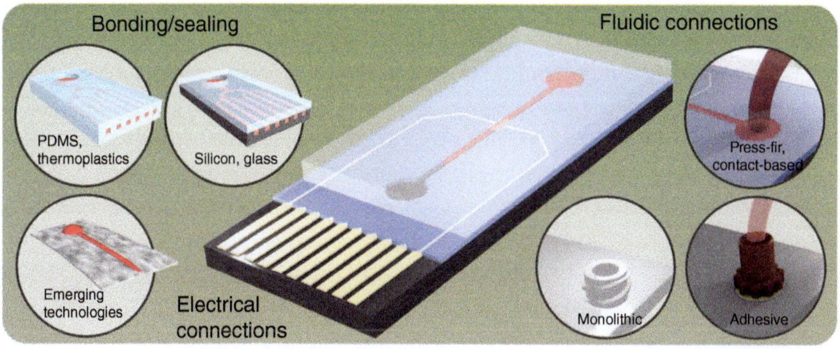

Figure 7.1 The schematic depiction of an LOC platform. Source: Temiz [73]. © 2014, Elsevier.

7.3.1 Paper-based LOC Platforms

In the preparation of paper-based LOC systems , hydrophobic, microchannels on hydrophilic paper that guide liquid from inlet to designated location are mainly printed or photoetched [74, 75]. This approach is quite cheap, disposable, facile to apply, and usually applied for the environmental testing and portable glucose detection [76, 77]. However, the capillary force-driven transportation of liquid shows that the precision of assay is usually limited and the hydrophobic barriers may not be stable after longtime store [78].

7.3.2 Centrifugal LOC Platforms

In centrifugal LOC systems, a centrifugal field is usually exploited to manipulate liquid [79]. In the analyses by using this type of miniaturized systems, a cartridge including liquid reagent is usually centrifuged as a disk during the liquid process. Centrifugal LOC systems are also called as "lab-on-a-disc systems" that exhibit great superiorities. Centrifugal LOC systems are composed of a motor as the pumping source, which is rather simple and compact. Furthermore, analyses can be easily processed since the centrifugal force exists everywhere on the disk-shaped cartridge [80].

7.3.3 Droplet-based LOC Platforms

Droplet-based LOC systems manipulate discrete volumes of fluids with the use of immiscible phases [81]. In these miniaturized systems, a large number of droplets are used as reactors to implement bioassays. A very striking application based on droplet-based LOC systems is digital PCR that exhibits great sensitivity toward nucleic acids [82, 83]. However, the generation and manipulation of a large number of tiny droplets require a careful fabrication technique and also a high precision manipulation strategy [84].

7.3.4 Digital LOC Platforms

Similar to droplet-based LOC systems, digital LOC systems also deal with the droplets. However, during the analysis by using digital LOC systems, droplets are usually manipulated on a set of insulated electrodes that are under software-driven electronic control [85]. In this type of LOC system, the most widely applied manipulation strategy is electrowetting-on-dielectric phenomenon which is based on the principle of modification of the interfacial tension with an applied electric field [86].

7.3.5 Surface Acoustic Wave-based LOC Platforms

Surface acoustic wave-based LOC systems, in which acoustic waves having frequency range between 10 and 1000 MHz are applied for the manipulation of microscale fluid, are extensively applied in drug delivery and point-of-care diagnostics. This manipulate strategy is of rapid fluid actuation, compact, and makes transfer a complete microfluidics solution at the microscale. However, it still currently has some drawbacks in understanding the physics of this manipulation approach, and the deformation of the fluid interface remains an unsolved issue [87, 88].

7.3.6 LOC Platforms for Biological Applications

In the past years, LOC technology has received great interest from the researchers for biological applications due to the superiorities of LOC platforms in biological sample processing such as rapid analysis, high throughput, reduced reagent, and sample consumption. [89–94].

LOC technology was successfully applied for the design and fabrication of miniaturized platforms for biological applications. For example, Fan et al. reported the development of an LOC platform having an integrated blood barcode chip for the sensitive detection of various proteins in plasma samples [95]. The successful combination of a DNA-encoded barcode chip with a microfluidic system provides a sensitive and reliable on-chip blood separation and in situ detection of different plasma proteins within a very short time (10 minutes). The results indicated that developed LOC platform can be successfully applied as a potential miniaturized device for point-of-care testing (POCT) applications since the device is a minimally invasive, low-cost, and an informative clinical diagnostic platform.

In another interesting work carried out by the research group of Lee [96], a self-powered integrated microfluidic blood analysis system based on LOC technology that efficiently separates blood plasma from a 5 μl droplet of undiluted whole blood followed by multiple protein binding assays was successfully designed and fabricated (Figure 7.2). For on-chip blood plasma separation, blood cells are effectively separated from undiluted whole blood by trapping them in an integral filter trench structure. This integrated device does not require any external networks or tubing to transport and analyze the introduced whole blood sample within 10 minutes because of the degas-driven flow powered by pre-evacuation of

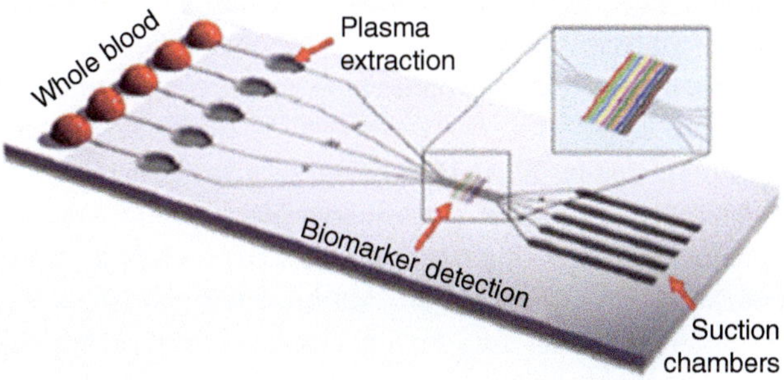

Figure 7.2 The schematic depiction of an LOC platform developed by Dimov et al. Source: Dimov et al. [96]. © 2011, Royal Chemistry of Society.

its polydimethylsiloxane (PDMS) substrate. The developed LOC-based microfluidic blood analysis system is very suitable for POCT applications because of it is a high-throughput, robust, portable, low-cost, low sample volume, simple-to-use disposable platform. In addition, the objective of the self-powered integrated microfluidic blood analysis system based on LOC technology for diagnostic applications is to minimize the number of elements while achieving the self-powered, tubing-free single-chip integration critical for maximum efficiency in performing bioassays.

In a study carried out by the research group of Shevkoplyas [97], a microfluidic paper-based LOC system with integrated blood plasma separation was designed and fabricated. The blood plasma was efficiently separated from finger-prick volumes of whole blood because of the red blood cell agglutination in the paper-fluidic platform.

In another interesting work [98], Nestorova et al. described a thermoelectric technique for the sensitive recognition of nucleic acid sequences which is based on the measurement of the heat which is generated when two DNA sequences hybridize. In this study, a microfluidic-based LOC system having an integrated thin-film antimony/bismuth thermopile was designed, prepared, and efficiently applied for the sensitive detection of the heat produced during the reaction of DNA hybridization.

In another interesting research [99], cardiac troponin, which is a highly sensitive and specific biomarker for myocardial damage, was successfully detected by an LOC system developed by Mohammed and Desmulliez. The developed LOC system was integrated to optical lenses for detection of obtained signals and an autonomous capillary microfluidics for the passive transport of the sample and reagents. The achieved results confirmed that the developed LOC system was successfully applied for the sensitive detection of the target biomarker cardiac troponin. The detection process was carried out in a very short time (7–9 minutes), and LOD was found to be 24 pg ml^{-1}.

In a study performed by Park et al. [100], an LOC platform was designed and developed for the detection of cardiac troponin in saliva and blood samples. A centrifugal force was applied for the transport of liquid samples in the developed LOC platform. The detection of the target cardiac troponin was carried out by using microbead-based sandwich ELISA approach. For this purpose, the beads conjugated with various biomarker detection reagents were loaded into the microfluidic channels in the LOC platform. The authors concluded that the developed LOC platform was also effectively applied for the sensitive recognition of other biomarkers for myocardial damage such as N-terminal pro-B type natriuretic peptide and C-reactive protein.

In another crucial study conducted by Zhou et al. [101], an aptamer-based LOC system was designed and fabricated for the recognition of living cells. For this purpose, LOC system was fabricated by using two layers of PDMS, glass substrate and Au electrodes. One of the PDMS layers contains semi-circular microcups and microchannels, while the other layer was used for pneumatic control during the analysis. Square wave voltammetry technique was applied for the investigation of the interactions between the specific aptamer and cell-secreted protein. The achieved data indicated that aptamer-based LOC system can be successfully applied for the sensitive recognition of living cells. The LOD value for the proteins secreted by cells was obtained as 5 ng ml^{-1}.

7.4 Biomedical Applications of NMs

NMs such as graphene oxide (GO) and reduced GO can be successfully used for biomedical applications because of their unique features including their excellent biocompatibility and tunable physicochemical features [102, 103]. The reported studies on the application of these unique nanomaterials in drug delivery, and photothermal cancer therapy, were well documented [104–107]. The photothermal features of GO and reduced GO confer these nanomaterials excellent potentials in cancer therapy applications [108–110]. Figure 7.3 shows the schematic demonstration of the drug delivery and photothermal cancer therapy applications of graphene-based nanomaterials.

The presence of the functional groups in the structures of GO and reduced GO provides them a high feasibility to conjugate with pharmaceutical compounds, polymers, and other functional molecules. Many nanomaterials such as polymeric nanoparticles [112, 113], gold nanoparticles [114, 115], and liposomes [116, 117] were designed and prepared as carriers for the drug delivery applications. Among these materials, GO and reduced GO are excellent candidates for the chemical modifications and loading of pharmaceutical compounds in drug delivery applications.

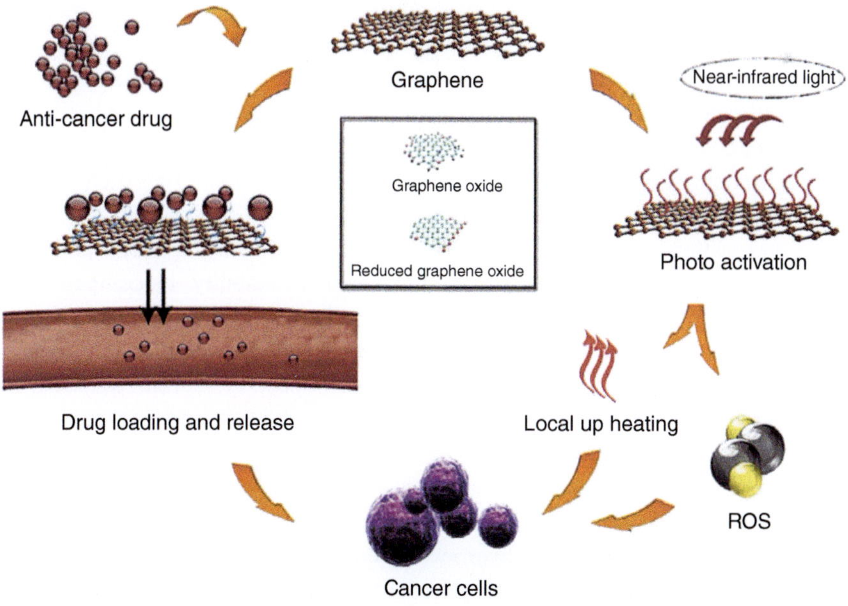

Figure 7.3 The schematic demonstration of the drug delivery and photothermal cancer therapy applications of graphene-based nanomaterials. Source: Liu et al. [111]. © 2018, Elsevier.

One of the crucial techniques for the tumor-targeted drug delivery applications is the conjugation of certain targeting ligands to the GO and reduced GO. In a study carried out by Tan and colleagues [118], hydroxypropyl-β-cyclodextrin was conjugated to the carboxylated GO and successfully employed for the delivery of paclitaxel. The developed targeted paclitaxel delivery system showed great stability and relatively high loading capacity of the drug (c. 29.93%). The obtained results indicated that the release of the paclitaxel from the developed GO-based nanocarrier depends on the medium pH.

In a study reported by Pan et al. [119], lactobionic acid and carboxymethyl chitosan were successfully grafted on the surface of GO and efficiently applied for the release of doxorubicin (DOX). In their study, GO grafted with lactobionic acid exhibited a great ability of targeting cancer cells through the specific recognition of asialoglycoprotein receptors which were overexpressed by the cancer cells.

In another important work conducted by Wojtoniszak et al. [120], methotrexate that is a widely used anticancer drug was successfully loaded to the structure of GO. The obtained results confirmed that the release of the methotrexate from the prepared methotrexate-loaded GO was faster under physiological conditions than that in biocompatible polymers such as polyethylene glycol and polysodium-4-styrene sulfonate.

Reduced GO was also efficiently used for the design and development of nanosystems for the effective release of anticancer compounds. For example, Miao et al. reported the preparation of reduced GO functionalized with cholesteryl hyaluronic acid for the delivery of doxorubicin [121]. In their study, the prepared cholesteryl hyaluronic acid-functionalized reduced GO displayed a higher stability in saline buffer and exhibited a fourfold higher drug loading capacity compared to the plain reduced GO. In addition, doxorubicin-loaded reduced GO functionalized with cholesteryl hyaluronic acid showed an excellent ability to kill KB cells under in vitro conditions.

The combination of GO and reduced GO with various nanomaterials such as magnetic Fe_3O_4 nanoparticles and carbon nanotubes (CNTs) is an efficient approach for the preparation of nanocomposite materials for drug delivery applications [122]. Magnetic Fe_3O_4 nanoparticles exhibit high chemical stability, biocompatibility, and feasibility for combining with GO and reduced GO. Therefore, they are great candidates for the preparation of GO-based nanocomposite materials. In a study reported by Gupta et al. [123], a magnetic nanocomposite material composed of Fe_3O_4 nanoparticles and reduced GO was designed and developed for the efficient release of doxorubicin. The results showed that the best drug release was achieved at pH 4.3 which is the pH value of microenvironment of cancer cells.

Yang et al. designed and developed magnetic Fe_3O_4/GO nanocomposite for the effective delivery of doxorubicin hydrochloride [124]. In their study, it was found that magnetic nanocomposite exhibits excellent potential for the delivery of the anticancer drug doxorubicin hydrochloride because it can aggregate under acidic conditions and transport by applying an external magnetic field. The achieved loading capacity of the drug was 1.08 mg mg^{-1}.

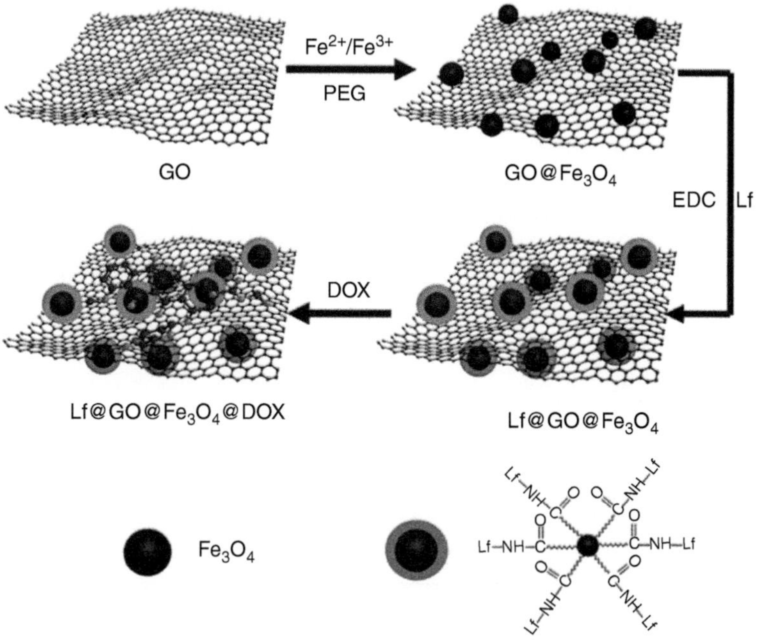

Figure 7.4 The schematic depiction of the preparation of magnetic nanocomposite composed of Fe_3O_4 nanoparticles and GO functionalized with lactoferrin (Lf) for the delivery of doxorubicin (DOX). Source: Song et al. [125]. © 2017, Elsevier.

In another interesting study carried out by Song et al. [125], a magnetic nanocomposite composed of Fe_3O_4 nanoparticles and GO functionalized with lactoferrin (Lf) was developed for the delivery of DOX to the glioma cells (Figure 7.4). In this study, Lf was chosen as the targeted ligand since it is an iron-transporting glycoprotein in serum and can be effectively bound to the surface of the receptors overexpressed at the surface of glioma cells. The results indicated that the delivery of the DOX depends on the medium pH. The loading capacity of the DOX was obtained as 0.8 mg mg^{-1}.

7.5 Sensor Applications of NMs

Graphene and graphene-based nanocomposite materials were also successfully applied in sensor applications. For example, Kong et al. reported the preparation of a nanocomposite material composed of reduced GO and Au/Ag bimetallic nanoparticles for the sensitive detection of daunorubicin in human serum samples [126]. In this study, the prepared nanosensor showed high recognition behavior toward the target compound daunorubicin in a linear concentration range between 0.01 and 15 μg ml^{-1}. The detection limit was obtained as 3 ng ml^{-1}.

In another important study conducted by Arumugam and Kim [127], sensitive recognition of ascorbic acid in aqueous samples was successfully carried out by using the prepared GO modified with CdSe and ZnS QDs. In their study, the prepared GO-based nanosensor displayed great affinity and sensitivity for the ascorbic acid, and the detection limit was achieved as 568 pM.

In an interesting work performed by and Gevaerd et al. [128], an electrochemical sensor composed of GO functionalized with imidazole groups was developed for the efficient detection of progesterone in in pharmaceutical formulations. The voltammetric detection of the target compound progesterone was successfully carried out by using the developed GO-based electrochemical sensor. The response of the developed electrochemical sensor toward progesterone was linear in the range from 0.22 to 14.0 $\mu mol\, l^{-1}$. The detection limit and quantification limit values were found to be 68 and 210 $nmol\, l^{-1}$, respectively.

In another interesting study [129], Ensafi et al. designed and prepared an electrochemical sensor having graphene nanoplatelets, graphene quantum dots (GQDs) , and molecularly imprinted polymers (MIPs) toward metronidazole. The achieved results confirmed that the prepared electrochemical sensor was successfully applied for the detection of metronidazole in human plasma samples with an excellent sensitivity. The obtained two linear response range values varied from 0.005 to 0.75 $\mu mol\, l^{-1}$ and 0.75 to 10.0 $\mu mol\, l^{-1}$. The detection limit was achieved as 0.52 $nmol\, l^{-1}$.

In a study carried out by Adumitrachioaie et al. [130], an electrochemical sensor having nanocomposite material composed of GO and chitosan was developed for the sensitive detection of serotonin in complex biological samples including tears, saliva, urine, and serum (Figure 7.5). The developed electrochemical sensor exhibited great sensitivity and selectivity toward the target compound serotonin. The achieved sensor response was linear in the range between 10 nM and 100 mM, and the detection limit was obtained as 3.2 nM.

In another important work [131], a nanosensor composed of antimorphine-functionalized graphene QDs was prepared by Masteri-Farahani and Mosleh. The prepared nanosensor was successfully applied for the sensitive recognition of morphine. The achieved results indicated that the prepared nanosensor having antimorphine-functionalized graphene QDs displayed great affinity and selectivity toward the target compound morphine in the existence of other competing compounds such as codeine and methamphetamine. The obtained detection limit was 0.06 μM.

Prongmanee et al. reported the preparation of an electrochemical sensor having nanocomposite material composed of GO and hydroxyapatite for the efficient detection of L-tryptophan [132]. For this purpose, screen-printed electrode was modified with GO and hydroxyapatite. The preparation of GO-hydroxyapatite nanocomposite was carried out by the hydroxyapatite on the GO in the simulated body fluid. The developed electrochemical sensor showed a linear response toward L-tryptophan in the range of 7 μM to 1000 μm, and the detection limit was found to be 5.5 μM.

Ai et al. reported the design and preparation of graphene aerogels doped with nitrogen for the sensitive detection of dopamine in serum samples [133]. The obtained results confirmed that the prepared graphene aerogels were successfully employed for the efficient recognition of dopamine in serum samples. The linear

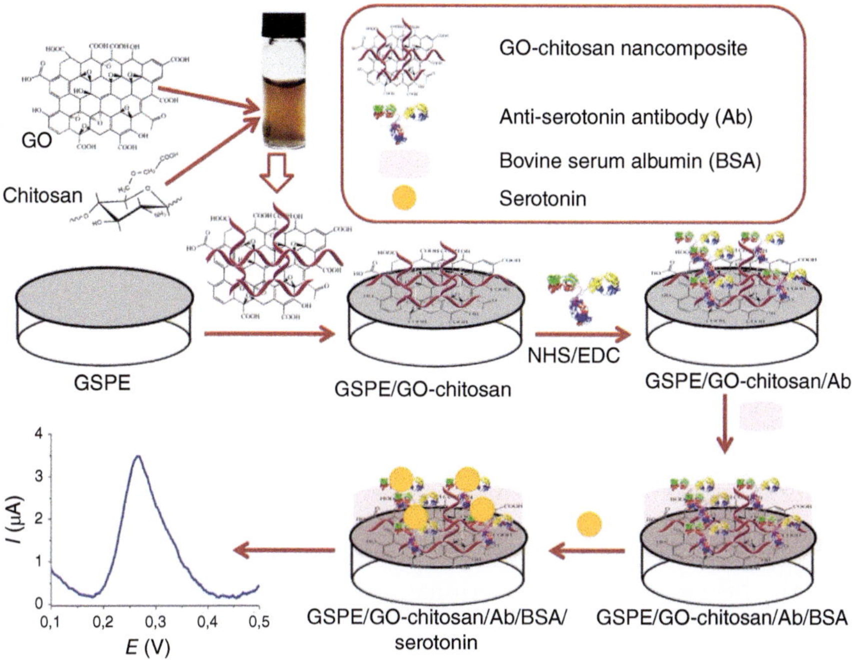

Figure 7.5 The schematic demonstration of the preparation of GO-chitosan nanocomposite-based electrochemical sensor toward serotonin. Source: Republished with permission from Adumitrachioaie et al. [130]. © 2019, Elsevier.

response was in the range from 1 to 250 μM. The detection limit was obtained as 0.1 μM.

Basiruddin and Swain developed a GO-based nanosensor for the sensitive detection of glucose in aqueous samples. In their study, reduced GO was functionalized with phenylboronic acid (PBA) for the preparation of fluorescence nanosensor [134]. Figure 7.6 shows the schematic demonstration of the fluorescence-based detection of glucose by using nanosensor composed of GO and PBA. The prepared GO-based fluorescence nanosensor exhibited high sensitivity and selectivity toward the target compound glucose in a linear concentration range between 2 and 75 $mg\,ml^{-1}$.

On the other hand, molecular imprinting of polymers is a process where the target analyte molecule is mixed together during the co-polymerization of functional and cross-linking monomers. The analyte molecule serves as a template for the monomers to polymerize around forming a complex that are held together by the interacting functional groups present between the analyte and the monomers. The template analyte molecule is subsequently removed which leaves binding sites that are specifically complementary to the size, shape, and functionality of the analyte. These binding sites are capable or rebinding with the analyte at a very high specificity [135–138]. The overall process is schematically summarized in Figure 7.7. The imprinting of the target molecule can be done either covalently or noncovalently. Covalent bonding interactions are indisputably more stable and stronger that noncovalent interactions, thus covalent imprinting produces a

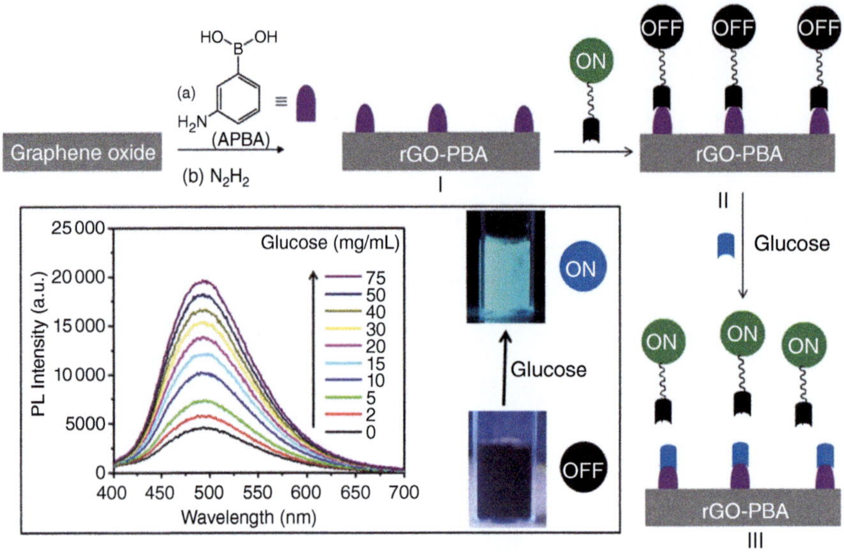

Figure 7.6 The schematic demonstration of fluorescence-based detection of glucose by using nanosensor composed of GO and PBA. Source: Republished with permission from Basiruddin and Swain [134]. © 2016, Elsevier.

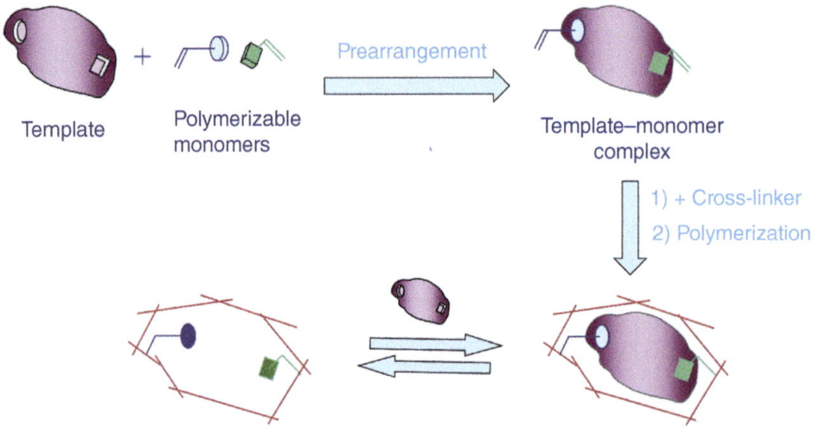

Figure 7.7 The schematic demonstration of molecular imprinting technique. Source: Turiel and Martín-Esteban [141]. © 2019, Elsevier.

more homogenous population of binding sites. In addition, these binding sites are much more specific yielding a much higher efficiency of successfully generated binding sites relative to the amount of imprint molecule introduced. However, the stronger covalent bond requires effective protocols for the removal of the template. On the other hand, noncovalent imprinting is much more flexible with more available functional monomers for possible target molecules. Most natural binding processes rely on noncovalent interactions; thus, noncovalent imprinting is a more appropriate biomimetic system [139, 140].

In a research reported by Masteri-Farahani et al. [142], a molecular imprinting-based fluorescence nanosensor was developed for the efficient detection of methamphetamine which is a quite addictive and illegal psychostimulant drug. For this purpose, they embedded GQDs into the MIP particles. The obtained outputs showed that the developed molecular imprinting-based fluorescence nanosensor can sensitively and selectively recognize methamphetamine even in the existence of other similar compounds (i.e. ibuprofen, amphetamine, morphine, and codeine). The detection limit was achieved as 1.7 $\mu g\,l^{-1}$.

Liu et al. prepared a sensitive molecular imprinting-based fluorescence nanosensor for the recognition of metronidazole in serum samples [143]. In this research, first, carbon dots (CDs) were synthesized using longan peels by applying high pressure microwave approach. Then, the preparation of CDs-embedded MIPs was carried out by using the functional monomers acrylamide (AM) and glycidyl methacrylate (GMA), cross-linker ethylene glycol dimethylacrylate (EGDMA), and initiator azobisisobutyronitrile (AIBN). The developed molecular imprinting-based fluorescence nanosensor was employed for the sensitive and selective recognition of metronidazole in serum samples. The achieved sensor response toward metronidazole was linear in the range from 50 to 1200 $ng\,ml^{-1}$, and the limit of detection was obtained as 17.4 $ng\,ml^{-1}$.

In a work performed by Jalili et al. [144], penicillin G in milk samples was successfully detected by using the molecular imprinting-based fluorescence nanosensor having CDs. In their work, first, the synthesis of silica nanoparticles embedded with blue-emissive CDs was performed. Then, MIP film was formed on the surface of the prepared nanocomposite (CDs@silica nanoparticles) by using 3-aminopropyltriethoxysilane (APTES) and tetraethoxysilane (TEOS) as functional monomers. The prepared molecular imprinting-based fluorescence nanosensor having CDs displayed excellent sensitivity toward penicillin G. The obtained sensor response was linear in the range between 1 and 32 nM. The limit of detection was found as 0.34 nM.

In another research conducted by Kumar et al. [145], a molecular imprinting-based fluorescence nanosensor was fabricated for the recognition of a fluoroquinolone antibiotic ciprofloxacin. In this research, magnetic molecularly imprinted nanoparticles having fluorescent bismuth phosphate@graphene oxide were prepared by using *N*-vinyl caprolactam and *N*,*N*-methylene bis-acrylamide as the biocompatible functional monomer and cross-linker, respectively. The obtained results from the fluorescence studies showed that the fabricated molecular imprinting-based fluorescence nanosensor can be effectively employed for the detection of ciprofloxacin in various complex matrices including milk, whole blood, and serum samples. The fluorescence sensor response was linear in the range between 39.0 and 328.0 $\mu g\,l^{-1}$. The limit of detection was achieved as 0.40 $\mu g\,l^{-1}$.

In an interesting research performed by Bai et al. [146], olaquindox that is an antimicrobial drug was successfully detected by the developed molecular imprinting-based electrochemical sensor. In this research, dopamine@graphene (DGr) was coated on the surface of GCE in the first step. Graphene was used for increasing sensitivity and conductivity. Dopamine was introduced for effective

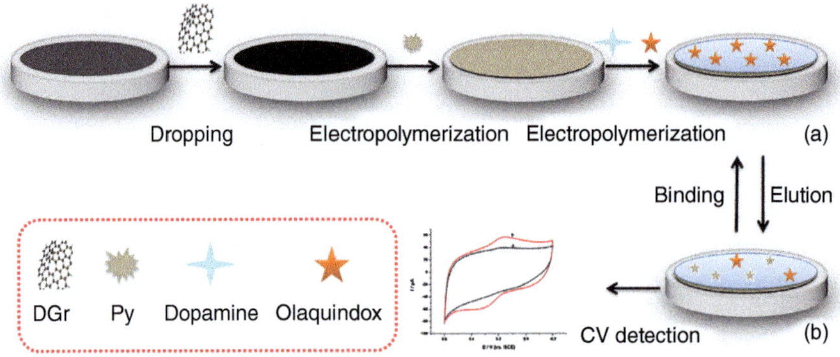

Figure 7.8 The illustration of the fabrication of molecular imprinting-based electrochemical sensor for olaquindox. Source: Bai et al. [146]. © 2020, Elsevier.

dispersion and adhesion of graphene on the electrode surface. Figure 7.8 shows the schematic illustration of the preparation of molecular imprinting-based electrochemical sensor toward olaquindox. The developed molecular imprinting-based electrochemical sensor was effectively employed for the selective recognition of olaquindox. The achieved linear sensor response was in the range between 50 and 500 $nmol\,l^{-1}$. The detection limit was obtained as 7.5 $nmol\,l^{-1}$.

7.6 Conclusions

The rapidly increasing number of studies in which NMs were successfully employed biological sample preparation processes exhibited that these unique materials having excellent properties are promising for the design and fabrication of novel sensing platforms (i.e. LOC, nanosensors etc.) in biological applications. The reported interesting studies briefly demonstrated in this chapter highlight the recent advancements of NMs in LOC platforms and biomedical and sensor applications over the past years.

On the other hand, one of the important goals in this area is the design and development of new approaches and techniques for the functionalization of the NMs and developing their physical and chemical stabilities, as well as the sensitivity and selectivity toward the target compound/s in complex matrices such as biological samples.

References

1 Niemeyer, C.M. and Mirkin, C.A. (2004). *Nanobiotechnology: Concepts, Applications and Perspectives*. Weinheim, Germany: Wiley-VCH.

2 Muller, D.J. and Dufrene, Y.F. (2008). Atomic force microscopy as a multifunctional molecular toolbox in nanobiotechnology. *Nat. Nanotechnol.* 3 (5): 261–269.

3 Rosi, N.L. and Mirkin, C.A. (2005). Nanostructures in biodiagnostics. *Chem. Rev.* 105 (4): 1547–1562.

4 Giljohann, D.A. and Mirkin, C.A. (2009). Drivers of biodiagnostic development. *Nature* 462 (7272): 461–464.

5 Chen, J., Saeki, F., Wiley, B.J. et al. (2005). Gold nanocages: bioconjugation and their potential use as optical imaging contrast agents. *Nano Lett.* 5 (3): 473–477.

6 De La Zerda, A., Zavaleta, C., Keren, S. et al. (2008). Carbon nanotubes as photoacoustic molecular imaging agents in living mice. *Nat. Nanotechnol.* 3 (9): 557–562.

7 Na, H.B., Song, I.C., and Hyeon, T. (2009). Inorganic nanoparticles for MRI contrast agents. *Adv. Mater.* 21 (21): 2133–2148.

8 Peer, D., Karp, J.M., Hong, S. et al. (2007). Nanocarriers as an emerging platform for cancer therapy. *Nat. Nanotechnol.* 2 (12): 751–760.

9 Dreaden, E.C. and El-Sayed, M.A. (2012). Detecting and destroying cancer cells in more than one way with noble metals and different confinement properties on the nanoscale. *Acc. Chem. Res.* 45 (11): 1854–1865.

10 Sailor, M.J. and Park, J.-H. (2012). Hybrid nanoparticles for detection and treatment of cancer. *Adv. Mater.* 24 (28): 3779–3802.

11 de la Rica, R., Aili, D., and Stevens, M.M. (2012). Enzyme-responsive nanoparticles for drug release and diagnostics. *Adv. Drug Deliv. Rev.* 64 (11): 967–978.

12 Lutolf, M.P. and Hubbell, J.A. (2005). Synthetic biomaterials as instructive extracellular microenvironments for morphogenesis in tissue engineering. *Nat. Biotechnol.* 23 (1): 47–55.

13 Patterson, J., Martino, M.M., and Hubbell, J.A. (2010). Biomimetic materials in tissue engineering. *Mater. Today* 13 (1–2): 14–22.

14 Dvir, T., Timko, B.P., Kohane, D.S., and Langer, R. (2011). Nanotechnological strategies for engineering complex tissues. *Nat. Nanotechnol.* 6 (1): 13–22.

15 Harrison, R.H., St-Pierre, J.-P., and Stevens, M.M. (2014). Tissue engineering and regenerative medicine: a year in review. *Tissue Eng. B Rev.* 20 (1): 1–16.

16 Sahoo, S.K., Parveen, S., and Panda, J.J. (2007). The present and future of nanotechnology in human health care. *Nanomedicine* 3 (1): 20–31.

17 Jain, K.K. (2008). Nanomedicine: application of nanobiotechnology in medical practice. *Med. Princ. Pract.* 17 (2): 89–101.

18 Lowe, S.B., Dick, J.A.G., Cohen, B.E., and Stevens, M.M. (2012). Multiplex sensing of protease and kinase enzyme activity via orthogonal coupling of quantum dot–peptide conjugates. *ACS Nano* 6 (1): 851–857.

19 Bae, S.W., Tan, W., and Hong, J.-I. (2012). Fluorescent dye-doped silica nanoparticles: new tools for bioapplications. *Chem. Commun.* 48 (17): 2270–2282.

20 Howes, P., Green, M., Levitt, J. et al. (2010). Phospholipid encapsulated semiconducting polymer nanoparticles: their use in cell imaging and protein attachment. *J. Am. Chem. Soc.* 132 (11): 3989–3996.

21 Ao, L., Gao, F., Pan, B. et al. (2006). Fluoroimmunoassay for antigen based on fluorescence quenching signal of gold nanoparticles. *Anal. Chem.* 78 (4): 1104–1106.

22 Wu, S., Han, G., Milliron, D.J. et al. (2009). Non-blinking and photostable upconverted luminescence from single lanthanide-doped nanocrystals. *Proc. Natl. Acad. Sci.* 106 (27): 10917–10921.

23 Lin, Y., Chapman, R., and Stevens, M.M. (2015). Integrative self-assembly of graphene quantum dots and biopolymers into a versatile biosensing toolkit. *Adv. Funct. Mater.* 25 (21): 3183–3192.

24 Palit, S. and Hussain, C.M. (2020). Green nanomaterials: a sustainable perspective. In: *Green Nanomaterials: Advanced Structured Materials*, vol. 126 (eds. S. Ahmed and W. Ali), 23–41. Singapore: Springer.

25 Singh, S.B. and Hussain, C.M. (2020). Functionalized nanographene for catalysis, Chapter 5. In: *Micro and Nano Technologies, Handbook of Functionalized Nanomaterials for Industrial Applications* (ed. C.M. Hussain), 111–129. Elsevier.

26 Palit, S. and Hussain, C.M. (2020). Functionalization of nanomaterials for industrial applications: recent and future perspectives, Chapter 1. In: *Micro and Nano Technologies, Handbook of Functionalized Nanomaterials for Industrial Applications* (ed. C.M. Hussain), 3–14. Elsevier.

27 Hussain, C.M. and Keçili, R. (2020). *Modern Environmental Analysis Techniques for Pollutants*. Netherlands: Elsevier.

28 Büyüktiryaki, S., Keçili, R., and Hussain, C.M. (2020). Functionalized nanomaterials in dispersive solid phase extraction: advances and prospects. *Trends Anal. Chem.* 127: 115893.

29 Keçili, R., Büyüktiryaki, S., and Hussain, C.M. (2019). Advancement in bioanalytical science through nanotechnology: past, present and future. *Trends Anal. Chem.* 110: 259–276.

30 Hussain, C.M. and Kharisov, B. (2017). *Advanced Environmental Analysis-Application of Nanomaterials*. The Royal Society of Chemistry.

31 Hussain, C.M. (2017). Magnetic nanomaterials for environmental analysis. In: *Advanced Environmental Analysis-Application of Nanomaterials* (eds. C.M. Hussain and B. Kharisov). The Royal Society of Chemistry.

32 Keçili, R., Büyüktiryaki, S., Dolak, İ., and Hussain, C.M. (2020). The use of magnetic nanoparticles in sample preparation devices and tools, Chapter 5. In: *Handbook of Nanomaterials in Analytical Chemistry* (ed. C.M. Hussain), 75–95. Elsevier.

33 Keçili, R., Büyüktiryaki, S., and Hussain, C.M. (2020). Membrane applications of nanomaterials, Chapter 7. In: *Handbook of Nanomaterials in Analytical Chemistry* (ed. C.M. Hussain), 159–182. Elsevier.

34 Büyüktiryaki, S., Keçili, R., and Hussain, C.M. (2020). Modern age of analytical chemistry: nanomaterials, Chater 2. In: *Handbook of Nanomaterials in Analytical Chemistry* (ed. C.M. Hussain), 29–40. Elsevier.

35 Keçili, R., Büyüktiryaki, S., and Hussain, C.M. (2020). Micro total analysis systems with nanomaterials, Chapter 8. In: *Handbook of Nanomaterials in Analytical Chemistry* (ed. C.M. Hussain), 185–198. Elsevier.

36 Büyüktiryaki, S., Sümbelli, Y., Keçili, R., and Hussain, C.M. (2019). Lab-on-chip platforms for environmental analysis. In: *Encyclopedia of Analytical Science*, 3e (eds. P. Worsfold, C. Poole, A. Townshend and M. Miró), 267–273. Academic Press.

37 Keçili, R. and Hussain, C.M. (2018). Recent progress of imprinted nanomaterials in analytical chemistry. *Int J. Anal. Chem.*: 8503853.

38 Keçili, R. and Hussain, C.M. (2018). Mechanism of adsorption on nanomaterials, Chapter 4. In: *Nanomaterials Chromatography*, 89–115. Elsevier.

39 Keçili, R., Büyüktiryaki, S., and Hussain, C.M. (2018). Engineered nanosensors based on molecular imprinting technology, Chapter 57. In: *Micro and Nano Technologies, Handbook of Nanomaterials for Industrial Applications* (ed. C.M. Hussain), 1031–1046. Elsevier.

40 Keçili, R., Özcan, A.A., Ersöz, A. et al. (2011). Superparamagnetic nanotraps containing MIP based mimic lipase for biotransformations uses. *J. Nanopart. Res.* 13 (5): 2073–2079.

41 Say, R., Keçili, R., and Ersöz, A. (2016). Molecularly imprinted polymer-based micro- and nanotraps for solid-phase extraction. In: *Advanced Molecularly Imprinting Materials* (eds. A. Tiwari and L. Uzun), 129–163. Wiley.

42 Cecchin, D., de la Rica, R., Bain, R.E.S. et al. (2014). Plasmonic ELISA for the detection of gp120 at ultralow concentrations with the naked eye. *Nanoscale* 6 (16): 9559–9562.

43 Gliddon, H.D., Howes, P.D., Kaforou, M. et al. (2016). A nucleic acid strand displacement system for the multiplexed detection of tuberculosis-specific mRNA using quantum dots. *Nanoscale* 8: 10087–10095.

44 Lu, C.H., Yang, H.H., Zhu, C.L. et al. (2009). A graphene platform for sensing biomolecules. *Angew. Chem. Int. Ed.* 48 (26): 4785–4787.

45 Zhang, C.Y., Yeh, H.C., Kuroki, M.T., and Wang, T.H. (2005). Single-quantum-dot-based DNA nanosensor. *Nat. Mater.* 4 (11): 826–831.

46 Kalluri, J.R., Arbneshi, T., Khan, S.A. et al. (2009). Use of gold nanoparticles in a simple colorimetric and ultrasensitive dynamic light scattering assay: selective detection of arsenic in groundwater. *Angew. Chem. Int. Ed.* 48 (51): 9668–9671.

47 Wang, J., Wang, L.H., Liu, X.F. et al. (2007). A gold nanoparticle-based aptamer target binding readout for ATP assay. *Adv. Mater.* 19 (22): 3943–3946.

48 Cai, D., Ren, L., Zhao, H.Z. et al. (2010). A molecular-imprint nanosensor for ultrasensitive detection of proteins. *Nat. Nano* 5 (8): 597–601.

49 de la Rica, R. and Stevens, M.M. (2013). Plasmonic ELISA for the detection of analytes at ultralow concentrations with the naked eye. *Nat. Protoc.* 8 (9): 1759–1764.

50 Kurkina, T. and Balasubramanian, K. (2012). Towards in vitro molecular diagnostics using nanostructures. *Cell. Mol. Life Sci.* 69 (3): 373–388.

51 Peng, F., Su, Y., Zhong, Y. et al. (2014). Silicon nanomaterials platform for bioimaging, biosensing, and cancer therapy. *Acc. Chem. Res.* 47 (2): 612–623.

52 Ding, C., Zhu, A., and Tian, Y. (2014). Functional surface engineering of C-dots for fluorescent biosensing and in vivo bioimaging. *Acc. Chem. Res.* 47 (1): 20–30.

53 Nel, A.E., Madler, L., Velegol, D. et al. (2009). Understanding biophysicochemical interactions at the nano-bio interface. *Nat. Mater.* 8 (7): 543–557.

54 Zhan, N., Palui, G., Safi, M. et al. (2013). Multidentate zwitterionic ligands provide compact and highly biocompatible quantum dots. *J. Am. Chem. Soc.* 135 (37): 13786–13795.

55 Niemeyer, C.M. (2001). Nanoparticles, proteins, and nucleic acids: biotechnology meets materials science. *Angew. Chem. Int. Ed.* 40 (22): 4128–4158.

56 Han, H.S., Devaraj, N.K., Lee, J. et al. (2010). Development of a bioorthogonal and highly efficient conjugation method for quantum dots using tetrazine-norbornene cycloaddition. *J. Am. Chem. Soc.* 132 (23): 7838.

57 Petershans, A., Wedlich, D., and Fruk, L. (2011). Bioconjugation of CdSe/ZnS nanoparticles with SNAP tagged proteins. *Chem. Commun.* 47 (38): 10671–10673.

58 Bao, M. and Wang, W. (1996). Future of microelectromechanical systems (MEMS). *Sensor. Actuat. A* 56: 135–141.

59 Terry, S.C., Hermanand, J.H., and B Angell J. (1979). A gas chromatographic air analyzer fabricated on a silicon wafer. *IEEE T Electron Dev.* 26 (12): 1880–1886.

60 Manz, A., Graber, N., and Widmer, H.M. (1990). Miniaturized total chemical analysis systems: a novel concept for chemical sensing. *Sensor. Actuat. B Chem.* 1 (1–6): 244–248.

61 Seiler, K., Harrison, D.J., and Manz, A. (1993). Planar glass chips for capillary electrophoresis: repetitive sample injection, quantitation, and separation efficiency. *Anal. Chem.* 15: 1481–1488.

62 Woolley, A.T. and Mathies, R.A. (1994). Ultra-high-speed DNA fragment separations using microfabricated capillary array electrophoresis chips. *Proc. Natl. Acad. Sci. U. S. A.* 91: 11348–11352.

63 Adamski, K., Kubicki, W., and Walczak, R. (2016). 3D printed electrophoretic lab-on-chip for DNA separation. *Procedia Eng.* 168: 1454–1457.

64 Figeys, D., Ning, Y., and Aebersold, R. (1997). A microfabricated device for rapid protein identification by microelectrospray ion trap mass spectrometry. *Anal. Chem.* 69 (16): 3153–3160.

65 Weigl, B.H. and Yager, P. (1997). Silicon-microfabricated diffusion-based optical chemical sensor. *Sensor. Actuat. B Chem.* 38–39: 452–457.

66 Xie, P., Cao, X., Lin, Z., and Javanmard, M. (2017). Top-down fabrication meets bottom-up synthesis for nanoelectronic barcoding of microparticles. *Lab Chip* 17 (11): 1939–1947.

67 Harrison, D.J., Fluri, K., Chiem, N. et al. (1996). Micromachinng chemical and biochemical analysis and reaction systems on glass substrates. *Sensor. Actuat. B Chem.* 33 (1–3): 105–109.

68 Daniel, J.H., Iqbal, S., Millingston, R.B. et al. (1998). Silicon microchambers for DNA amplification. *Sensor. Actuat. A Phys.* 81: 81–88.

69 Pang, X., Lewis, A.C., and Ródenas-García, M. (2013). Microfluidic lab-on-a-chip derivatization for gaseous carbonyl analysis. *J. Chromatogr. A* 1296: 93–103.
70 Hür, D., Say, M.G., Diltemiz, S. et al. (2018). 3D micropatterned all-flexible microfluidic platform for microwave-assisted flow organic synthesis. *ChemPlusChem* 83 (1): 42–46.
71 Becker, H. (2009). It's the economy. *Lab Chip* 9: 2759–2762.
72 Tantra, R. and van Heeren, H. (2013). Product qualification: a barrier to point-of-care microfluidic-based diagnostics? *Lab Chip* 13: 2199–2201.
73 Temiz, Y., Lovchik, R.D., Kaigala, G.V., and Delamarche, E. (2015). Lab-on-a-chip devices: how to close and plug the lab? *Microelectron. Eng.* 132: 156–175.
74 Yamada, K., Shibata, H., Suzuki, K., and Citterio, D. (2017). Toward practical application of paper-based microfluidics for medical diagnostics: state-of-the-art and challenges. *Lab Chip* 17: 1206–1249.
75 Carrilho, E., Martinez, A.W., and Whiteside, G.M. (2009). Understanding wax printing: a simple micropatterning process for paper-based microfluidics. *Anal. Chem.* 81: 7091–7095.
76 Li, X., Ballerini, D.R., and Shen, W. (2012). A perspective on paper-based microfluidics: current status and future trends. *Biomicrofluidics* 6: 11301–1130113.
77 Mao, X. and Huang, T.J. (2012). Microfluidic diagnostics for the developing world. *Lab Chip* 12: 1412–1416.
78 Haeberle, S. and Zengerle, R. (2007). Microfluidic platforms for lab-on-a-chip applications. *Lab Chip* 7: 1094–1110.
79 Gorkin, R., Park, J., Siegrist, J. et al. (2010). Centrifugal microfluidics for biomedical applications. *Lab Chip* 10: 1758–1773.
80 Burger, R., Kirby, D., Glynn, M. et al. (2012). Centrifugal microfluidics for cell analysis. *Curr. Opin. Chem. Biol.* 16: 409–414.
81 Teh, S.Y., Lin, R., Hung, L.H., and Lee, A.P. (2008). Droplet microfluidics. *Lab Chip* 8: 198–220.
82 Ottesen, E.A., Hong, J.W., Quake, S.R., and Leadbetter, J.R. (2006). Microfluidic digital PCR enables multigene analysis of individual environmental bacteria. *Science* 314: 1464–1467.
83 Hindson, B.J., Ness, K.D., Masquelier, D.A. et al. (2011). High-throughput droplet digital PCR system for absolute quantitation of DNA copy number. *Anal. Chem.* 83: 8604–8610.
84 Seemann, R., Brinkmann, M., Pfohl, T., and Herminghaus, S. (2012). Droplet based microfluidics. *Rep. Prog. Phys.* 75: 016601.
85 Barbulovic-Nad, I., Yang, H., Park, P.S., and Wheeler, A.R. (2008). Digital microfluidics for cell-based assays. *Lab Chip* 8: 519–526.
86 Abdelgawad, M. and Wheeler, A.R. (2009). The digital revolution: a new paradigm for microfluidics. *Adv. Mater.* 21: 920–925.
87 Yeo, L.Y. and Friend, J.R. (2014). Surface acoustic wave microfluidics. *Annu. Rev. Fluid Mech.* 46: 379–406.

88 Ding, X., Li, P., Lin, S.C. et al. (2013). Surface acoustic wave microfluidics. *Lab Chip* 13: 3626–3649.

89 Yager, P., Edwards, T., Fu, E. et al. (2006). Microfluidic diagnostic technologies for global public health. *Nature* 442: 412–418.

90 Whitesides, G.M. (2006). The origins and the future of microfluidics. *Nature* 442: 368–373.

91 Madhavan, B., Yue, S., Galli, U. et al. (2015). Combined evaluation of a panel of protein and miRNA serum-exosome biomarkers for pancreatic cancer diagnosis increases sensitivity and specificity. *Int. J. Cancer* 136: 2616–2627.

92 Sackmann, E.K., Fulton, A.L., and Beebe, D.J. (2014). The present and future role of microfluidics in biomedical research. *Nature* 507: 181–189.

93 Jung, W., Han, J., Choi, J.-W., and Ahn, C.H. (2015). Point-of-care testing (POCT) diagnostic systems using microfluidic lab-on-a-chip technologies. *Microelectron. Eng.* 132: 46–57.

94 Gebauer, A., Schmidt, S., and Hoffmann, W. (2016). Status and perspective of lab-on-a-chip systems for common diseases: a systematic review from 2003 to 2013. *Pers. Med.* 13: 71–91.

95 Fan, R., Vermesh, O., Srivastava, A. et al. (2008). Integrated barcode chips for rapid, multiplexed analysis of proteins in microliter quantities of blood. *Nat. Biotechnol.* 26: 1373–1378.

96 Dimov, I.K., Basabe-Desmonts, L., Garcia-Cordero, J.L. et al. (2011). Stand-alone self-powered integrated microfluidic blood analysis system (SIMBAS). *Lab Chip* 11: 845–850.

97 Yang, X., Forouzan, O., Brown, T.P., and Shevkoplyas, S.S. (2012). Integrated separation of blood plasma from whole blood for microfluidic paper-based analytical devices. *Lab Chip* 12: 274–280.

98 Nestorova, G.G., Adapa, B.S., Kopparthy, V.L., and Guilbeau, E.J. (2016). Lab-on-a-chip thermoelectric DNA biosensor for label-free detectionof nucleic acid sequences. *Sensor. Actuat. B Chem.* 225: 174–180.

99 Mohammed, M.I. and Desmulliez, M.P.Y. (2014). Autonomous capillary microfluidic system with embedded optics for improved troponin I cardiac biomarker detection. *Biosens. Bioelectron.* 61: 478–484.

100 Park, J., Sunkara, V., Kim, T.-H. et al. (2012). Lab-on-a-disc for fully integrated multiplex immunoassays. *Anal. Chem.* 84: 2133–2140.

101 Zhou, Q., Kwa, T., Gao, Y. et al. (2014). On-chip regeneration of aptasensors for monitoring cell secretion. *Lab Chip* 14: 276–279.

102 Yao, J., Sun, Y., Yang, M., and Duan, Y. (2012). Chemistry, physics and biology of graphene based nanomaterials: new horizons for sensing, imaging and medicine. *J. Mater. Chem.* 22: 14313.

103 Zhu, Y., Murali, S., Cai, W. et al. (2010). Graphene and graphene oxide: synthesis, properties, and applications. *Adv. Mater.* 22: 3906–3924.

104 Sun, X., Liu, Z., Welsher, K. et al. (2008). Nano-graphene oxide for cellular imaging and drug delivery. *Nano Res.* 1: 203–212.

105 Dong, H., Dong, C., Ren, T. et al. (2014). Surface-engineered graphene-based nanomaterials for drug delivery. *J. Biomed. Nanotechnol.* 10: 2086–2106.

106 Shim, G., Kim, M.G., Park, J.Y., and Oh, Y.K. (2016, 2016). Graphene-based nanosheets for delivery of chemotherapeutics and biological drugs. *Adv. Drug Deliv. Rev.* 105: 205–227.

107 Lin, J., Chen, X., and Huang, P. (2016). Graphene-based nanomaterials for bioimaging. *Adv. Drug Deliv. Rev.* 105: 242–254.

108 Shanmugam, V., Selvakumar, S., and Yeh, C.S. (2014). Near-infrared light-responsive nanomaterials in cancer therapeutics. *Chem. Soc. Rev.* 43: 6254–6287.

109 Rahman, M., Akhter, S., Ahmad, M.Z. et al. (2015). Emerging advances in cancer nanotheranostics with graphene nanocomposites: opportunities and challenges. *Nanomedicine (London)* 10: 2405–2422.

110 Jaleel, J.A., Sruthi, S., and Pramod, K. (2017). Reinforcing nanomedicine using graphene family nanomaterials. *J. Control. Release* 255: 218–230.

111 Liu, J., Dong, J., Zhang, T., and Peng, Q. (2018). Graphene-based nanomaterials and their potentials in advanced drug delivery and cancer therapy. *J. Control. Release* 286: 64–73.

112 Miao, L., Guo, S., Lin, C.M. et al. (2017). Nanoformulations for combination or cascade anticancer therapy. *Adv. Drug Deliv. Rev.* 115: 3–22.

113 Tyagi, P. and Subramony, J.A. (2018). Nanotherapeutics in oral and parenteral drug delivery: key learnings and future outlooks as we think small. *J. Control. Release* 272: 159–168.

114 Dreaden, E.C., Alkilany, A.M., Huang, X. et al. (2012). The golden age: gold nanoparticles for biomedicine. *Chem. Soc. Rev.* 41: 2740–2779.

115 Mazur, J., Roy, K., and Kanwar, J.R. (2018). Recent advances in nanomedicine and survivin targeting in brain cancers. *Nanomedicine (London)* 13: 105–137.

116 Cao, Z., Tong, R., Mishra, A. et al. (2009). Reversible cellspecific drug delivery with aptamer-functionalized liposomes. *Angew. Chem.* 48: 6616–6620.

117 Lu, Y., Wu, P., Yin, Y. et al. (2014). Aptamer-functionalized graphene oxide for highly efficient loading and cancer cell-specific delivery of antitumor drug. *J. Mater. Chem. B* 2: 3849–3859.

118 Tan, J., Meng, N., Fan, Y. et al. (2016). Hydroxypropyl-betacyclodextrin-graphene oxide conjugates: carriers for anti-cancer drugs. *Mater. Sci. Eng. C* 61: 681–687.

119 Pan, Q., Lv, Y., Williams, G.R. et al. (2016). Lactobionic acid and carboxymethyl chitosan functionalized graphene oxide nanocomposites as targeted anticancer drug delivery systems. *Carbohydr. Polym.* 151: 812–820.

120 Wojtoniszak, M., Urbas, K., Perużyńska, M. et al. (2013, 2013). Covalent conjugation of graphene oxide with methotrexate and its antitumor activity. *Chem. Phys. Lett.* 568-569: 151–156.

121 Miao, W., Shim, G., Kang, C.M. et al. (2013). Cholesteryl hyaluronic acid-coated, reduced graphene oxide nanosheets for anti-cancer drug delivery. *Biomaterials* 34: 9638–9647.

122 Lerra, L., Farfalla, A., Sanz, B. et al. (2019). Graphene oxide functional nanohybrids with magnetic nanoparticles for improved vectorization of doxorubicin to neuroblastoma cells. *Pharmaceutics* 11: 1–17.

123 Gupta, J., Prakash, A., Jaiswal, M.K. et al. (2018). Superparamagnetic iron oxide-reduced graphene oxide nanohybrid-a vehicle for targeted drug delivery and hyperthermia treatment of cancer. *J. Magn. Magn. Mater.* 448: 332–338.

124 Yang, X., Zhang, X., Ma, Y. et al. (2009). Superparamagnetic graphene oxide–Fe_3O_4 nanoparticles hybrid for controlled targeted drug carriers. *J. Mater. Chem.* 19: 2710–2714.

125 Song, M.M., Xu, H.L., Liang, J.X. et al. (2017). Lactoferrin modified graphene oxide iron oxide nanocomposite for glioma-targeted drug delivery. *Mater. Sci. Eng. C Mater. Biol. Appl.* 77: 904–911.

126 Kong, F.Y., Yao, L., Li, R.F. et al. (2019). Synthesis of nitrogen-doped reduced graphene oxide loading with Au–Ag bimetallic nanoparticles for electrochemical detection of daunorubicin. *J. Alloys Compd.* 797: 413–420.

127 Arumugam, N. and Kim, J.S. (2018). Quantum dots attached to graphene oxide for sensitive detection of ascorbic acid in aqueous solutions. *Mater. Sci. Eng. C* 92: 720–725.

128 Gevaerd, A., Blaskievicz, S.F., Zarbin, A.J.G. et al. (2018). Nonenzymatic electrochemical sensor based on imidazole-functionalized graphene oxide for progesterone detection. *Biosens. Bioelectron.* 112: 108–113.

129 Ensafi, A.A., Nasr-Esfahani, P., and Rezaei, B. (2018). Metronidazole determination with an extremely sensitive ands elective electrochemical sensor based on graphene nanoplatelets and molecularly imprinted polymers on graphene quantum dots. *Sens. Actuat. B Chem.* 270: 192–199.

130 Adumitrachioaie, A., Tertiş, M., Suciu, M. et al. (2019). A novel immunosensing platform for serotonin detection in complex real samples based on graphene oxide and chitosan. *Electrochim. Acta* 311: 50–61.

131 Masteri-Farahani, M. and Mosleh, N. (2019). Functionalization of graphene quantum dots with antimorphine: design of selective nanosensor for detection of morphine. *Mater. Lett.* 241: 206–209.

132 Prongmanee, W., Alam, I., and Asanithi, P. (2019). Hydroxyapatite/graphene oxide composite for electrochemical detection of L-tryptophan. *J. Taiwan Inst. Chem. Eng.* 102: 415–523.

133 Ai, S., Chen, Y., Liu, Y. et al. (2018). Facile synthesis of nitrogen-doped graphene aerogels for electrochemical detection of dopamine. *Solid State Sci.* 86: 6–11.

134 Basiruddin, S.K. and Swain, S.K. (2016). Phenylboronic acid functionalized reduced graphene oxide based fluorescence nano sensor for glucose sensing. *Mater. Sci. Eng. C* 58: 103–109.

135 Kryscioa, D.R. and Peppas, N.A. (2012). Critical review and perspective of macromolecularly imprinted polymers. *Acta Biomater.* 8: 461–473.

136 Cheong, W.J., Yang, S.H., and Ali, F. (2013). Molecular imprinted polymers for separation science: a review of reviews. *J. Sep. Sci.* 36: 609–628.

137 Vasapollo, G., Del Sole, R., Mergola, L. et al. (2011). Molecularly imprinted polymers: present and future prospective. *Int. J. Mol. Sci.* 12: 5908–5945.

138 Kupai, J., Razali, M., Büyüktiryaki, S. et al. (2017). Long-term stability and reusability of molecularly imprinted polymers. *Polym. Chem.* 8 (4): 666–673.

139 Dolak, İ., Keçili, R., Hür, D. et al. (2015). Ion-imprinted polymers for selective recognition of neodymium(III) in environmental samples. *Ind. Eng. Chem. Res.* 54: 5328–5335.

140 Keçili, R. (2018). Selective recognition of myoglobin in biological samples using molecularly ımprinted polymer-based affinity traps. *Int J. Anal. Chem.*: 4359892.

141 Turiel, E. and Martín-Esteban, A. (2019). Molecularly imprinted polymers-based microextraction techniques. *TrAC Trends Anal. Chem.* 118: 574–586.

142 Masteri-Farahani, M., Mashhadi-Ramezani, S., and Mosleh, N. (2020). Molecularly imprinted polymer containing fluorescent graphene quantum dots as a new fluorescent nanosensor for detection of methamphetamine. *Spectrochim. Acta A* 229: 118021.

143 Liu, H., Ding, J., Zhang, K., and Ding, L. (2020). Fabrication of carbon dots@restricted access molecularly imprinted polymers for selective detection of metronidazole in serum. *Talanta* 209: 120508.

144 Jalili, R., Khataee, A., Rashidi, M.R., and Razmjou, A. (2020). Detection of penicillin G residues in milk based on dual-emission carbon dots and molecularly imprinted polymers. *Food Chem.* 314: 126172.

145 Kumar, S., Karfa, P., Majhi, K.C., and Madhuri, R. (2020). Photocatalytic, fluorescent $BiPO_4$@graphene oxide based magnetic molecularly imprinted polymer for detection, removal and degradation of ciprofloxacin. *Mater. Sci. Eng. C* 111: 110777.

146 Bai, X., Zhang, B., Liu, M. et al. (2020). Molecularly imprinted electrochemical sensor based on polypyrrole/dopamine@graphene incorporated with surface molecularly imprinted polymers thin film for recognition of olaquindox. *Bioelectrochemistry* 132: 107398.

8

Magnetic Nanomaterials for Sample Preparation

8.1 Introduction

The sampling and sample preparation stages take almost two-thirds of the total time of the analytical processes [1]. These stages are also the main source of the uncertainties in the analyses [2]. Thus, researchers put so much effort in the development of facile, fast, low-cost, efficient, selective, and environmentally friendly sample preparation techniques.

The conventional sample preparation approaches such as liquid–liquid extraction have some drawbacks (i.e. time-consuming, show low selectivity and efficiency toward the target compound/s and high consumption of toxic organic solvents, etc.) To overcome these disadvantages of conventional techniques, new sample preparation techniques have been developed and successfully employed for the analysis of environmental samples [3–10]. These techniques have many superiorities such as high selectivity for the target compound/s, facile automation, low cost, environmentally friendly, rapid processing of samples, high reproducibility, and the need of low volumes of organic toxic solvents.

Nanoparticles are extensively applied in sample preparation processes [11–16]. They exhibit high surface-area-to-volume ratio, which promises a much higher capacity for extraction of target compound/s from complex matrices and efficiency compared with other conventional adsorbents. In addition, the surface functionality of the nanoparticles can successfully be modified to achieve effective and selective extraction. Among various types of nanoparticles, magnetic nanoparticles are interesting advanced materials and have gained great attention among researchers from a wide range of disciplines in science [17–20]. Magnetic nanoparticles were successfully applied for different applications such as magnetic fluids [21], catalysis [22], magnetic resonance imaging [23], and environmental remediation [24, 25]. Magnetic nanoparticles exhibit superparamagnetic feature, which can be attracted to a magnetic field, but retain no residual magnetism after the field is removed. Thus, it is very easy to separate these magnetic nanoparticles adhered with target compound/s from complex matrices such as environmental, food, and biological samples by using an external magnetic field. There is no centrifugation or filtration step needed during this process. However, an unavoidable drawback associated with these nanoparticles in this range of size is instability of these magnetic

Sample Preparation with Nanomaterials: Next Generation Techniques and Applications,
First Edition. Chaudhery Mustansar Hussain, Rüstem Keçili, and Chaudhery Ghazanfar Hussain.

nanoparticles, which tends to cause formation of agglomerates in the sample solution. Moreover, naked metallic nanoparticles are highly chemically active and can easily be oxidized in air, which leads to loss of their dispersibility and magnetism feature. Therefore, it is important to design and develop protection approaches for the chemical stabilization of the naked magnetic nanoparticles. These approaches for the modification of the surfaces of magnetic nanoparticle consist of grafting technique or coating with an inorganic layer such as carbon or silica or coating with organic compounds including polymers and surfactants [26].

This chapter provides an overview of the recent progresses on the design and development of new magnetic nanoparticles for the sample preparation processes. The chapter starts with the brief explanation of various techniques commonly used for the synthesis of magnetic nanoparticles. Then, solid-phase extraction (SPE) as the sample preparation tool is described. Finally, the recent SPE applications of magnetic nanoparticles for environmental, biological, food, and beverage samples are demonstrated.

8.2 Synthesis of Magnetic Nanoparticles

There are different approaches for the synthesis of monodisperse, shape-controllable and stable magnetic nanoparticles such as thermal decomposition, coprecipitation, sol–gel synthesis, hydrothermal synthesis, microemulsion, flow injection, and aerosol/vapor-phase-based synthesis techniques. To date, thermal decomposition and coprecipitation approaches to synthesize magnetic nanoparticles are the best studied types of the synthesis techniques of the magnetic nanoparticles.

8.2.1 Thermal Decomposition Technique

Magnetic nanoparticles can be successfully synthesized by thermal decomposition of metal precursors including metal carbonyls such as $Fe(CO)_5$, $Co_2(CO)_8$, and $Ni(CO)_4$ and metal oleates in autoclaves or reactors by applying at high pressure and temperature during the synthesis process [27]. Thermal decomposition technique yields products having great monodispersity, excellent control in particle size, narrow size distribution, and excellent crystallinity of individual and dispersible iron oxide magnetic nanoparticles. The main drawbacks of this approach are the need for high temperature during the synthesis process, high cost, use of toxic organic solvents/chemicals, and the need for further modification to increase the biocompatibility of the synthesized magnetic nanoparticles (i.e. Fe_3O_4 nanoparticles) [28]. A number of magnetic metal nanoparticles were synthesized by applying this approach with the use of metal carbonyl precursors (i.e. $Co_2(CO)_8$ and $Fe(CO)_5$) [29, 30]. Magnetic metal nanoparticles (i.e. Ni and Co) can also easily be synthesized by decomposition of low-valent alkene or polyene metal complexes with H_2 at room temperature [31, 32]. Recently, water-soluble magnetic nanoparticles with controllable particle size and narrow size distribution were successfully synthesized. In a study carried out by Lee et al. [33], high-quality magnetism-engineered iron oxide nanoparticles were prepared in the organic medium at high temperature. In this study, the authors successfully used the prepared high-quality magnetism-engineered iron oxide nanoparticles as contrast agents for the magnetic resonance applications and tested their applicability as molecular imaging probes for the sensitive recognition of target biomolecules. In another interesting work [34], Hu et al. developed a method of one-pot preparation of water-soluble magnetic nanoparticles at 245 °C. The prepared magnetic nanoparticles were efficiently applied as magnetic resonance imaging contrast agents for cancer diagnosis.

8.2.2 Coprecipitation Technique

Coprecipitation technique is a facile and effective approach for the preparation of iron oxides with either γ-Fe_2O_3 or Fe_3O_4 structure. This technique is based on the chemical reactions occurring in an aqueous solution with the process of both the nucleation and growth of iron hydroxide nuclei [35]. The coprecipitation process of Fe^{2+}/Fe^{3+} salts is carried out at ambient temperature under alkaline conditions. The synthesis of magnetic iron oxide nanoparticles with suitable diameter, magnetic responsiveness, and surface properties can successfully be achieved by the optimization of experimental conditions (i.e. medium pH, temperature, ionic strength, the molar ratio of reactants, etc.) [36]. The chemical reaction of the Fe_3O_4 synthesis can be expressed by Eq. (8.1).

$$Fe^{2+} + 2Fe^{3+} + 8OH^- \rightarrow Fe_3O_4 + 4H_2O \tag{8.1}$$

One of the main superiorities of the coprecipitation approach is that magnetic iron oxide nanoparticles can successfully be obtained in large scale [37]. However, it needs the careful optimization of the reaction conditions and usually results in a rather broad size distribution of magnetic nanoparticles. The first successful synthesis of magnetic iron oxide nanoparticles was conducted by Massart and Cabuil under alkaline conditions by coprecipitation of $FeCl_2$ and $FeCl_3$ [38]. In their study, the prepared magnetic nanoparticles have around 8 nm average diameter and were roughly spherical.

8.2.3 Sol–Gel Synthesis

In the sol–gel synthesis of the magnetic nanoparticles, the hydroxylation and condensation of the reactants in solutions are used to obtain a sol of nanoparticles. In the next step, a gel with a three-dimensional network of metal oxide is prepared by the condensation as well as inorganic polymers of aerosol. Then, crystalline state is achieved by heating of the prepared gel [39]. The gel structure and its performance can easily be controlled by optimization of the conditions of hydroxylation and condensation process such as medium pH, temperature, solvent and concentration of the salt precursors, etc. [40]. This approach is facile, effective and provides various advantages such as high reaction activity, good mixing uniformity, low process temperature. However, the synthesized magnetic nanoparticles can be coagulated during posttreatment process.

8.2.4 Hydrothermal Synthesis

In the hydrothermal synthesis technique, the synthesis of magnetic nanoparticles is performed by using aqueous medium at high temperature and pressure [41]. Applying high temperature during the synthesis of magnetic nanoparticles is crucial and beneficial for the improvement of the nucleation rate and to speed up the growth of the new nanoparticles, which ensures obtaining small-size particles. Hydrolysis and oxidation are two of the main ways for the synthesis of magnetic nanoparticles under hydrothermal conditions, and another way is neutralization of hybrid metal hydroxides.

8.2.5 Microemulsion-Based Synthesis

A microemulsion is defined as a thermodynamically stable isotropic dispersion of two immiscible liquids in which the microdomain of either or both liquids was stabilized by an interfacial film of surface active compounds [42]. Microemulsion-based synthesis is usually carried out by accomplishing the aforementioned chemical reactions such as coprecipitation or reduction in microemulsions to carefully control the size of the prepared magnetic nanoparticles. It is achieved by the reaction of either two reactants lying in two separate micellar systems through mixing and coalescence or reactants inside a single micelle through thermal initiation [43]. Monodisperse magnetic nanoparticles having various morphologies can be synthesized by applying microemulsion-based approach. The main disadvantages of this technique are high consumption of organic solvent during the production process of the magnetic nanoparticles and low production yield [44].

8.2.6 Flow Injection Synthesis

Flow injection synthesis technique is another effective approach for the synthesis of magnetic nanoparticles having narrow size distribution. This technique is based on the segmented or continuous mixing of the chemical reagents by applying a laminar flow in a capillary reactor during the synthesis process of the magnetic nanoparticles. In a study reported by Alvarez and colleagues [45], magnetic Fe_3O_4 nanoparticles having a narrow size distribution and varying in the range between 2 and 7 nm were successfully synthesized by applying this technique. Flow injection synthesis approach has various superiorities such as excellent mixing homogeneity and great reproducibility because of the laminar plug-flow conditions.

8.2.7 Aerosol/Vapor-Phase-Based Synthesis

Aerosol-based techniques such as laser pyrolysis and spray approach are efficient techniques because of high rate synthesis in continuous chemical processes [46]. In spray pyrolysis technique, a solution composed of an organic reducing agent and ferric salts is sprayed into a series of reactors followed by condensation of the aerosol solute and evaporation of the solvent. The obtained dried residues are transformed into magnetic nanoparticles. The size of these nanoparticles depends on the initial size of the droplet. On the other hand, laser pyrolysis technique is another promising approach that is applied for the preparation of magnetic nanoparticles. Laser heating of a gaseous mixture of iron precursor such as iron pentacarbonyl is used for decreasing the reaction time during the production process. The synthesis reaction is continuously carried out into generators to obtain high-yield magnetic nanoparticles. Laser pyrolysis technique is successfully applied for the synthesis of narrow size distributed, small, and non-aggregated magnetic nanoparticles. Magnetic nanoparticles having the crystal size with a quite narrow size distribution in the range between 2 and 7 nm can be effectively prepared under optimum process conditions [47].

8.3 Solid-Phase Extraction (SPE)

SPE is an efficient tool for the sample preparation in chemistry. SPE was first applied in 1940s [48], and the rapid progresses in various SPE applications were initiated in the 1970s. It is now successfully applied in different fields of science. Conventional SPE materials such as silica-based [49], carbon-based [50], and clay-based [51] resins were used in various SPE examples. Many adsorbents for SPE applications now commercially exist in different format on the market such as SPE tubes and pipet tip formats such as Oasis hydrophilic–lipophilic balance (HLB; produced by Waters), Omix (produced by Agilent), MonoTips (produced by GL Sciences), etc.

The basic principles of SPE are similar to liquid–liquid extraction. Both methods involve the distribution of dissolved species between two phases. However, SPE involves the dispersion of the analyte between a liquid (sample medium) and a solid (adsorbent) phase instead of the two liquid phases, which are not mixed together as in liquid–liquid extraction. This technique allows the enrichment and purification of the analytes on a solid adsorbent through adsorption from the solution.

8.4 Magnetic Solid-Phase Extraction (MSPE)

Magnetic nanoparticles are interesting materials that are widely used in many applications such as catalysis [52], separation [53, 54], and drug delivery [55]. Magnetic nanoparticles can easily be separated from the complex matrices such as biological and environmental samples by using an external magnetic field. Thus, no filtration or centrifugation steps are required. Although magnetic nanoparticles exhibit excellent features, there are some drawbacks such as agglomeration of the particles and stability issues. Bare metallic nanoparticles are chemically very active and easily oxidized, which lead to loss of their magnetic features. Thus, these nanoparticles are need to be coated with a layer (e.g. carbon, silica, or polymers) to protect their stability.

Magnetic solid-phase extraction (MSPE) was proposed for the facile and fast sample preparation processes [56]. The magnetic materials are generally directly dispersed in the sample solutions for the rapid extraction process because they can be readily recovered by a magnet. It also overcomes the drawbacks with traditional SPE process such as eliminating the packing of columns and avoiding the time-consuming process of loading large-volume of samples. Typically, the magnetic SPE material is composed of a core material such as Fe_3O_4 nanoparticles to collect by using a magnet and a supported nanoparticle for the extraction of various target compounds from complex matrices (i.e. environmental samples, food and beverage samples, and biological samples).

8.4.1 MSPE for Environmental Samples

In recent years, several studies were reported on the application of MSPE for environmental samples using magnetic nanoparticles. For example, Haeri and Abbasi developed magnetic nanoparticles for the pre-concentration of atrazine from water samples [57]. The results indicated that the analytical response of the prepared nanoparticles was linear over the range 0.1–50 $\mu g\,l^{-1}$ with LOD of 0.033 $\mu g\,l^{-1}$ by using high-performance liquid chromatography under optimized conditions. They obtained the enrichment factor as 268 for atrazine in their study.

In another interesting study carried out by Dai et al. [58], magnetic $Fe_3O_4/SiO_2/$dithiocarbamate nanoparticles were prepared for the efficient removal of Cu^{2+} and Ni^{2+} ions from aqueous solutions. The obtained results confirmed that the prepared magnetic $Fe_3O_4/SiO_2/$dithiocarbamate nanoparticles exhibited excellent binding affinity toward the target ions. The binding capacity values for Cu^{2+} and Ni^{2+} ions were achieved as 230.49 and 235.23 $mg\,g^{-1}$, respectively.

Sun et al. developed magnetic Fe_3O_4/ZrO_2 nanoparticles for the effective removal of As(III) ions from water samples via photocatalytic oxidation process [59]. In their study, As(III) ions were completely oxidized to As(V) ions, which have less toxic feature by using the prepared magnetic Fe_3O_4/ZrO_2 nanoparticles. As(V) ions were simultaneously adsorbed on the surface of the magnetic nanoparticles. The achieved maximum binding capacity was 133.48 $mg\,g^{-1}$.

In an important study conducted by Ma et al. [60], magnetic lignin-based carbon nanoparticles were prepared by applying precipitation/carbonization approach. The prepared magnetic nanoparticles were successfully used for the efficient removal of methyl orange dye from aqueous solutions. The obtained maximum binding capacity was 113.0 $mg\,g^{-1}$.

In another study [61], Golmohammadi et al. successfully removed reactive yellow 15 dye pollutant from water samples by using magnetic Fe_3O_4/SiO_2 nanoparticles functionalized with ionic liquid [3-(trimethoxysilylpropyl)]ammonium chloride. The obtained results indicated that the prepared magnetic nanoparticles exhibited high affinity toward the target dye compound. The binding capacity was achieved as 63.69 $mg\,g^{-1}$.

In an interesting work conducted by Tang et al. [62], methylene blue was efficiently removed from wastewater samples by using magnetic Fe_3O_4 nanoparticles having flower-like $MoSe_2$ nanosheets. The schematic demonstration of the preparation of magnetic $Fe_3O_4/MoSe_2$ nanocomposite toward methylene blue is shown in Figure 8.1. In their study, the highest binding capacity of the prepared magnetic $Fe_3O_4/MoSe_2$ nanocomposite for methylene blue was achieved as 88.1 $mg\,g^{-1}$.

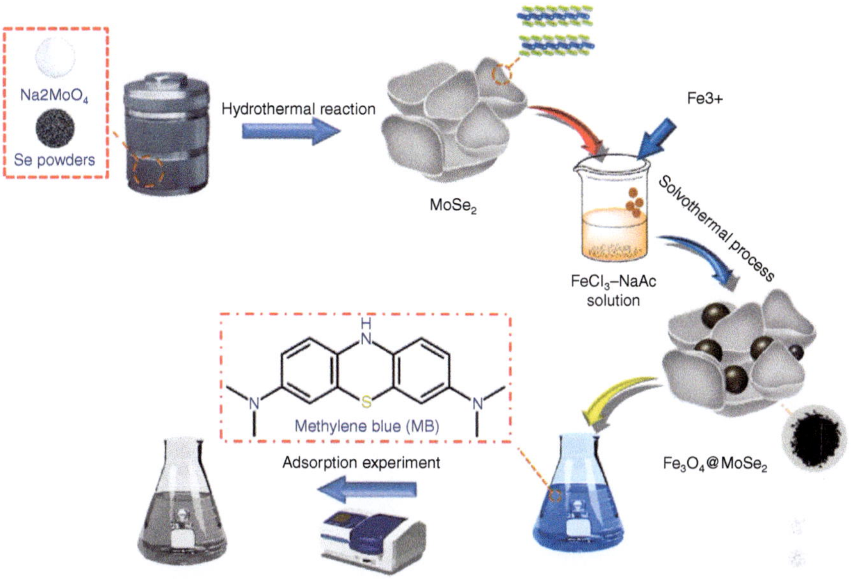

Figure 8.1 The schematic demonstration of the preparation of magnetic $Fe_3O_4/MoSe_2$ nanocomposite toward methylene blue. Source: Tang, G. et al., (2020) Construction of magnetic Fe_3O_4 nanoparticles coupled with flower-like $MoSe_2$ nanosheets for efficient adsorptive removal of methylene blue. Colloids and Surfaces A587, 124291. © 2019, Elsevier.

On the other hand, molecularly imprinted polymers (MIPs) are efficiently utilized as molecularly imprinted solid-phase extraction (MISPE) to decrease cumbersome sample preparation steps and facilitate selective isolation, sample purification, and pre-concentration of chemical species. Application of MIPs, especially as MISPE extraction material, has been well researched for compounds including mycotoxins [63], sterols [64], various drugs [65], amide anesthetics [66], and organic dyes [67]. The majority of the MISPE studies indicated that MIPs are stable under strong elution conditions including a wide range of pH, different solvents, and various organic modifiers. In addition, a study by Pichon et al. confirmed that MIPs are more selective than C18 or ion exchange materials, which are traditionally used in sample treatment methods such as SPE, and are also more selective and stable than immunoextraction matrices [68].

Among the various physical forms of MIPs such as monoliths, gels, membranes, spherical particles in micro- and nanoscale, molecularly imprinted nanoparticles have received great attention from the researchers due to their facile preparation, rapid recognition kinetics toward target compound/s [69].

Molecularly imprinted magnetic nanoparticles were successfully applied for the sensitive detection of pollutants in environmental samples. For example, in an interesting work carried out by Jiang et al. [70], ion-imprinted magnetic Fe_3O_4/SiO_2 core–shell nanoparticles were prepared for the rapid and sensitive detection of methyl mercury (MeHg) in natural water samples. For this purpose, methacrylic acid (MAA) and trimethylolpropane trimethacrylate (TMPTM) were preferred as the functional monomer and cross-linker, respectively. In this study, excellent detection limit was achieved (0.084 pg ml^{-1}), and the binding capacity of the prepared magnetic ion-imprinted magnetic nanoparticles for MeHg was obtained as 25 mg g^{-1}.

In a study conducted by Hu et al. [71], molecularly imprinted magnetic carboxylated cellulose nanocrystals were used for the selective separation of ofloxacin. The prepared MIP-based nanocrystals exhibited very fast binding kinetics toward the target compound ofloxacin, and 34.09 mg g^{-1} binding capacity was obtained in a short time (two minutes). The selectivity studies were also performed using lomefloxacin, marbofloxacin, orbifloxacin, sulfamethoxazole, gatifloxacin, sarafloxacin, and difloxacin as competing fluoroquinolones. The maximum binding capacity was obtained as 40.65 mg g^{-1} after 20 minutes.

Patiño-Ropero et al. reported the application of magnetic molecularly imprinted core–shell nanoparticles for the efficient extraction of various triazines (i.e. atrazine, desethylatrazine, simazine, desisopropylatrazine, and propazine) from soil samples [72]. In this study, firstly, magnetic Fe_3O_4 nanoparticles having vinyl functional groups were synthesized. Then, propazine-imprinted thin layer was synthesized on the surface of magnetic nanoparticles using MAA and EGDMA as the functional monomer and cross-linker, respectively. The prepared magnetic molecularly imprinted core–shell nanoparticles were successfully applied for the extraction of target five different triazine pollutants from soil samples. Zhou et al. prepared magnetic ion-imprinted polymer (IIP) for the effective removal of Cr(VI) ions from aqueous solutions [73]. In this study, 4-vinyl pyridine (4-VP) was used as the functional monomer for the preparation of magnetic IIP for Cr(VI) ions (Figure 8.2).

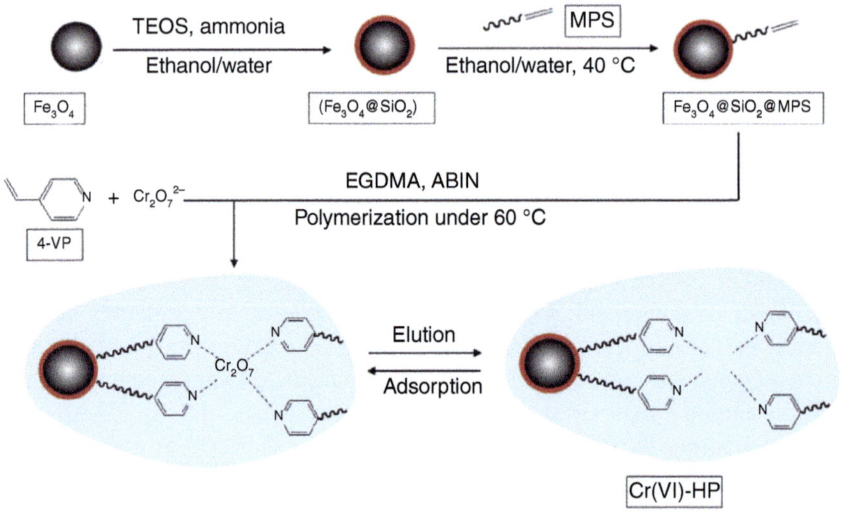

Figure 8.2 The schematic depiction of preparation of magnetic IIPs toward Cr(VI) ions. Source: Zhou, Z. et al., (2020) Preparation of highly efficient ion-imprinted polymers with Fe_3O_4 nanoparticles as carrier for removal of Cr(VI) from aqueous solution. Science of the Total Environment, 699, 134334. © 2019, Elsevier.

The obtained results confirmed that the prepared magnetic IIPs displayed great affinity and selectivity toward the target Cr(VI) ions in the competitive ion systems including $Cr_2O_7^{2-}/AlF_4^-$ and $Cr_2O_7^{2-}/CrO_4^{2-}$. In their study, the maximum binding capacity was found as 201.55 mg g^{-1}.

Barati et al. developed MIP-based magnetic graphene oxide (GO)/chitosan (CS) nanocomposite toward fluoxetine in biological and environmental samples [74]. In their study, MIP-based magnetic GO/CS nanocomposite was prepared. The prepared MIP-based magnetic nanocomposite was successfully used for the selective separation of fluoxetine from urine and water samples. The maximum binding capacity was calculated as 66 mg g^{-1}.

In another research conducted by Lu et al. [75], triclosan (TCS) was efficiently removed from aqueous samples by using magnetic Fe_3O_4/SiO_2 nanocomposites. For this purpose, MAA and EGDMA were chosen as the functional monomer and cross-linker, respectively (Figure 8.3). The prepared magnetic nanocomposites displayed excellent binding efficiency for the target compound TCS. In their research, the achieved maximum binding capacity was 53.12 mg g^{-1}.

Huang et al. reported design and preparation of magnetic silica nanoparticles coated with CS and decorated with 2-phosphonobutane-1,2,4-tricarboxylic acid (PBTCA) ($CoFe_2O_4$@SiO_2@CS-PBTCA) for the efficient removal of U(VI) ions from aqueous samples [76]. The schematic demonstration of the preparation of magnetic $CoFe_2O_4$@SiO_2@CS-PBTCA nanocomposites toward the target U(VI) ions is shown in Figure 8.4. In this work, the maximum binding capacity of the prepared magnetic nanocomposite toward U(VI) ions was obtained as 83.16 mg g^{-1}.

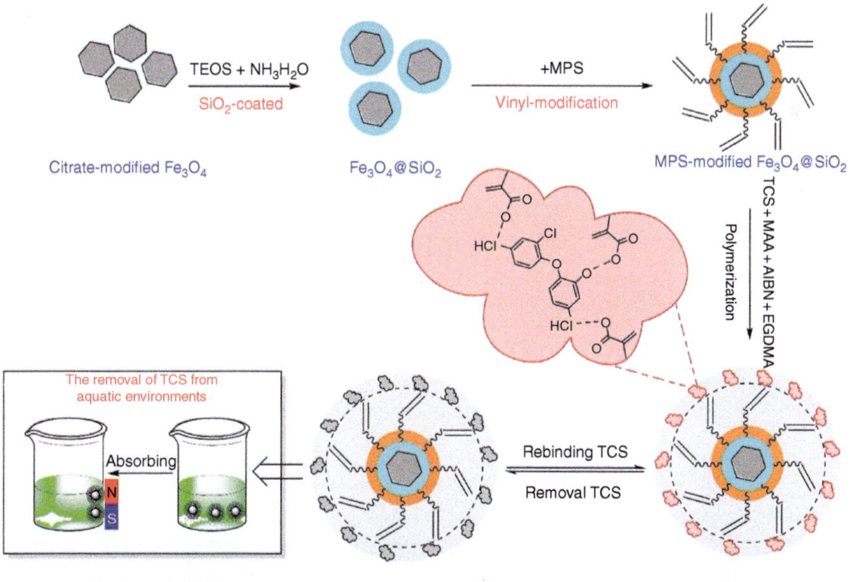

Figure 8.3 The schematic depiction of the preparation of magnetic Fe_3O_4/SiO_2 nanocomposites for triclosan (TCS). Source: Lu, Y-C. et al., (2020) A novel strategy for selective removal and rapid collection of triclosan from aquatic environment using magnetic molecularly imprinted nano-polymers, Chemosphere238, 124640. © 2019, Elsevier.

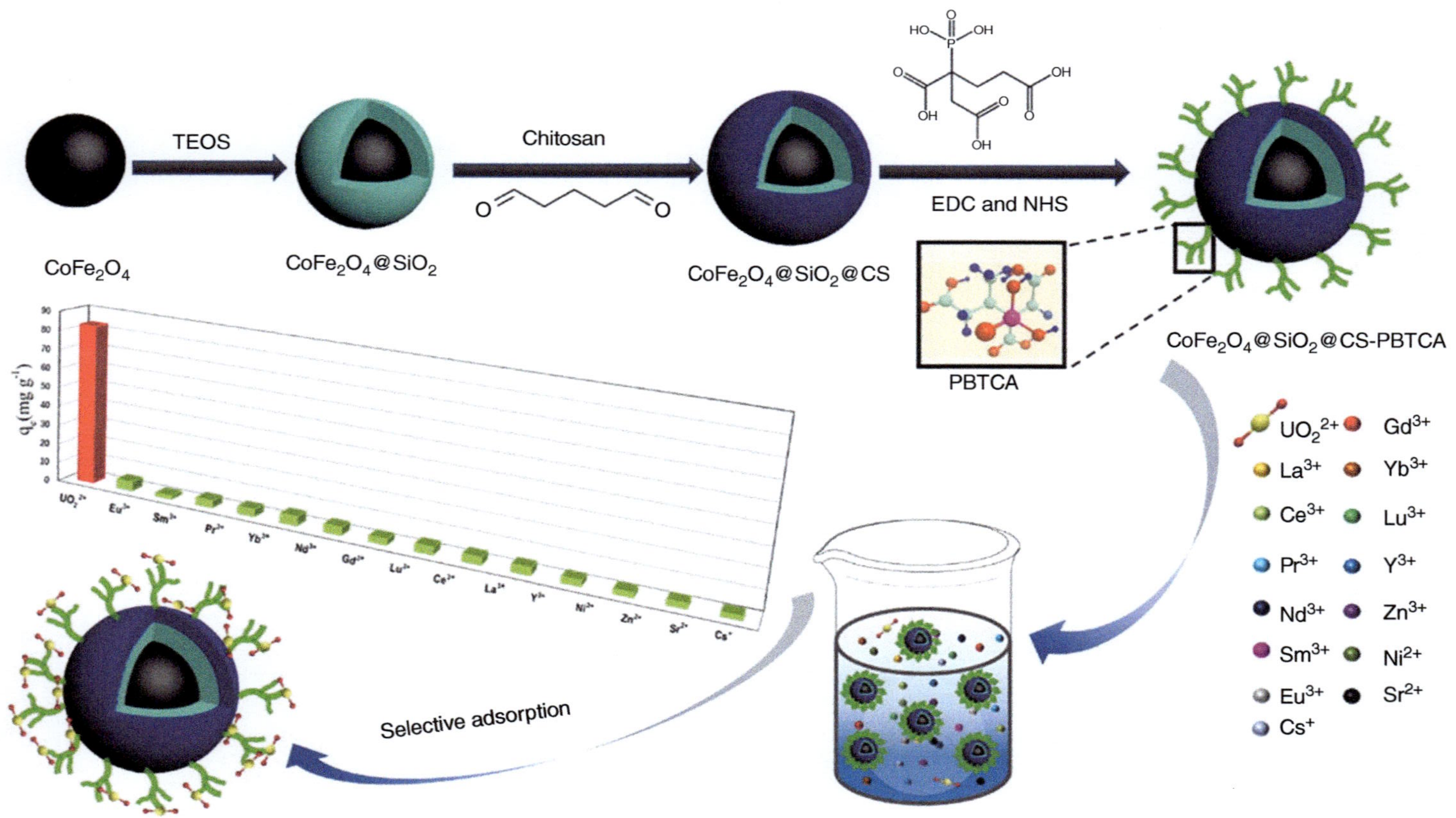

Figure 8.4 The schematic depiction of the preparation of magnetic $CoFe_2O_4@SiO_2@CS$-PBTCA nanocomposites toward U(VI) ions. Source: Huang, Y. et al., (2020) Highly selective uranium adsorption on 2-phosphonobutane-1,2,4- tricarboxylic acid-decorated chitosan-coated magnetic silica nanoparticles, Chemical Engineering Journal, 388, 124349. © 2020, Elsevier

In another interesting study conducted by Shahriman et al. [77], magnetic Fe_3O_4 nanoparticles coated with polyaniline–dicationic ionic liquid were prepared for the selective extraction of polycyclic aromatic hydrocarbons (PAHs) from soil, sludge, and environmental water samples. The achieved results confirmed that the prepared magnetic nanoparticles exhibited excellent extraction performance toward the target PAHs in environmental samples. The detection limit values varied in the range between 0.0008 and 0.2086 $\mu g\, l^{-1}$.

Al-Harahsheh et al. reported the preparation of magnetic Fe_3O_4 nanoparticles functionalized with polyamine/amide for the efficient extraction of U(VI) ions from aqueous solutions [78]. In their study, the binding capacity of the prepared magnetic nanoparticles was obtained as 127.5 $mg\, g^{-1}$.

8.4.2 MSPE for Food and Beverage Samples

Magnetic nanoparticles were also successfully applied for the sample preparation step of the analysis of food and beverage samples. Some examples are briefly described as follows.

In a study carried out by Li et al. [79], magnetic Fe_3O_4 nanoparticles coated with polyvinyl alcohol (PVA) layer were synthesized for the selective extraction of aminoglycoside antibiotics including streptomycin, dihydrostreptomycin, and kanamycin in honey samples. The prepared magnetic nanoparticles successfully extracted the target aminoglycoside antibiotics in a very short time (30 seconds). The detection limit values for streptomycin, dihydrostreptomycin, and kanamycin were found as 0.993, 0.913, and 1.23 $\mu g\, kg^{-1}$, respectively.

In another interesting study [80], Zhao et al. prepared magnetic Fe_3O_4 nanoparticles grafted with alendronate sodium for the selective extraction of trans-resveratrol in peanut oils. The achieved results showed that the whole extraction process was accomplished in 10 minutes. The obtained recovery values varied in the range between 78.6% and 118.9%. The detection limit was calculated as 0.3 $ng\, g^{-1}$.

Khodadadi et al. reported the design and development of magnetic $Fe_3O_4@SiO_2$ nanoparticles functionalized with aptamer for the efficient extraction of aflatoxin M1 from milk samples [81]. The detection limit was found as 0.2 $ng\, l^{-1}$. Authors reported that this is the achieved lowest value, which has been reported in the literature up to now.

In another important work conducted by Wang et al. [82], various estrogens including 17-β-estradiol, estrone, and diethylstilbestrol were successfully extracted from pork samples by using magnetic Fe_3O_4@caprylic acid nanoparticles coated with cetyltrimethylammonium bromide. In their study, the obtained results showed that extraction equilibrium for the target estrogens was obtained in a very short time (two minutes). The achieved detection limit values varied in the range from 0.021 to 0.033 $ng\, ml^{-1}$.

In a study carried out by Turan et al. [83], magnetic Fe_3O_4 nanoparticles functionalized with oleic acid were designed and developed for the efficient extraction of a pesticide fenazaquin from almond samples. Dispersive extraction process by using oleic acid–modified magnetic Fe_3O_4 nanoparticles is schematically demonstrated in Figure 8.5. The achieved results confirmed that the prepared magnetic nanoparticles were successfully applied for the extraction of target pesticide fenazaquin from

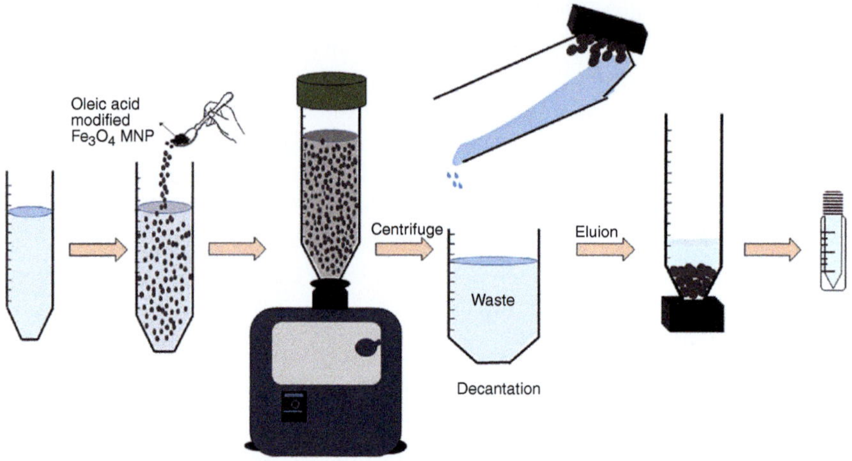

Figure 8.5 The schematic demonstration of the dispersive extraction process by using oleic acid–modified magnetic Fe_3O_4 nanoparticles. Source: Turan, N.B. et al., (2020) Application of oleic acid functionalized magnetic nanoparticles for a highlysensitive and efficient dispersive magnetic solid phase extraction of fenazaquin in almond samples for determination by gas chromatrography mass spectrometry. Microchemical Journal, 153, 104329. © 2019, Elsevier.

almond samples. The limit of detection and limit of quantification values achieved are 0.06 and 0.21 $\mu g\, l^{-1}$, respectively.

In another interesting study carried out by Zhao et al. [84], chlorogenic acid was successfully extracted from honeysuckle tea samples by using MIP-based magnetic nanoparticles. In their study, magnetic Fe_3O_4 nanoparticles were first modified with ethanediamine and methyl acrylate to provide the high number of functional amino groups. Then, chlorogenic acid–selective MIP layer was synthesized on the surface of magnetic Fe_3O_4 nanoparticles using dopamine as the functional monomer. The prepared MIP-based magnetic nanoparticles were successfully applied for the efficient extraction of chlorogenic acid from honeysuckle tea samples. The detection limit and quantification limit values were achieved as 0.01 and 0.038 $\mu g\, ml^{-1}$, respectively.

Guo et al. prepared magnetic $Fe_3O_4@SiO_2$ nanoparticles modified with octadecyl (C_{18}) functional groups for the extraction of phthalic acid esters from Chinese herbs [85]. In this study, the achieved detection limit was <1.89 $ng\, ml^{-1}$.

In another interesting work reported by Liang et al. [86], magnetic $Fe_3O_4@SiO_2$/GO nanocomposites functionalized with metal-organic framework MIL-101 (Cr) were successfully prepared and applied for the efficient extraction of various triazine herbicides including secbumeton, terbuthylazine, trietazine, atrazine, terbumeton, atraton, and prometon from rice samples. The obtained detection limit values for the target triazine herbicides were in the range between 0.010 and 0.080 $\mu g\, kg^{-1}$.

González-Sálamo et al. reported the preparation of magnetic Fe_3O_4 nanoparticles having poly(dopamine) layer for the effective extraction of various mycotoxins including zearalenone, α-zearalanol, β-zearalanol, α-zearalenol, β-zearalenol, and zearalanone from yogurt and milk samples [87]. In their study, the achieved detection limit values for yogurt samples and milk samples were in the range from 0.29 to 4.54 $\mu g\, kg^{-1}$ and 0.21 to 4.77 $\mu g\, l^{-1}$.

8.4.3 MSPE for Biological Samples

MSPE for biological samples (i.e. blood plasma, urine, etc.) is another crucial application area of magnetic nanoparticles. In a study carried out by Heidari and Limouei-Khosrowshahi [88], magnetic Fe_3O_4 nanoparticles modified with carbon were prepared for the extraction of various pharmaceutical compounds including carvedilol, losartan, and amlodipine besylate from plasma samples. In this study, the detection limit values varied in the range between 0.09 and 0.69 ng ml^{-1}. In another crucial study reported by Zhao et al. [89], a magnetic nanocomposite composed of Fe_3O_4, SiO_2 nanoparticles and polydopamine (Fe_3O_4@SiO_2@PDA) was developed. The prepared Fe_3O_4@SiO_2@PDA nanoparticles were successfully used for the enrichment of ginsenosides from rat plasma samples. In this study, the obtained results confirmed that the proposed approach exhibited excellent enrichment efficiency and could successfully determine the target 23 ginsenosides prototypes and metabolites in rat plasma samples. In addition, the reusability of the prepared Fe_3O_4@SiO_2@PDA nanoparticles was also investigated. The achieved recovery rate of the target ginsenosides exceeded 80% after recycling six times. Feng et al. reported the design and preparation of Fe_3O_4 nanoparticles-modified multiwalled carbon nanotubes (MWCNTs) [90]. The prepared MWCNTs-based nanocomposite was successfully used for the efficient extraction of brucine (a neurotoxic alkaloid exists in the nux-vomica tree) from human urine samples. In their study, the detection limit and quantification limit values were achieved as 6 and 21 ng ml^{-1}, respectively.

In another interesting conducted by Chen et al. [91], magnetic Fe_3O_4 nanoparticles having a layer composed of 1,3,5-triformylbenzene (Tb) and benzidine (Bd) (Fe_3O_4@TbBd) were prepared and successfully applied for the selective extraction of seven estrogens from human urine samples. The schematic demonstration of the preparation of magnetic Fe_3O_4@TbBd nanocomposite and extraction process for the target estrogens is shown in Figure 8.6. In this study, the achieved detection limit values varied in the range from 0.2 to 7.7 ng l^{-1}.

Mirzapour et al. reported the preparation of organic dendrimer–modified magnetic Fe_3O_4 nanoparticles for the efficient extraction of rosuvastatin from human urine, blood plasma, and tablet samples [92]. For this purpose, magnetic Fe_3O_4 nanoparticles were firstly synthesized by using $FeCl_2$ and $FeCl_3$. Then, the surface of the synthesized magnetic Fe_3O_4 nanoparticles was modified with organic dendrimers containing ethylene diamine and methyl methacrylate. The prepared organic dendrimer–modified magnetic Fe_3O_4 nanoparticles were successfully used for the extraction of the target pharmaceutical compound rosuvastatin from human urine, blood plasma, and tablet samples. In this study, the highest extraction capacity of the prepared organic dendrimer–modified magnetic Fe_3O_4 nanoparticles was achieved as 61 mg g^{-1}.

In another study reported by Ji et al. [93], magnetic Fe_3O_4 nanoparticles having MIP layer were successfully prepared and applied for the selective extraction of 9-hydroxyrisperidone and risperidone from human urine samples. The obtained detection limit values for 9-hydroxyrisperidone and risperidone were 0.24 and 0.21 ng ml^{-1}, respectively.

Azodi-Deilami et al. prepared magnetic Fe_3O_4 nanoparticles coated with molecularly imprinted polymeric shell for the selective extraction of paracetamol from

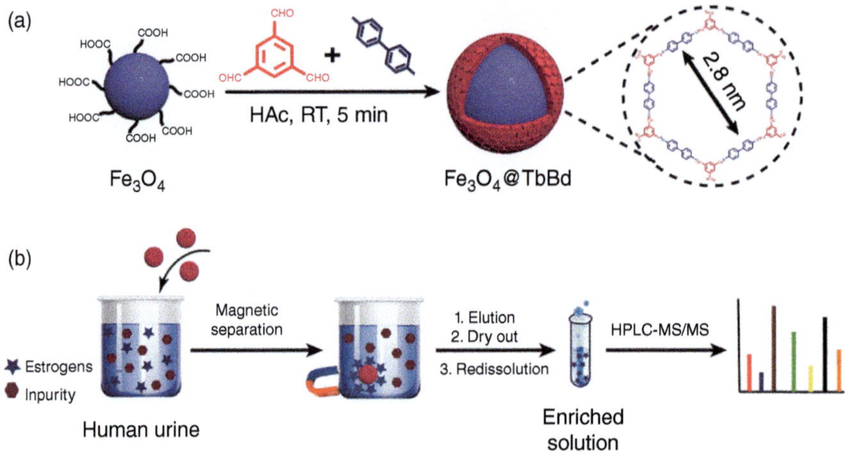

Figure 8.6 The schematic demonstration of the preparation of magnetic Fe_3O_4@TbBd nanocomposite (a) and extraction process for the target estrogens (b). Source: L. Chen et al., Facile synthesis of magnetic covalent organic framework nanobeads and application to magnetic solid-phase extraction of trace estrogens from human urine, J Chromatogr A., 2018, 1567, 136–146. © 2018, Elsevier.

human plasma samples [94]. In their study, the detection limit and quantification limit values for the target pharmaceutical compound paracetamol in human plasma were achieved as 0.17 and 0.4 $\mu g\, l^{-1}$, respectively.

In another interesting study [95], Jiang et al. reported the development of magnetic $Fe_3O_4@SiO_2$ nanoparticles modified with octadecyl (C_{18}) functional groups for the efficient extraction of aromatic amines including 1-aminonaphthalene, 4-aminobiphenyl, 4,4′-diaminodiphenylmethane, and 4-aminophenylthioether from human urine samples. The detection limit values for 1-aminonaphthalene, 4-aminobiphenyl, 4,4′-diaminodiphenylmethane, and 4-aminophenylthioether were obtained as 1.3, 0.88, 1.1, and 1.1 $ng\, ml^{-1}$, respectively.

Zhang et al. prepared PDA-coated magnetic Fe_3O_4 nanoparticles modified with MWCNTs for the extraction of extraction of antiepileptic drugs including phenytoin, oxcarbazepine, and carbamazepine from human urine, plasma, and cerebrospinal fluid samples [96]. In their study, the achieved detection limit values for the target drug compounds were in the range between 0.4 and 3.1 $ng\, ml^{-1}$.

8.5 Conclusions and Future Trends

The growing number of research in which magnetic nanoparticles were successfully applied in MSPE processes confirmed that these NMs with unique features are promising for the selective extraction of target compound/s from complex matrices (i.e. environmental, biological, food, and beverage samples). The reported examples briefly demonstrated in this chapter highlight the latest progresses in sample preparation applications by using magnetic nanoparticles over the past years. Magnetic nanoparticles exhibit great properties (i.e. small size, high surface area, active surface that can easily be modified, low toxicity, superparamagnetism, etc.) These great properties of magnetic nanoparticles make them unique candidates for the design and preparation of new functional NMs for the SPE processes. Magnetic nanoparticles are directly dispersed in the sample solutions for the quick extraction of the target compound/s since they can be readily recovered by a magnet, which overcomes drawbacks of the conventional extraction materials such as time consumption and large volume of samples.

Efficient extraction of the target compound/s by using magnetic nanoparticles can also eliminate the matrix effects usually observed in conventional SPE processes. In addition, high extraction capacity provides successful applications of the magnetic nanoparticles for efficient extraction of trace analytes from complex samples. The automation possibilities of the MSPE process also provide high-throughput analysis by applying this approach. Moreover, a promising and crucial goal in this area is the design and development of new strategies and techniques for the preparation and functionalization of the magnetic nanoparticles and increasing their chemical and physical stability, selectivity toward the target compound/s in complex samples and lifespan.

References

1 Wasik, A., Kot-Wasik, A., and Namiesnik, J. (2016). New trends in sample preparation techniques for the analysis of the residues of pharmaceuticals in environmental samples. *Curr. Anal. Chem.* 12: 280–302.

2 Ribeiro, C., Ribeiro, A.R., Maia, A.S. et al. (2014). *Crit. Rev. Anal. Chem.* 44: 142–185.

3 Calderilla, C., Maya, F., Leal, L.O., and Cerda, V. (2018). Recent advances in flow-based automated solid-phase extraction. *Trends Anal. Chem.* 108: 370–380.

4 Płotka-Wasylka, J., Szczepanska, N., de la Guardia, M., and Namiesnik, J. (2016). Modern trends in solid phase extraction: new sorbent media. *Trends Anal. Chem.* 77: 23–43.

5 Xu, J., Zheng, J., Tian, J. et al. (2013). New materials in solid-phase microextraction. *Trends Anal. Chem.* 47: 68–83.

6 Buszewski, B. and Szultka, M. (2012). Past, present, and future of solid phase extraction: a review. *Crit. Rev. Anal. Chem.* 42: 198–213.

7 Azzouz, A., Kailasa, S.K., Lee, S.S. et al. (2018). Review of nanomaterials as sorbents in solid-phase extraction for environmental samples. *Trends Anal. Chem.* 108: 347–369.

8 Reyes-Garcés, N., Gionfriddo, E., Gómez-Ríos, G.A. et al. (2018). Advances in solid phase microextraction and perspective on future directions. *Anal. Chem.* 90: 302–360.

9 Khezeli, T. and Daneshfar, A. (2017). Development of dispersive micro-solid phase extraction based on micro and nano sorbents. *Trends Anal. Chem.* 89: 99–118.

10 Wierucka, M. and Biziuk, M. (2014). Application of magnetic nanoparticles for magnetic solid-phase extraction in preparing biological, environmental and food samples. *Trends Anal. Chem.* 59: 50–58.

11 Hussain, C.M. and Kharisov, B. (2017). *Advanced Environmental Analysis-Application of Nanomaterials.* The Royal Society of Chemistry.

12 Hussain, C.M. (2018). *Nanomaterials in Chromatography: Current Trends in Chromatographic Research Technology and Techniques.* Elsevier.

13 Hussain, C.M. (2019). *Handbook of Nanomaterials in Analytical Chemistry: Modern Trends in Analysis.* Elsevier.

14 Hussain, C.M. and Keçili, R. (2020). *Modern Environmental Analysis Techniques for Pollutants.* Amsterdam, The Netherlands.: Elsevier.

15 Keçili, R., Büyüktiryaki, S., and Hussain, C.M. (2020). Micro total analysis systems with nanomaterials, Chapter 8. In: *Handbook of Nanomaterials in Analytical Chemistry* (ed. C.M. Hussain), 185–198. Elsevier.

16 Murtada, K. (2020). Trends in nanomaterial-based solid-phase microextraction with a focus on environmental applications – a review. *Trends in Environmental Analytical Chemistry* 25: e00077.

17 Hussain, C.M. (2017). Magnetic nanomaterials for environmental analysis. In: *Advanced Environmental Analysis-Application of Nanomaterials* (eds. C.M. Hussain and B. Kharisov). The Royal Society of Chemistry.

18 Keçili, R. and Hussain, C.M. (2018). Recent progress of imprinted nanomaterials in analytical chemistry. *Int. J. Anal. Chem.*: 8503853.

19 Büyüktiryaki, S., Sümbelli, Y., Keçili, R., and Hussain, C.M. (2019). Lab-on-chip platforms for environmental analysis. In: *Encyclopedia of Analytical Science*, 3e (eds. P. Worsfold, C. Poole, A. Townshend and M. Miró), 267–273. Academic Press.

20 Keçili, R., Büyüktiryaki, S., and Hussain, C.M. (2019). Advancement in bioanalytical science through nanotechnology: past, present and future. *TrAC Trends Anal. Chem.* 110: 259–276.

21 Chikazumi, S., Taketomi, S., Ukita, M. et al. (1987). Physics of magnetic fluids. *J. Magn. Magn. Mater.* 65: 245–251.

22 Lu, A.H., Schmidt, W., Matoussevitch, N. et al. (2004). Nanoengineering of a magnetically separable hydrogenation catalyst. *Angew. Chem. Int. Ed.* 43: 4303–4306.

23 Li, Z., Wei, L., Gao, M.Y., and Lei, H. (2005). One-pot reaction to synthesize biocompatible magnetite nanoparticles. *Adv. Mater.* 17: 1001–1005.

24 Farrukh, A., Akram, A., Ghaffar, A. et al. (2013). Design of polymer-brush-grafted magnetic nanoparticles for highly efficient water remediation. *ACS Appl. Mater. Interfaces* 5: 3784–3793.

25 Lin, P.C., Yu, C.C., Wu, H.T. et al. (2013). A chemically functionalized magnetic nanoplatform for rapid and specific biomolecular recognition and separation. *Biomacromolecules* 14: 160–168.

26 Lu, A.H., Salabas, E.L., and Schuth, F. (2007). Magnetic nanoparticles: synthesis, protection, functionalization, and application. *Angew. Chem. Int. Ed.* 46: 1222–1244.

27 Gao, L.Z., Wu, J.M., Lyle, S. et al. (2008). Magnetite nanoparticle-linked immunosorbent assay. *J. Phys. Chem. C* 112: 17357–17361.

28 Behrens, S. (2011). Preparation of functional magnetic nanocomposites and hybridmaterials: recent progress and future directions. *Nanoscale* 3: 877–892.

29 Ahniyaz, A., Sakamoto, Y., and Bergström, L. (2007). Magnetic field-induced assembly oforiented superlattices from maghemite nanocubes. *Proc. Natl. Acad. Sci. U. S. A.* 104: 17570–17574.

30 Redl, F.X., Cho, K.S., Murray, C.B., and O'Brien, S. (2003). Three-dimensional binary super-lattices of magnetic nanocrystals and semiconductor quantum dots. *Nature* 423: 968–971.

31 Ely, T.O., Amiens, C., Chaudret, B. et al. (1999). Synthesis of nickel nanoparticles. Influence of aggregation induced by modification of poly(vinylpyrrolidone) chain length on their magnetic properties. *Chem. Mater.* 11: 526–529.

32 Osuna, J., de Caro, D., Amiens, C. et al. (1996). Synthesis, characterization, and magnetic properties of cobaltnanoparticles from an organometallic precursor. *J. Phys. Chem. C* 100: 14571–14574.

33 Lee, J.H., Huh, Y.M., Jun, Y. et al. (2006). Artificially engineered magnetic nanoparticles for ultra-sensitive molecular imaging. *Nat. Med.* 13: 95–99.

34 Hu, F.Q., Wei, L., Zhou, Z. et al. (2006). Preparation of biocompatiblemagnetite nanocrystals for in vivo magnetic resonance detection of cancer. *Adv. Mater.* 18: 2553–2556.

35 Reddy, L.H., Arias, J.L., Nicolas, J., and Couvreur, P. (2012). Magnetic nanoparticles: designand characterization, toxicity and biocompatibility, pharmaceutical andbiomedical applications. *Chem. Rev.* 112: 5818–5878.

36 Massart, R. (1981). Preparation of aqueous magnetic liquids in alkaline and acidicmedia. *IEEE Trans. Magn.* 17: 1247–1248.

37 Liu, J., Qiao, S.Z., Hu, Q.H., and Lu, G.Q. (2011). Magnetic nanocomposites with mesoporousstructures: synthesis and applications. *Small* 7: 425–443.

38 Massart, R. and Cabuil, V. (1987). Effect of some parameters on the formation of colloidalmagnetite in alkaline medium: yield and particle size control. *J. Chim. Phys. PCB* 84: 967–973.

39 Itoh, H. and Sugimoto, T. (2003). Systematic control of size, shape, structure, and magnetic properties of uniform magnetite and maghemite particles. *J. Colloid Interface Sci.* 265: 283–295.

40 Cannas, C., Gatteschi, D., Musinu, A. et al. (1998). Structural and magnetic properties of Fe_2O_3 nanoparticles dispersed over a silica matrix. *J. Phys. Chem. B* 102: 7721–7726.

41 Chen, F., Gao, Q., Hong, G., and Ni, J. (2008). Synthesis and characterization of magnetite dodecahedron nanostructure by hydrothermal method. *J. Magn. Magn. Mater.* 320: 1775–1780.

42 Bee, A., Massart, R., and Neveu, S. (1995). Synthesis of very fine maghemite particles. *J. Magn. Magn. Mater.* 149: 6–9.

43 Petit, C., Taleb, A., and Pileni, M.P. (1999). Cobalt nanosized particles organized in a 2D superlattice: synthesis, characterization, and magnetic properties. *J. Phys. Chem. B* 103: 1805–1810.

44 Chen, D.H. and Wu, S.H. (2000). Synthesis of nickel nanoparticles in water-in-oilmicroemulsion. *Chem. Mater.* 12: 1354–1360.

45 Salazar-Alvarez, G., Muhammed, M., and Zagorodni, A.A. (2006). Novel flow injection synthesis of iron oxide nanoparticles with narrow size distribution. *Chem. Eng. Sci.* 61: 4625–4633.

46 Strobel, R. and Pratsinis, S.E. (2009). Direct synthesis of maghemite, magnetite and wustite nanoparticles by flame spray pyrolysis. *Adv. Powder Technol.* 20: 190–194.

47 Veintemillas-Verdaguer, S., Leconte, Y., Costo, R. et al. (2007). Continuous production of inorganic magnetic nanocomposites for biomedical applications by laser pyrolysis. *J. Magn. Magn. Mater.* 311: 120–124.

48 Liska, I. (2000). Fifty years of solid-phase extraction in water analysis – historical development and overview. *J. Chromatogr. A* 885: 3–16.

49 Spivakov, B.Y., Malofeeva, G.I., and Petrukhin, O.M. (2006). Solid-phase extraction on alkyl-bonded silica gels in inorganic analysis. *Anal. Sci.* 22: 503–519.

50 Rodríguez, J.A., Escamilla-Lara, K.A., Guevara-Lara, A. et al. (2015). Application of an activated carbon-based support for magnetic solid phase extraction followed by spectrophotometric determination of tartrazine in commercial beverages. *Int. J. Anal. Chem.* 2015: 1–8.

51 Valdés, M.G., Pérez-Cordoves, A.I., and Díaz-García, M.E. (2006). Zeolites and zeolite-based materials in analytical chemistry. *TrAC Trends Anal. Chem.* 25: 24–30.

52 Keçili, R., Özcan, A.A., Ersöz, A. et al. (2011). Superparamagnetic nanotraps containing MIP based mimic lipase for biotransformations uses. *J. Nanopart. Res.* 13: 2073–2079.

53 Bucak, S., Jones, D.A., Laibinis, P.E., and Hatton, T.A. (2003). Protein separations using colloidal magnetic nanoparticles. *Biotechnol. Progr.* 19: 477–484.

54 Büyüktiryaki, S., Uzun, L., Denizli, A. et al. (2016). Simultaneous depletion of albumin and immunoglobulin G by using twin affinity magnetic nanotraps. *Sep. Sci. Technol.* 51: 2080–2089.

55 Prabha, G. and Raj, V. (2016). Preparation and characterization of polymer nanocomposites coated magnetic nanoparticles for drug delivery applications. *J. Magn. Magn. Mater.* 408: 26–34.

56 Safarikova, M. and Safarik, I. (1995). Magnetic separations in natural sciences and biotechnology. *Chem. Listy* 89: 280–287.

57 Haeri, S.A. and Abbasi, S. (2016). New strategy for the biosorption of atrazine after magnetic solid-phase extraction from water followed by high-performance liquid chromatography analysis. *J. Sep. Sci.* 39: 2839–2845.

58 Dai, Y., Niu, L., Zou, J. et al. (2018). Preparation of core-shell magnetic Fe_3O_4@SiO_2-dithiocarbamate nanoparticle and its application for the Ni^{2+}, Cu^{2+} removal. *Chin. Chem. Lett.* 29: 887–891.

59 Sun, T., Zhao, Z., Liang, Z. et al. (2017). Efficient removal of arsenite through photocatalytic oxidation and adsorption by ZrO_2–Fe_3O_4 magnetic nanoparticles. *Appl. Surf. Sci.* 416: 656–665.

60 Ma, Y.-Z., Zheng, D.-F., Mo, Z.-Y. et al. (2018). Magnetic lignin-based carbon nanoparticles and the adsorption for removal of methyl orange. *Colloids Surf., A* 559: 226–234.

61 Golmohammadi, F., Hazrati, M., and Safari, M. (2019). Removal of reactive yellow 15 from water sample using a magnetite nanoparticles coated with covalently immobilized dimethyl octadecyl [3-(trimethoxysilylpropyl)]ammonium chloride ionic liquid. *Microchem. J.* 144: 64–72.

62 Tang, G., Chen, W., Wan, X. et al. (2020). Construction of magnetic Fe_3O_4 nanoparticles coupled with flower-like $MoSe_2$ nanosheets for efficient adsorptive removal of methylene blue. *Colloid. Surface. A* 587: 124291.

63 Pascale, M., De Girolamo, A., Visconti, A. et al. (2008). Use of itaconic acid-based polymers for solid-phase extraction of deoxynivalenol and application to pasta analysis. *Anal. Chim. Acta* 609: 131–138.

64 Puoci, F., Curcio, M., Cirillo, G. et al. (2008). Molecularly imprinted solid-phase extraction for cholesterol determination in cheese products. *Food Chem.* 106: 836–842.

65 Beltran, A., Marcé, R., Cormack, P., and Borrull, F. (2009). Synthesis by precipitation polymerisation of molecularly imprinted polymer microspheres for the selective extraction of carbamazepine and oxcarbazepine from human urine. *J. Chromatogr. A* 1216: 2248–2253.

66 Andersson, L., Abdel-Rehim, M., Nicklasson, L. et al. (2002). Towards molecular-imprint based SPE of local anaesthetics. *Chromatographia* 55: S65–S69.

67 Li, Y.H., Yang, T., Qi, X.L. et al. (2008). Development of a group selective molecularly imprinted polymers based solid phase extraction of malachite green from fish water and fish feed samples. *Anal. Chim. Acta* 624: 317–325.

68 Pichon, V., Bouzige, M., Miège, C., and Hennion, M.-C. (1999). Immunosorbents: natural molecular recognition materials for sample preparation of complex environmental matrices. *TrAC Trend. Anal. Chem.* 18: 219–235.

69 Wackerlig, J. and Lieberzeit, P.A. (2015). Molecularly imprinted polymer nanoparticles in chemical sensing – synthesis, characterisation and application. *Sens. Actuators, B* 207: 144–157.

70 Jiang, W., Jin, X., Yu, X. et al. (2017). Ion-imprinted magnetic nanoparticles for specific separation and concentration of ultra-trace methyl mercury from aqueous sample. *J. Chromatogr. A* 1496: 167–173.

71 Hu, Z.H., Wang, Y.F., Omer, A.M., and Ouyang, X.K. (2018). Fabrication of ofloxacin imprinted polymer on the surface of magnetic carboxylated cellulose nanocrystals for highly selective adsorption of fluoroquinolones from water. *Int. J. Biol. Macromol.* 107: 453–462.

72 Patiño-Ropero, M.J., Díaz-Álvarez, M., and Martín-Esteban, A. (2017). Molecularly imprinted core-shell magnetic nanoparticles for selective extraction of triazines in soils. *J. Mol. Recognit.* 30: e2593.

73 Zhou, Z., Liu, X., Zhang, M. et al. (2020). Preparation of highly efficient ion-imprinted polymers with Fe_3O_4 nanoparticles as carrier for removal of Cr(VI) from aqueous solution. *Sci. Total Environ.* 699: 134334.

74 Barati, A., Kazemi, E., Dadfarnia, S., and Shabani, A.M.H. (2017). Synthesis/characterization of molecular imprinted polymer based on magnetic chitosan/graphene oxide for selective separation/preconcentration of fluoxetine from environmental and biological samples. *J. Ind. Eng. Chem.* 46: 212–221.

75 Lu, Y.-C., Mao, J.-H., Zhang, W. et al. (2020). A novel strategy for selective removal and rapid collection of triclosan from aquatic environment using magnetic molecularly imprinted nano-polymers. *Chemosphere* 238: 124640.

76 Huang, Y., Zheng, H., Li, H. et al. (2020). Highly selective uranium adsorption on 2-phosphonobutane-1,2,4-tricarboxylic acid-decorated chitosan-coated magnetic silica nanoparticles. *Chem. Eng. J.* 388: 124349.

77 Shahriman, M.S., Ramachandran, M.R., Zain, N.N.M. et al. (2018). Polyaniline-dicationic ionic liquid coated with magnetic nanoparticles composite for magnetic solid phase extraction of polycyclic aromatic hydrocarbons in environmental samples. *Talanta* 178: 211–221.

78 Al-Harahsheh, M., AlJarrah, M., Mayyas, M., and Alrebaki, M. (2018). High-stability polyamine/amide-functionalized magnetic nanoparticles for enhanced extraction of uranium from aqueous solutions. *J. Taiwan Inst. Chem. E* 86: 148–157.

79 Li, D., Li, T., Wang, L., and Ji, S. (2018). A polyvinyl alcohol-coated core-shell magnetic nanoparticle for the extraction of aminoglycoside antibiotics residues from honey samples. *J. Chromatogr. A* 1581–1582: 1–7.

80 Zhao, Q., Cheng, D.-Q., Tao, M. et al. (2019). Rapid magnetic solid-phase extraction based on alendronate sodium grafted mesoporous magnetic nanoparticle for the determination of trans-resveratrol in peanut oils. *Food Chem.* 279: 187–193.

81 Khodadadi, M., Malekpour, A., and Mehrgardi, M.A. (2018). Aptamer functionalized magnetic nanoparticles for effectiveextraction of ultratrace amounts of aflatoxin M1 prior itsdetermination by HPLC. *J. Chromatogr. A* 1564: 85–93.

82 Wang, J., Chen, Z., Li, Z., and Yang, Y. (2016). Magnetic nanoparticles based dispersive micro-solid-phase extraction as a novel technique for the determination of estrogens in pork samples. *Food Chem.* 204: 135–140.

83 Turan, N.B., Zaman, B.T., and Bakırdere, S. (2020). Application of oleic acid functionalized magnetic nanoparticles for a highlysensitive and efficient dispersive magnetic solid phase extraction of fenazaquin in almond samples for determination by gas chromatrography mass spectrometry. *Microchem. J.* 153: 104329.

84 Zhao, Y., Tang, Y., He, J. et al. (2018). Surface imprinted polymers based on amino-hyperbranched magnetic nanoparticles for selective extraction and detection of chlorogenic acid in Honeysuckle tea. *Talanta* 181: 271–277.

85 Guo, B., Ji, S., Zhang, F. et al. (2014). Preparation of C18-functionalized Fe_3O_4@SiO_2 core–shell magnetic nanoparticles for extraction and determination of phthalic acid esters in Chinese herb preparations. *J. Pharm. Biomed. Anal.* 100: 365–368.

86 Liang, L., Wang, X., Sun, Y. et al. (2018). Magnetic solid-phase extraction of triazine herbicides from rice using metal-organic framework MIL-101(Cr) functionalized magnetic particles. *Talanta* 179: 512–519.

87 González-Sálamo, J., Socas-Rodríguez, B., Hernández-Borges, J., and Ángel Rodríguez-Delgado, M. (2017). Core-shell poly(dopamine) magnetic nanoparticles for the extraction of estrogenic mycotoxins from milk and yogurt prior to LC–MS analysis. *Food Chem.* 215: 362–368.

88 Heidari, H. and Limouei-Khosrowshahi, B. (2019). Magnetic solid phase extraction with carbon-coated Fe_3O_4 nanoparticles coupled to HPLC-UV for the simultaneous determination of losartan, carvedilol, and amlodipine besylate in plasma samples. *J. Chromatogr. B* 1114–1115: 24–30.

89 Zhao, N., Liu, S., Xing, J. et al. (2020). Trace determination and characterization of ginsenosides in rat plasma through magnetic dispersive solid-phase extraction based on core-shell polydopamine-coated magnetic nanoparticles. *J. Pharm. Anal.* 10: 86–95.

90 Feng, Z., Yang, R., and Du, B. (2017). Fast SPE and HPLC–DAD determination of brucine in human urine using multi-walled carbon nanotubes modified with magnetic nanoparticles. *J. Anal. Chem.* 72 (8): 862–869.

91 Chen, L., Zhang, M., Fu, F. et al. (2018). Facile synthesis of magnetic covalent organic framework nanobeadsand application to magnetic solid-phase extraction of trace estrogensfrom human urine. *J. Chromatogr. A* 1567: 136–146.

92 Mirzapour, H., Panahi, H.A., Moniri, E., and Feizbakhsh, A. (2018). Magnetic nanoparticles modified with organic dendrimers containing methyl methacrylate and ethylene diamine for the microextraction of rosuvastatin. *Microchim. Acta* 185: 440.

93 Ji, W.-H., Guo, Y.-S., Wang, X., and Guo, D.-S. (2018). A water-compatible magnetic molecularly imprinted polymer for the selective extraction of risperidone and 9-hydroxyrisperidone from human urine. *Talanta* 181: 392–400.

94 Azodi-Deilami, S., Najafabadi, A.H., Asadi, E. et al. (2014). Magnetic molecularly imprinted polymer nanoparticles for the solid-phase extraction of paracetamol from plasma samples, followed its determination by HPLC. *Microchim. Acta* 181: 1823–1832.

95 Jiang, C., Sun, Y., Yu, X. et al. (2014). Application of C18-functional magnetic nanoparticles for extraction of aromatic amines from human urine. *J. Chromatogr. B* 947–948: 49–56.

96 Zhang, R., Wang, S., Yang, Y. et al. (2018). Modification of polydopamine-coated Fe_3O_4 nanoparticles with multi-walled carbon nanotubes for magnetic-μ-dispersive solid-phase extraction of antiepileptic drugs in biological matrices. *Anal. Bioanal.Chem.* 410: 3779–3788.

9

Lab-on-a-Chip with Nanomaterials (NMs)

9.1 Introduction

Miniaturization is considered to be the trend to downscale biological and chemical processes, offering various superiorities compared to the macroscopic systems [1–10]. The time for the mass transfer and heat are reduced in microfluidic channels because of the shorter distances in the miniaturized systems. Lab-on-a-chip (LOC) systems are defined as miniaturized platforms that process through many specific analyses (i.e. biochemical detection or DNA sequencing) on a chip platform [11–24]. Most of the researches on the LOC platforms basically focus on the design and development of diagnostic systems and analysis of DNA. Less often, researches also put effort on the organic synthesis of chemicals using LOC technology [25, 26].

In the recent years, so much effort was put on the miniaturization, which is a widely applied biological and chemical process, offering various superiorities compared to the macroscopic systems [3, 4]. On the other hand, the high surface-to-volume ratio also provides homogeneous heat distribution. Miniaturization of various functional units (i.e. reaction, separation, detection, etc.) enables the serial processing of samples as well as the parallelization of the process for high throughput. The term "miniaturization" is not only associated with the decrease in size but also linked to the various challenges in analytical systems including automation, rapid analysis, operational cost, portability, etc.

On the other hand, there is a strong relation between nanotechnology and miniaturized or biosensing and (bio)analytical platforms. The advantages of miniaturized (bio)systems include the reduction of the volumes of samples and reagents consumed, detection facility, minimal handling of hazardous materials, and parallel multiple sample detection. It is well known that nanomaterials (NMs) are bringing a number of superiorities for the improvement of biosensing platforms [27, 28]. Given the advantages of NMs in the design of novel biosensing systems or improvements of the existing ones, it is logical to see a strong interaction between these novel materials and LOC platforms. NMs are being used in microfluidic platforms as detectors (or detectors modifiers), tools for microreactors, and others.

This chapter provides a comprehensive overview of LOC technology. The chapter starts with the description and brief history of LOC technology. Then, the latest

Sample Preparation with Nanomaterials: Next Generation Techniques and Applications,
First Edition. Chaudhery Mustansar Hussain, Rüstem Keçili, and Chaudhery Ghazanfar Hussain.

advancements and studies in the field of miniaturized systems based on LOC technologies are presented.

The integration of NMs into analytical devices for both detection and in-chip process improvement will also be discussed.

9.2 Lab-on-a-Chip (LOC) Concept

In the recent years, there has been a great effort in the development of novel approaches for the sensitive detection and treatment of diseases and ailments. Currently, the detection of various diseases is carried out by using expensive instruments. There are various issues that should be considered when performing these experiments. First of all, bulky equipments that can be pretty expensive to buy and/or repair are used in the experiments. Second, with these equipments having high dimensions, it can be certain that large amounts of samples and/or reagents have to be used. The third issue is the time needed to perform each experiment. Because of these issues, innovative techniques and approaches need to be addressed. Basically, a system for the sensitive detection has to be designed, which not only is light and portable but also should be rapid in detection and must take as little fluid as possible.

On the other hand, fast advancements of the microelectronic technology since 1960s established semiconductor production techniques to enhance the density of transistors in integrated circuits. In 1980s, these techniques led to the first design and fabrication of the microelectromechanical systems (MEMS) [29]. Further refinement of the fabrication technologies of silicon-based MEMS made it straightforward to design and prepare simple microfluidic platforms. The first demonstration of microfluidic systems was carried out by Terry et al. [30]. In their study, a microfluidic system was prepared by using a silicon wafer, and the prepared systems were successfully used as a miniaturized gas analysis system in terms of gas chromatography (GC) with a thermal conductivity detector. The microfluidic system was fabricated by applying photolithography technique and chemical etching. The efforts on the design and fabrication of microfluidic platforms did not pick up momentum until the early 1990s, when Manz first introduced the concept of micro total analysis system (μTAS) [31]. According to this approach, the injection, separation, and detection systems were found in the same interface. This research was not only defining the μTASs, but it was a realization at the same time for scientists to use the technique in life sciences. Since then, the developed miniaturized systems were successfully applied for different applications such as separation [32–34] and detection [35–37] or used as a reactor [38–41]. Then, scientists realized the potential of these unique systems having great advantages in other fields of science besides analytical chemistry.

Microfluidic techniques can be successfully applied for the design and fabrication of LOC platforms, which enables to produce a number of microchannels existing on the surface of single chip which fits in your hand. The microchannels enable the handling of fluids in quantities as low as a few picoliters as well as the manipulation of biochemical reactions at very small volumes. In addition to microchannels, LOC

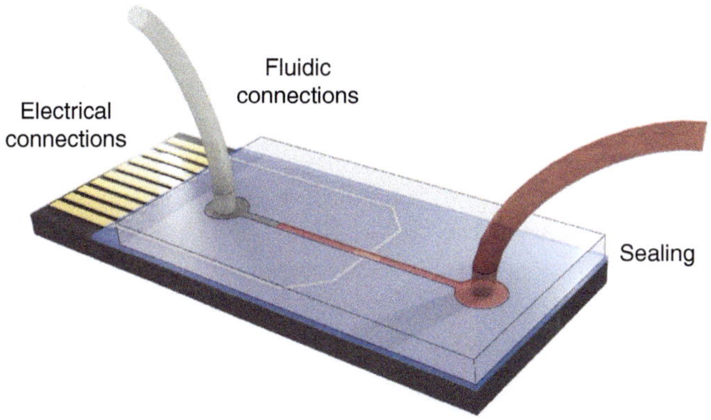

Figure 9.1 The schematic demonstration of an LOC system. Source: Y. Temiz, et al., Lab-on-a-chip devices: How to close and plug the lab?, Microelectron Eng, 2015, 132, 156–175. © 2014, Elsevier.

platforms also contain various components such as electrodes, integrated pumps, electrical field valves, and electronics. Figure 9.1 shows a schematic depiction of an LOC system having various electrical and fluidic connections. During the fabrication process of the system, electrodes, microfluidic structures, and various back-end processes are needed to obtain a fully functional LOC system [42, 43]. As can be seen from Figure 9.1, sealing of microfluidic structures can be performed by using a cover layer that is patterned or drilled for the formation of openings for electrical/fluidic interfaces. In addition, the system can be integrated to the peripheral equipments (i.e. pumps and valves) using inlet and outlet ports.

In an interesting study reported by Whitesides and coworkers in 2000, a technique for the fabrication of microfluidic systems using polydimethylsiloxane (PDMS) as the substrate material which is called soft lithography was developed[45]. This technique enabled a rapid and low-cost route for the fabrication of microfluidic cartridges. In 2002, S. Quake and coworkers reported the design and fabrication of a large-scale integration microfluidic device with the use of a number of micro-pumps and micro-valves [46].

Today, the concept of microfluidics is defined as the manipulation and precise control of the fluids on LOC platforms [47]. The main characteristics of microfluidic devices include miniaturization and automation. Miniaturization means that bioassays can be implemented with very little reagent use in a very short time, while automation means that bioassays can be implemented requiring no skilled researchers, which eliminates the errors caused by the man-made operation to a large extent [48]. Microfluidics technology integrates sampling, dilution, separation, detection, and other experimental operations on a chip surface of several square centimeters or less, thereby reducing sample reagent consumption, improving detection sensitivity, shortening reaction time, and reducing the average cost. The microfluidic chips enable a very broad prospect in the fields of biomedical research [49, 50], drug synthesis screening [51], environmental monitoring or protection, quarantine, biomaterials synthesis, forensic identification, and many other fields [52]. Especially, in clinical medical diagnostics, the miniaturization, integration, and automation of microfluidics-based devices are highly relevant to the development needs of point-of-care testing (POCT), which is defined as medical diagnostic testing at the time and place of patient care. In recent years, microfluidics has increasingly become a research hotspot in the field of POCT [53–57]. As stated above, microfluidic technologies can be successfully applied for the design and fabrication of LOC platforms. With the rapid progresses in the design and development of LOC systems, they can be categorized into different branches such as paper-based LOC systems, centrifugal LOC systems, droplet-based LOC systems, digital LOC systems, and surface acoustic wave-based LOC systems.

9.2.1 Paper-based LOC Systems

In the preparation of paper-based LOC systems, hydrophobic microchannels on hydrophilic paper, which guides liquid from inlet to designated location, are mainly printed or photoetched [58, 59]. This approach is quite cheap, disposable, facile to apply, and usually applied for the environmental testing and portable glucose detection [60, 61]. However, the capillary force-driven transportation of liquid makes that the precision of assay is usually limited and the hydrophobic barriers may not be stable after longtime store [62].

9.2.2 Centrifugal LOC Systems

In centrifugal LOC systems, a centrifugal field is usually exploited to manipulate liquid [63]. In the analyses by using this type of miniaturized systems, a cartridge including liquid reagent is usually centrifuged as a disk during the liquid process. Centrifugal LOC systems are also called as "lab-on-a-disk systems" that exhibit great superiorities. Centrifugal LOC systems are composed of a motor as the pumping source, which is rather simple and compact. Furthermore, analyses can be easily processed since the centrifugal force exists everywhere on the disk-shaped cartridge [64]. In addition, existence of centrifugal field also makes no residual volume transportation of fluid- and density-based sample preparation inherently available [65, 66]. However, due to the spin of cartridge in this LOC system, optical detection may be difficult to be implemented as it is a little difficult to make the final detection chamber and photodetector focused on one line.

9.2.3 Droplet-Based LOC Systems

Droplet-based LOC systems manipulate discrete volumes of fluids with the use of immiscible phases [67]. In these miniaturized systems, a large number of droplets are used as reactors to implement bioassays. A very striking application based on droplet-based LOC systems is digital polymerase chain reaction (PCR) that exhibits great sensitivity toward nucleic acids [68, 69]. However, the generation and manipulation of a large number of tiny droplets require a careful fabrication technique and also high-precision manipulation strategy [70].

9.2.4 Digital LOC Systems

Similar to droplet-based LOC systems, digital LOC systems also deal with the droplets. However, during the analysis by using digital LOC systems, droplets are usually manipulated on a set of insulated electrodes that are under software-driven electronic control [71]. In this type of LOC system, the most widely applied manipulation strategy is electrowetting-on-dielectric phenomenon, which is based on the principle of modification of the interfacial tension with an applied electric field [72].

9.2.5 Surface Acoustic Wave-Based LOC Systems

Surface acoustic wave-based LOC systems, in which acoustic waves having frequency range between 10 and 1000 MHz are applied for the manipulation of microscale fluid, are extensively applied in drug delivery and point-of-care (POC) diagnostics. This manipulate strategy is of rapid fluid actuation, is compact, and makes transfer a complete microfluidics solution at the microscale. However, it still currently has some drawbacks in understanding the physics of this manipulation approach, and the deformation of the fluid interface remains an unsolved issue [73, 74].

9.3 NM-Based LOC Platforms

9.3.1 NM-Based Transducers

The application of NMs in the development of sensors and biosensors represents one of the main focuses of current research in nanotechnology. Their interesting optical and electrical properties are making NMs new building blocks for the design of various kinds of sensing and biosensing devices with improved performance. In the past decade, NMs have been very attractive for the modification of transducers that are the most important part of the (bio)sensing systems to be integrated into LOCs. The transducer's modification through NMs can improve the electrochemical (amperometric, impedimetric, potentiometric), optical (light absorbance, dispersion, fluorescence), or other mixed signals (i.e. electrochemilumniscence, ECL) based measurements.

9.3.1.1 Electrochemical Detection Systems

Various NM-based electrochemical detectors integrated within the LOC systems have been reported. Among these, carbon nanotubes (CNTs) are the most extensively studied [75–78]. Indeed, they present interesting properties such as extraordinary thermal and electrical conductivity, as well as excellent mechanical resistance. CNTs can be easily deposited or grown onto different surfaces through simple dropping, chemical vapor deposition (CVD), or plasma-enhanced CVD. The CNTs modified transducers have a larger surface area, which increases the current density as well as offers an increased surface for biomolecule attachment. This can improve the current stability and increase the sensitivity of the detector. The electrical behavior of such platforms depends on CNTs chirality, the number of carbon layers, defects, and their functionalization, which must be carefully considered during their application. All these CNTs properties are useful in catalysis, enzyme immobilization, protein detection, and metal detection.

Okuno et al. developed a label-free immunosensor modified with CNTs to enhance the detection of total prostate-specific antigen (PSA) [79]. An array of platinum microelectrodes was modified by CVD with single-walled carbon nanotubes (SWCNTs). CNTs allow a better electron transfer, improving the limit of detection (LOD) in relation to the non-modified electrodes, obtaining an LOD of 0.25 ng ml^{-1}.

CNTs are widely used in the fabrication of field effect transistors (FETs); Javey et al. have obtained a CNT-based FET by combining ohmic metal-tube contacts, high dielectric constant films as gate insulators, and electrostatically doped nanotube segments as source/drain electrodes [80]. This approach looks like those of metal oxide semiconductor field effect transistors (MOSFETs) and can provide excellent ON–OFF states to nanotube-based FETs under aggressive vertical scaling.

Tey et al. described a liquid gate field effect transistor (LG-FET) that integrates SWCNTs [81]. The electrical detection was performed by using CNTs as contact electrodes and semiconductive channels. An LOD of 1 pM of poly-L-lysine was achieved. The SWCNTs were used as "channel modulation labels" to sense changes in their immediate environment as a result of specific interactions with biomolecules. In this design, the electrolyte replaces the dielectric in normal operation. Also, the electrical double-layer capacitance created by the electrolyte and high surface area of the SWCNT network yields a total gate capacitance, which is increased by an order of magnitude in comparison to typical Si/SiO_2 bottom gate configuration.

A nonenzymatic CNT sensor for sugar detection was developed by Vlandas et al. [82]. The sensor was also assembled as an LG-FET. The semiconductive channel is based on SWCNTs functionalized with boronic acid receptors. This sensor is capable of detecting glucose in a range of concentration from 5 to 30 mM by measuring the change of impedance at different frequencies and gate voltages, which is in the range of physiological glucose detection.

Graphene is another interesting NM that has recently attracted significant attention due to its interesting input for electrochemical detection. It offers higher electron transference and is easy to be functionalized with (bio)molecules. This material, which is easy to produce from graphite [83], seems to be much cheaper than CNTs and consequently a good alternative.

Chua et al. reported a microfluidic graphene oxide-based amperometric detector that showed higher peak sensitivity, resolution, and separation efficiency than the same device without modification [84]. Figure 9.2 shows the schematic depiction of the preparation of a graphene-based thin film transistor where the drain and source pads were made with graphene oxide and the channel with reduced graphene (rGO). The detection of fibronectin was demonstrated as proof of concept, with a detection limit of 0.5 nM [85].

Finally, nanoelectrodes and nanostructured electrodes are attracting the attention for biological sample analysis at very low concentrations where the reduction of the

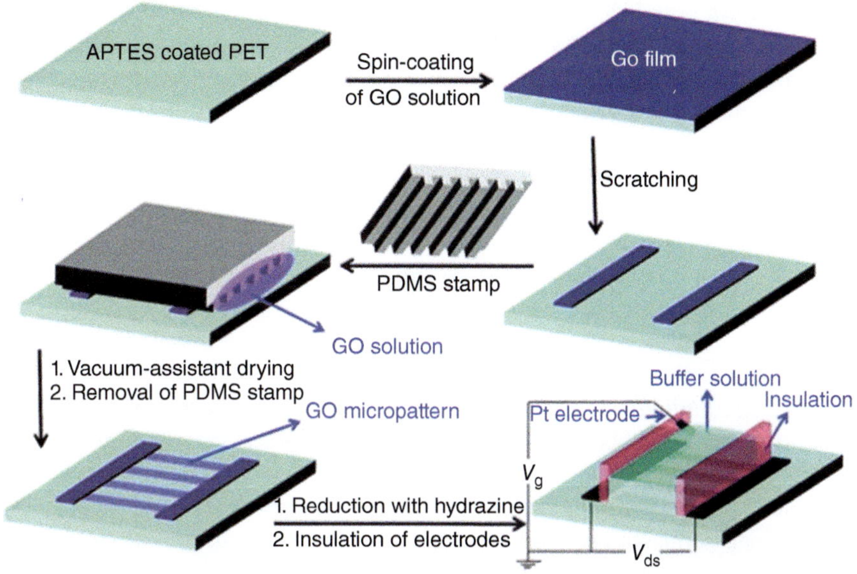

Figure 9.2 The schematic depiction of the preparation of a graphene-based thin film transistor. Source: Q. He et al., Transparent, Flexible, All-Reduced Graphene Oxide Thin Film Transistors, ACS Nano, 2011, 5(6), 5038–5044. © 2011, American Chemical Society.

sensor size seems to facilitate the detection of reaction products at a similar scale [86].

Ahn's group reported the functionalization of nano interdigitated electrode array (nIDA) fabricated onto cyclic olefin copolymer substrate for direct bio-affinity sensing (Figure 9.3) [87]. The protein immobilization was based on the self-assembly of alkanethiol groups on the electrode surface. Different concentrations of mouse monoclonal anti-rabbit immunoglobulin were used as the analyte for biomolecular

Figure 9.3 The schematic demonstration of the gold nIDA on polymer substrate with a polymer microfluidics chip. Source: Z. Zou et al., Functionalized nano interdigitated electrodes arrays on polymer with integrated microfluidics for direct bio-affinity sensing using impedimetric measurement, Sens. Actuators, A, 2007, 136, 518–526. © 2006, Elsevier.

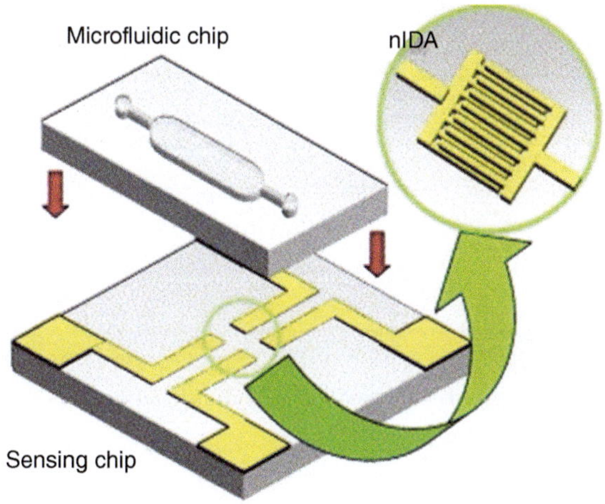

binding sensing tests. The protein was covalently immobilized on the alkanethiol self-assembled monolayers (SAMs) that recognize the amino groups of the protein. Finally, the variations of impedance to the electrode were measured for different protein concentrations. In this example, the scale and thickness of the electrode play an important role due to the possibility of being integrated in microfluidic platforms, as well as improving the sensitivity when the analyte is detected at the same scale as the detection platform.

These kinds of electrodes could also be modified with nanoparticles, as Lee et al. reported in 2011 [88]. Indium oxide and silica nanoparticles (SNPs) were selectively assembled onto multisite channel area of the resistors, using layer-by-layer self-assembly. Glucose oxidase was immobilized onto the SNPs layer, and glucose detection was performed by conductometry. A current dependence upon different glucose concentrations with a sensitivity that depends on the channel and resistors dimensions was obtained. The sensitivity for glucose detection was in the range between 4 and 12 nA mM^{-1}.

In general, modifying transducing platforms with nanoparticles allows generating really high-conductive surface area interfaces that enable the sensitive/catalytic detection of ionic, molecular, and biomolecular analytes. These highly sensitive sensors are easy to be integrated and become good candidates for LOCs.

9.3.1.2 Optical Detection Systems

The integration of photonics and microfluidics is a well-known strategy that provides enhanced sensing and simplification of microsystems. Various optofluidic detections such as those based on surface-enhanced Raman spectroscopy (SERS), fluorescence, and absorbance and related with the use of NMs modified detectors have been reported. SERS is a surface-sensitive technique that enhances Raman scattering by molecules adsorbed onto rough metal surfaces. For this reason, the use of NMs provides a wide range of possibilities for sensitive detection.

The use of gold nanowires (NWs) (diameter of 170 nm) synthesized by an electro crystallization method into a polycarbonate microchannel was reported [89]. Gold NWs produce large increases in a Raman cross-section. This kind of SERS-active dielectric polymer/NWs itself is a good candidate for in-microchip integration. An increased sensitivity was also observed by the detection of isonicotinic acid in perchloric acid due to the higher surface-to-volume ratio. Figure 9.4 shows the schematic depiction of the preparation of the microchannel network with embedded gold NW 3D arrays and SEM images of microchannel containing metallic NWs.

Different SERS spectra for a microfluidic separation chip with an integrated chromatographic column based on pillar arrays were evaluated [90]. Diffusive mixing of silver nanoparticles (AgNPs) solution with the eluent stream in a laminar flow to induce SERS signals due to their advantages as relatively easy preparation, manipulation, and possibility to have high SERS reproducibility has been reported. Interesting information about complex mixtures without complete resolution of chromatography bands was obtained. Analysis of the mixing kinetics effect on efficiency was also performed.

The combination of localized surface plasmon resonance (LSPR) and SERS has also been reported by Canpean and Astilean [91]. LSPR provides detection of molecular adsorption in both transmission and reflectivity modes, and SERS gives the

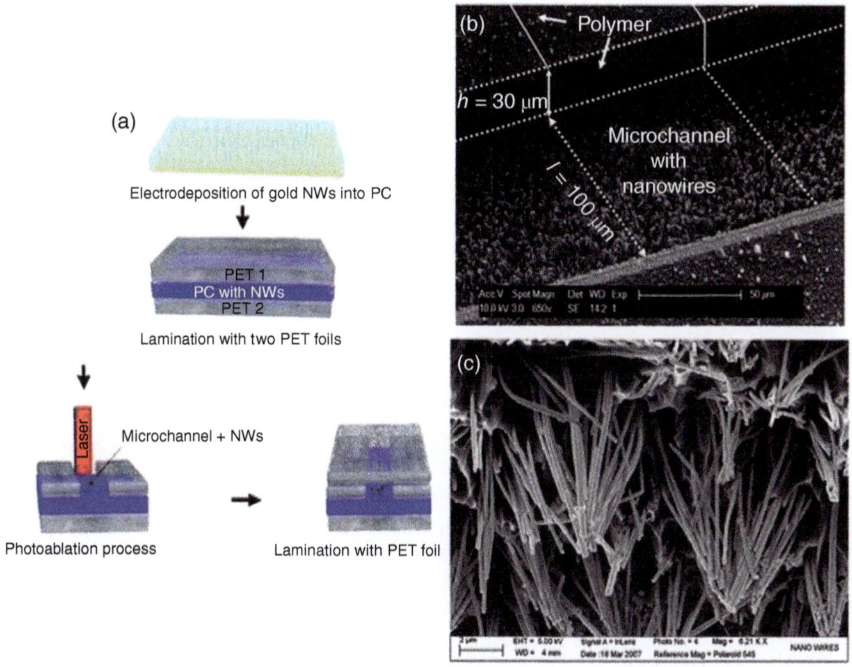

Figure 9.4 (a) The schematic depiction of the preparation of the microchannel network with embedded gold nanowires 3D platforms. (b) SEM image of microchannel containing metallic nanowires. (c) Field emission gun scanning electron microscopy (FEG-SEM) at higher magnification after selective plasma etching of polycarbonate. Source: J. Gamby et al., Polycarbonate microchannel network with carpet of Gold NanoWires as SERS-active device, Lab Chip, 2009, 9, 1806–1808. © 2009, Royal Society of Chemistry.

analysis of the adsorbed molecule and its structure and orientation on the metal surface. The optical and spectroscopy properties of regular arrays of subwavelength holes in a thin gold film by combining surface plasmon resonance (SPR) and SERS were demonstrated by the binding of the p-ATP monolayer, observing a shift in SPR. A change in SERS related to the orientation of the absorbed molecule, with a greater sensitivity of 300 nm/RIU (refractive index unit) and spectral shift of 16 nm, was also observed.

The glass surface of a glass–PDMS microfluidic channel was modified by immobilization of quantum dots (QDs) to develop a solid-phase assay for nucleic acid determination as schematically shown in Figure 9.5 [92]. QDs served as energy donors in fluorescence resonance energy transfer (FRET) for the transduction of nucleic acid hybridization. A selective detection of femtomole quantities of the target using 20% formamide was reported.

A novel alternative for fluorescence detection by using on-chip super localization and super-resolution imaging of single molecules and nanoparticles under different microscopy modes was reported by Luo et al. [93]. The detection was performed by using epi-fluorescence and differential interference contrast (DIC) modes at the same time. The performance was validated by precisely mapping micro necklaces made of fluorescent microtubules and gold nanoparticles (AuNPs) (40 nm of

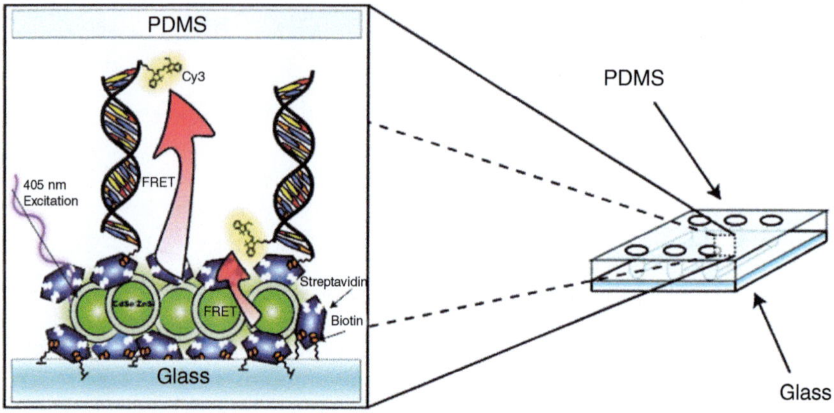

Figure 9.5 The schematic demonstration of the solid-phase approach for the transduction of nucleic acid hybridization using immobilized QDs and FRET on microfluidic platform. Source: A. J. Tavares et al., On-Chip Transduction of Nucleic Acid Hybridization Using Spatial Profiles of Immobilized Quantum Dots and Fluorescence Resonance Energy Transfer, Anal. Chem., 2012, 84(1), 312–319. © 2012, American Chemical Society.

diameter) and by demonstrating the activation and excitation cycles of single Alexa Fluor 647 dyes for direct stochastic optical. In-chip measurements of light absorbance have also been reported by Demming et al. They demonstrate that nanoparticles can be used to modify the index of reflection of the material, thus increasing the obtained signal. [94]. A PDMS matrix involving copper nanoparticles (CuNPs) and AgNPs for microfluidic devices was also performed. These nanoparticles were suspended in PDMS to obtain a permanent modification. The antiseptic behavior of the different nanoparticles was studied by observing the absorption of *Saccharomyces cerevisiae* culture cells.

AuNPs embedded in a polymeric waveguide for LSPR, capable of capturing minor variations in the refractive index of its microenvironment, have been demonstrated [95]. A linear behavior of refractive index changes of SU8 waveguide bound to AuNPs was also observed. Optical characterization should also be performed in microfluidic platforms, mainly by SERS detection. Qu et al. reported the development of disposable screen-printed SERS arrays by using AgNPs ink. This nanostructured surface was characterized by detecting rhodamine 6G with an LOD of 1.6×10^{-13} M [96].

The combination between quantum dots-labeled nerve growth factor (QDs-NGF), microfluidic chambers for neuronal culture, and pseudo total internal reflection (TIRF) microscopy provided a single molecule detection and analysis [97]. Chu's group developed a platform to detect retrograde axonal transport of NGF signals which is critical for the survival, differentiation, and maintenance of peripheral sympathetic and sensory neurons and basal forebrain cholinergic neurons. QD-NGF was used to track the movement of NGF in real time in compartmentalized culture of rat dorsal root ganglion (DRG) neurons. They obtained quantitative analysis at the single molecule level demonstrating that the majority of NGF containing endosomes had only a single NGF dimer. Electron microscopic analysis of axonal vesicles carrying QD-NGF confirmed this finding. Kuyper et al. proposed real-time nanoparticles synthesis (QDs, gold colloids, fluorescent and nonfluorescent beads) allowing the possibility to optimize reactions based on real-time readout of particle dimensions [98]. Confocal correlation spectroscopy (CCS) for on-chip sizing of both fluorescent and nonfluorescent nanoparticles was used. This technique can be an advantageous powerful tool to characterize colloids, polymer beads, and biological particles such as vesicles, viruses, and DNA molecules.

Optical detection is the most common detection method in microfluidic analyses. Fluorescence-based detection can be considered more usual for its facility to be integrated onto microfluidic devices (i.e. fluorescence excitation and detection components) as well as its ability to detect low-volume samples. SERS technique is also a potential tool to carry out qualitative and quantitative characterization of analytes; however, its integration with microfluidic devices still needs more research. In the same way, fluorescence and absorbance detection are moving toward information-improving techniques, such as fluorescence lifetime imaging microscopy or multicolor analyses. NMs play an increasingly important role as labeling and surface modification materials in fluorescence imaging to improve the sensitivity and the sensibility of existing methods.

9.3.1.3 Other Detection Techniques

Other detection methods, such as ECL and mass detection using nanoresonators, are also applied in LOC technologies. ECL is a kind of luminescence produced during electrochemical reactions in which a material emits light in response to an electric current or electrical field.

Nanoparticles could be used to improve the sensitivity of ECL. The MWCNT-CdS can react with H_2O_2, generating strong and stable ECL emission in neutral solution. By following the same procedure, a biosensor based on CdS–graphene has shown promising results for H_2O_2 sensing with an acceptable linear response from 5 μM up to 1 mM and a detection limit of 1.7 μM [99].

Nanomechanical biosensors can also be integrated into microfluidics. A variation of the mass onto the cantilever will produce a change in the length by material deformation. This deformation can be detected by different transducers.

Waggoner et al. were able to detect concentrations of 1–100 fM of PSA in serum, using a AuNP-based mass-labeling sandwich assay [100]. A reduced time of analysis with microfluidic assay was observed. Such a platform using nanoparticles allows multiplexed protein analysis with a very high sensitivity, offering many new alternatives of applications in clinical and biological studies for the early detection of diseases. The schematic demonstration of the prepared interfacing nanomechanical resonant sensor systems having microfluidics is shown in Figure 9.6. In the developed sensor systems, modular design brings together chips of resonators with gasket-like microfluidic channels and an encapsulating top with ports for bringing fluids in and out (Figure 9.6a). A number of arrays of resonators exist in a channel-like layout on each chip (Figure 9.6b), and all resonators are conjugated with antibodies to specifically recognize target biomarkers or proteins as well as subsequently bound labels, including secondary antibodies as well as nanoparticle mass labels (Figure 9.6c).

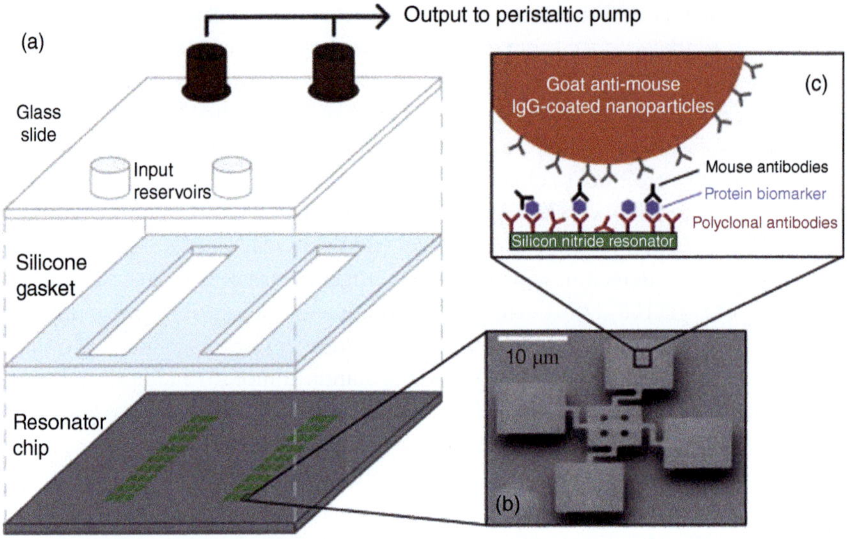

Figure 9.6 The schematic demonstration of the prepared interfacing nanomechanical resonant sensor systems having microfluidics. (a) The overall design, (b) individual resonators from on-chip arrays, and (c) a depiction of the specific surface chemistry used for biosensing. Source: P. S. Waggoner et al., Craighead, Microfluidic integration of nanomechanical resonators for protein analysis in serum, Sens. Actuators, B, 2010, 150, 550–555. © 2010, Elsevier.

NWs or nanotubes could also be employed as mass detectors. Hwang et al. have developed a nanocantilever-based sensor with an FET as a transducer. The signal from the gate changes when the cantilever is deflected due to specific binding of biomolecules on its surface [101].

9.3.2 Nanoparticles as Labels in Microfluidics

Biomarkers related to various diseases are present at very low concentrations. Therefore, enhancement strategies such as nanoparticle labeling, besides labeling with fluorescence dyes, enzymes, etc., have been demonstrated with interest for LOC diagnostic applications. Some examples related to nanoparticles used as labels in different LOC applications.

QDs are the most common nanoparticles used as labeling platforms in microfluidics. They could be attached to antibodies, aptamers, oligonucleotides, or peptides [102]. The increased and stable fluorescence properties are of special interest to develop ultrasensitive and robust LOCs. Furthermore, they also have interesting electrochemical/electrocatalytic properties that could be very useful for the integration of electrochemical detection for LOC applications with high sensitivity and selectivity.

A nanoparticle-labeling-based LOC platform involving different detection principles, including electrochemical detection by using QDs as biotracers/labels for tumor marker protein detection, is reported [103]. In a study conducted by Wang and colleagues [104], CdS QDs were synthesized based on a modification of the procedure with anti-carcinoembryonic antibody (αCEA) and detected by square wave anodic stripping voltammetry (SWASV). The electrode surface was modified with carbon nanoparticles enhancing the performance of the electrochemical reaction, while the QD labels allow the amplification of the signal, improving the sensitivity and selectivity of the detection system.

Another interesting application involving QDs as biomarkers is the development of nano-bio-microfluidic chips for multiplexed protein detection in serum and saliva [105]. Fluorescence detection of CEA, cancer antigen 125 (CA125), and Her-2/Neu (C-erbB-2) was achieved. The platform used to capture the antigen is a microporous agarose bead array where an antibody labeled with QDs is captured. An amplification of 30 times higher in comparison to the standard molecular fluorophores is observed. In addition, the use of different sized QDs with a wide range of wavelength emissions for multiple detection is also possible.

Zhang et al. reported the use of QDs as fluorescent labels for virus detection and genotyping [106]. The device could detect 1000 copies ml^{-1} of hepatitis B virus in clinical serum samples using in vitro transcribed RNA as the target molecules. Based on DNA hybridization with QDs labels, on-chip virus genotyping was also described.

Another interesting study for virus detection was performed by Liu et al. [107]. In their study, a microfluidic platform for immunofluorescent detection using QDs was employed. The LOD was improved from 360 (ELISA assay) to 22 ng ml^{-1} using these QD-based microfluidic system. An immunochromatographic strip based on nanoparticle probe was also demonstrated [108]. This platform allows the determination of cotinine in human serum. In this case, the QDs serve as signal vehicles for electrochemical readout with a linear range from 1 to 100 ng ml^{-1} of cotinine and a detection limit of 1 ng ml^{-1}. This sensor provides a rapid, accurate, and quantitative tool for cotinine detection and shows great promise for in-field and POC quantitative screening of tobacco smoke exposure.

AuNPs are also good label candidates. They are easy to produce in a variety of sizes and are also easy to functionalize. AuNP can be used for both fluorescence and electrochemical detections. PSA in the samples was detected by using AuNPs as labels at concentrations in the range between 40 pM and 40 fM [109]. The reported detection limit of this assay is 10 fM, with the entire assay, from sample injection to final data analysis, being completed in 80 minutes. Lei et al. also developed a biochip for protein detection involving AuNPs [110]. AuNPs indicate the presence of a protein. The concentration of protein was related to the resistance between interdigitated electrodes, achieving an LOD of 50 ng ml^{-1}.

Another detection principle applied in relation to labeling is the magnetoresistance, where the sensor is a magnetic field transducer that exhibits a linear change in electrical resistance under an external field [111, 112]. An integrated microfluidic system incorporating spin valve (SV) sensors has been fabricated, allowing cell detection and counting. This detection system can be used as a flow chip cytometer having the same efficiency for high concentration samples as the hemocytometer method, with approximately half the error (4.5% in comparison to 8.5%). Germano et al. reported a platform for biomolecular recognition detection using magnetic nanoparticles (MNPs, 250 nm), achieving a robust detection of up to 20 pM of DNA (20 mer single-stranded DNA oligonucleotides hybridization) [113]. The signal between complementary and non-complementary controls was possible to be discriminated.

Martins et al. developed a biochip portable platform for bioassays using a DNA–DNA oligonucleotides hybridization model [114], where 130 nm MNPs were used for their magnetic resonance contribution, allowing the detection in real time of 20 mer single-stranded DNA sequences. The same MNPs were used by Chen et al. for both capturing and labeling applications [115]. Their device enabled the determination of CEAs in human serum samples. Briefly, the immunoassay was performed by immobilizing the CEA onto MNPs modified with the first Ab-CEA. Then, second Ab-CEA labeled with PbS NPs was added. An LOD of 0.058 $\mu g\,l^{-1}$, suitable for CEA determination in human serum samples, was achieved.

NMs as labels give a wide range of possibilities to improve sensibility and selectivity. They have also been shown to be of particular interest in developing highly sensitive (bio)detection devices with multidetection capability and compatibility with LOCs. Various detection techniques ranging from optical, electrochemical, and mass changes can be applied in LOCs while using NM-based labeling technologies.

9.3.3 NMs for Process Improvement

NMs could also be employed to improve analytical processes. Electrophoresis separation, capture, and cell sorting between others are some of the reported examples improved by NMs. Concerning electrophoresis, a microchannel used for separation purposes could be modified with nanoparticles to control the electro-osmotic flow (EOF).

Wang and colleagues modified microfluidic channels by citrate-stabilized AuNPs after coating a layer of linear polyethylenimine (LPEI) [116]. This modification was successfully applied for the separation of pinephrine and dopamine that were difficult to separate from the baseline in native and hybrid PDMS microchannels. Measurements by in-channel amperometric detection with a single carbon fiber cylindrical electrode were performed. Linear responses from 25 to 600 μM for both compounds with detection limits of 2 μM for dopamine and 5 μM for epinephrine, respectively, were obtained.

The use of nanopore arrays is also extended for the capture and release of bacteria, as described by Zare's group [117]. In this work, conical nanopores on membranes inside a microfluidic device were developed, demonstrating the optimal capture of cyanobacteria (one bacteria per pore), in a defined orientation with viabilities as high as 100%. The platform can discriminate cyanobacteria from a mixed suspension of cyanobacteria and chlamydomonas with 90% selectivity.

For filtering and/or capturing processes, nanoporous elements have been integrated into microfluidics to manipulate bioparticles. Porous materials have been limited to membranes sandwiched between microchannel layers and polymer monoliths, but Chen et al. reported the use of micropatterned CNTs confined inside microfluidic channels for mechanically and/or capturing particles ranging over 3 orders of magnitude in size [118]. Nanoparticles with a size below the internanotube spacing can penetrate inside the forest and interact with the large surface area of CNTs. To validate this platform, specific biomolecular interactions were demonstrated by using cells, bacterial and viral-sized particles.

CNTs for LOC applications, especially their integration into microfluidic devices to improve super lubrifying, microchannels, filtering and separation processes, have also been studied by Choong et al. [75]. The use of nanoassembled graphene oxide appears as a novel capture alternative.

MNPs can also efficiently be used as a molecule transport controller to pre-concentrate the sample on the detector chamber in magnetoresistive sensors. Oscarsson's group reported a magnetic microchip for the controlled transport of an attomole level of proteins. The activity of proteins immobilized on the beads was also demonstrated by injecting antibodies into the channel. Parameters like the sticking forces between beads and substrate during transportation of proteins and the charge interaction in comparison with the hydrophobic forces were also optimized. The performance of the platform was tested by detecting a solution of fluorescein iso-thiocyanate (FITC)-labeled anti-HSA (human serum albumin), achieving an LOD of 0.02 mg ml^{-1} by fluorescence methods [119]. Another interesting application for process improvement is microfluidic propulsion by using magnetically actuated artificial cilia, which was carried out by Khaderi et al. [120]. In their study, artificial cilia was fabricated by using thin polymer films with embedded MNPs. These artificial cilia were fabricated using thin polymer films with embedded MNPs, and their deformation under different external magnetic fields and flows was studied.

9.4 Conclusions and Future Perspectives

NM-based LOC technology is shown to be an interesting alternative for a wide range of (bio)analytical device applications. The integration of NMs into LOC systems seems to bring additional advantages in relation to the system sensitivity and miniaturization, besides the versatility of the designs and application possibilities. NM application as detectors or detector modifiers is improving the performance of conventional LOCs. The use of CNTs and graphene is bringing significant advances in electrical or optical detections. NM applications as microchannel modifiers for separation enhancement are offering novel functionalization possibilities for improvements in sample pretreatment or separation processes.

Microfluidic platforms are also being used as a novel strategy for the production and characterization of NMs, opening new routes for encapsulation and homogenous shell–particle synthesis. Of notable progress are the synthesis of QDs and NWs using microfluidic devices. Very good reproducibility, reduction of the preparation time, and the reagents consumed are some of the advantages of in-chip NM preparation and applications. NMs are also involved in microanalytical process enhancement. NMs can be used for filtering, sorting, biomolecule capturing, and other actuations, with the possibility for various applications. The in-chip NM application spectra also include applications such as tissue nano-scaffolds and drug delivery systems. This is a challenging field for future nanodrug delivery platforms.

Finally, problems such as the application of complicated fabrication and measuring technologies are probably not suitable for mass production, and user-friendly technologies need to be designed to bring these devices to the end-users interested in microdevices.

References

1 Lindström, S. and Andersson-Svahn, H. (2011). Miniaturization of biological assays – overview on microwell devices for single-cell analyses. *Biochim. Biophys. Acta* 1810: 308–316.

2 Ríos, Á. and Zougagh, M. (2015). Modern qualitative analysis by miniaturized and microfluidic systems. *Trends Anal. Chem.* 69: 105–113.

3 Mitsudo, K., Kurimoto, Y., Yoshioka, K., and Suga, S. (2018). Miniaturization and combinatorial approach in organic electrochemistry. *Chem. Rev.* 118: 5985–5999.

4 Dixit, C.K., Kadimisetty, K., and Rusling, J. (2018). 3D-printed miniaturized fluidic tools in chemistry and biology. *Trends Anal. Chem.* 106: 37–52.

5 Liu, D., Wang, J., Wu, L. et al. (2020). Trends in miniaturized biosensors for point-of-care testing. *Trends Anal. Chem.* 122: 115701.

6 Katseli, V., Thomaidis, N., Economou, A., and Kokkinos, C. (2020). Miniature 3D-printed integrated electrochemical cell for trace voltammetric Hg(II) determination. *Sensors Actuators B Chem.* 308: 127715.

7 Lam, S.C., Coates, L.J., Hemida, M. et al. (2020). Miniature and fully portable gradient capillary liquid chromatograph. *Anal. Chim. Acta* 1101: 199–210.

8 Zhang, Y.-J., Chen, S., Yu, Y.-L., and Wang, J.-H. (2020). A miniaturized photoacoustic device with laptop readout for point-of-care testing of blood glucose. *Talanta* 209: 120527.

9 Kang, M., Zang, W., Dong, L. et al. (2020). On-site testing of multiple drugs of abuse in urine by a miniature dual-LIT mass spectrometer. *Anal. Chim. Acta* 1101: 74–80.

10 Gutiérrez-Serpa, A., Jiménez-Abizanda, A.I., Jiménez-Moreno, F. et al. (2020). Core-shell microparticles formed by the metal-organic framework CIM-80(Al) (Silica@CIM-80(Al)) as sorbent material in miniaturized dispersive solid-phase extraction. *Talanta* 211: 120723.

11 Tavakoli, H., Zhou, W., Ma, L. et al. (2019). Recent advances in microfluidic platforms for single-cell analysis in cancer biology, diagnosis and therapy. *Trends Anal. Chem.* 117: 13–26.

12 Sengupta, J. and Hussain, C.M. (2019). Graphene and its derivatives for analytical lab on chip platforms. *Trends Anal. Chem.* 114: 326–337.

13 Ai, Y., Zhang, F., Wang, C. et al. (2019). Recent progress in lab-on-a-chip for pharmaceutical analysis and pharmacological/toxicological test. *Trends Anal. Chem.* 117: 215–230.

14 Büyüktiryaki, S., Sümbelli, Y., Keçili, R., and Hussain, C.M. (2019). Lab-on-chip platforms for environmental analysis. In: *Encyclopedia of Analytical Science*, 3e

(eds. P. Worsfold, C. Poole, A. Townshend and M. Miró), 267–273. Academic Press.

15 Keçili, R., Büyüktiryaki, S., and Hussain, C.M. (2019). Advancement in bioanalytical science through nanotechnology: past, present and future. *Trends Anal. Chem.* 110: 259–276.

16 Hussain, C.M. and Keçili, R. (2020). *Modern Environmental Analysis Techniques for Pollutants*. The Netherlands: Elsevier.

17 Keçili, R., Büyüktiryaki, S., and Hussain, C.M. (2020). Micro total analysis systems with nanomaterials, Chapter 8. In: *Handbook of Nanomaterials in Analytical Chemistry* (ed. C.M. Hussain), 185–198. Elsevier.

18 Pravinraj, T. and Patrikar, R.M. (2019). Design, fabrication, and modeling of a low-cost droplet on demand device for lab-on-chip applications. *Microelectron. Eng.* 216: 111030.

19 Sengupta, P., Khanra, K., Chowdhury, A.R., and Datta, P. (2019). Lab-on-a-chip sensing devices for biomedical applications, Chapter 4. In: *Woodhead Publishing Series in Electronic and Optical Materials, Bioelectronics and Medical Devices* (eds. K. Pal, H.-B. Kraatz, A. Khasnobish, et al.), 47–95. Woodhead Publishing.

20 Francesko, A., Cardoso, V.F., and Lanceros-Méndez, S. (2019). Lab-on-a-chip technology and microfluidics, Chapter 1. In: *Micro and Nano Technologies, Microfluidics for Pharmaceutical Applications* (eds. H.A. Santos, D. Liu and H. Zhang), 3–36. William Andrew Publishing.

21 Hussain, C.M. and Kharisov, B. (2017). *Advanced Environmental Analysis-Application of Nanomaterials*. The Royal Society of Chemistry.

22 Hussain, C.M. (2017). Magnetic nanomaterials for environmental analysis. In: *Advanced Environmental Analysis-Application of Nanomaterials* (eds. C.M. Hussain and B. Kharisov). The Royal Society of Chemistry.

23 Hussain, C.M. (2018). *Nanomaterials in Chromatography: Current Trends in Chromatographic Research Technology and Techniques*. Elsevier.

24 Hussain, C.M. (2019). *Handbook of Nanomaterials in Analytical Chemistry: Modern Trends in Analysis*. Elsevier.

25 Belder, D. (2006). Integrating chemical synthesis and analysis on a chip. *Anal. Bioanal. Chem.* 385: 416–418.

26 Heiland, J.J., Warias, R., Lotter, C. et al. (2017). On-chip integration of organic synthesis and HPLC/MS analysis for monitoring stereoselective transformations at the micro-scale. *Lab Chip* 17: 76–81.

27 Dimcheva, N. (2020). Nanostructures of noble metals as functional materials in biosensors. *Curr. Opin. Electrochem.* 19: 35–41.

28 Shao, B. and Xiao, Z. (2020). Recent achievements in exosomal biomarkers detection by nanomaterials-based optical biosensors – a review. *Anal. Chim. Acta* https://doi.org/10.1016/j.aca.2020.02.041.

29 Bao, M. and Wang, W. (1996). Future of microelectromechanical systems (MEMS). *Sensor. Actuat. A-Phys.* 56: 135–141.

30 Terry, S.C., Hermanand, J.H., and Angell, J.B. (1979). A gas chromatographic air analyzer fabricated on a silicon wafer. *IEEE Trans. Electron Devices* 26 (12): 1880–1886.

31 Manz, A., Graber, N., and Widmer, H.M. (1990). Miniaturized total chemical analysis systems: a novel concept for chemical sensing. *Sensor. Actuat. B-Chem.* 1 (1–6): 244–248.

32 Seiler, K., Harrison, D.J., and Manz, A. (1993). Planar glass chips for capillary electrophoresis: repetitive sample injection, quantitation, and separation efficiency. *Anal. Chem.* 15: 1481–1488.

33 Woolley, A.T. and Mathies, R.A. (1994). Ultra-high-speed DNA fragment separations using microfabricated capillary array electrophoresis chips. *Proc. Natl. Acad. Sci. U. S. A.* 91: 11348–11352.

34 Adamski, K., Kubicki, W., and Walczak, R. (2016). 3D printed electrophoretic lab-on-chip for DNA separation. *Procedia Eng.* 168: 1454–1457.

35 Figeys, D., Ning, Y., and Aebersold, R. (1997). A microfabricated device for rapid protein identification by microelectrospray ion trap mass spectrometry. *Anal. Chem.* 69 (16): 3153–3160.

36 Weigl, B.H. and Yager, P. (1997). Silicon-microfabricated diffusion-based optical chemical sensor. *Sensor Actuat. B Chem.* 38–39: 452–457.

37 Xie, P., Cao, X., Lin, Z., and Javanmard, M. (2017). Top-down fabrication meets bottom-up synthesis for nanoelectronic barcoding of microparticles. *Lab Chip* 17 (11): 1939–1947.

38 Harrison, D.J., Fluri, K., Chiem, N. et al. (1996). Micromachinng chemical and biochemical analysis and reaction systems on glass substrates. *Sensors Actuators B Chem.* 33 (1–3): 105–109.

39 Daniel, J.H., Iqbal, S., Millingston, R.B. et al. (1998). Silicon microchambers for DNA amplification. *Sensor. Actuat. A-Phys.* 71: 81–88.

40 Pang, X., Lewis, A.C., and Ródenas-García, M. (2013). Microfluidic lab-on-a-chip derivatization for gaseous carbonyl analysis. *J. Chromatogr. A* 1296: 93–103.

41 Hür, D., Say, M.G., Diltemiz, S. et al. (2018). 3D micropatterned all-flexible microfluidic platform for microwave-assisted flow organic synthesis. *ChemPlusChem* 83 (1): 42–46.

42 Becker, H. (2009). It's the economy. *Lab Chip* 9: 2759–2762.

43 Tantra, R. and van Heeren, H. (2013). Product qualification: a barrier to point-of-care microfluidic-based diagnostics? *Lab Chip* 13: 2199–2201.

44 Temiz, Y., Lovchik, R.D., Kaigala, G.V., and Delamarche, E. (2015). Lab-on-a-chip devices: How to close and plug the lab? *Microelectron. Eng.* 132: 156–175.

45 Anderson, J.R., Chiu, D.T., Wu, H. et al. (2000). Fabrication of microfluidic systems in poly (dimethylsiloxane). *Electrophoresis* 21: 27–40.

46 Thorsen, T., Maerkl, S.J., and Quake, S.R. (2002). Microfluidic large-scale integration. *Science* 298: 580–584.

47 Whitesides, G.M. (2006). The origins and the future of microfluidics. *Nature* 442: 368–373.

48 Tang, M., Wang, G., Kong, S.-K., and Ho, H.-P. (2016). A review of biomedical centrifugal microfluidic platforms. *Micromachines* 7: 26.

49 Beebe, D.J., Mensing, G.A., and Walker, G.M. (2002). Physics and applications of microfluidics in biology. *Annu. Rev. Biomed. Eng.* 4: 261–286.

50 Fair, R.B. (2007). Digital microfluidics: is a true lab-on-a-chip possible? *Microfluid. Nanofluid.* 3: 245–281.

51 Dittrich, P.S. and Manz, A. (2006). Lab-on-a-chip: microfluidics in drug discovery. *Nat. Rev. Drug Discov.* 5: 210–218.

52 Günther, A. and Jensen, K.F. (2006). Multiphase microfluidics: from flow characteristics to chemical and materials synthesis. *Lab Chip* 6: 1487–1503.

53 Jung, W., Han, J., Choi, J.-W., and Ahn, C.H. (2015). Point-of-care testing (POCT) diagnostic systems using microfluidic lab-on-a-chip technologies. *Microelectron. Eng.* 132: 46–57.

54 Tian, T., Wei, X., Jia, S. et al. (2016). Integration of target responsive hydrogel with cascaded enzymatic reactions and microfluidic paper-based analytic devices (μPADs) for point-of-care testing (POCT). *Biosens. Bioelectron.* 77: 537–542.

55 Sista, R., Hua, Z., Thwar, P. et al. (2008). Development of a digital microfluidic platform for point of care testing. *Lab Chip* 8: 2091–2104.

56 Moon, S., Keles, H.O., Ozcan, A. et al. (2009). Integrating microfluidics and lensless imaging for point-of-care testing. *Biosens. Bioelectron.* 24: 3208–3214.

57 Wang, S., Ge, L., Yan, M. et al. (2013). 3D microfluidic origami electrochemiluminescence immunodevice for sensitive point-of-care testing of carcinoma antigen 125. *Sens. Actuat. B-Chem.* 176: 1–8.

58 Yamada, K., Shibata, H., Suzuki, K., and Citterio, D. (2017). Toward practical application of paper-based microfluidics for medical diagnostics: state-of-the-art and challenges. *Lab Chip* 17: 1206–1249.

59 Carrilho, E., Martinez, A.W., and Whitesides, G.M. (2009). Understanding wax printing: a simple micropatterning process for paper-based microfluidics. *Anal. Chem.* 81: 7091–7095.

60 Li, X., Ballerini, D.R., and Shen, W. (2012). A perspective on paper-based microfluidics: current status and future trends. *Biomicrofluidics* 6: 11301–1130113.

61 Mao, X. and Huang, T.J. (2012). Microfluidic diagnostics for the developing world. *Lab Chip* 12: 1412–1416.

62 Haeberle, S. and Zengerle, R. (2007). Microfluidic platforms for lab-on-a-chip applications. *Lab Chip* 7: 1094–1110.

63 Gorkin, R., Park, J., Siegrist, J. et al. (2010). Centrifugal microfluidics for biomedical applications. *Lab Chip* 10: 1758–1773.

64 Burger, R., Kirby, D., Glynn, M. et al. (2012). Centrifugal microfluidics for cell analysis. *Curr. Opin. Chem. Biol.* 16: 409–414.

65 Strohmeier, O., Keller, M., Schwemmer, F. et al. (2015). Centrifugal microfluidic platforms: advanced unit operations and applications. *Chem. Soc. Rev.* 44: 6187–6229.

66 Ducrée, J., Haeberle, S., Lutz, S. et al. (2007). The centrifugal microfluidic bio-disk platform. *J. Micromech. Microeng.* 17: S103.

67 Teh, S.Y., Lin, R., Hung, L.H., and Lee, A.P. (2008). Droplet microfluidics. *Lab Chip* 8: 198–220.

68 Ottesen, E.A., Hong, J.W., Quake, S.R., and Leadbetter, J.R. (2006). Microfluidic digital PCR enables multigene analysis of individual environmental bacteria. *Science* 314: 1464–1467.

69 Hindson, B.J., Ness, K.D., Masquelier, D.A. et al. (2011). High-throughput droplet digital pcr system for absolute quantitation of DNA copy number. *Anal. Chem.* 83: 8604–8610.

70 Seemann, R., Brinkmann, M., Pfohl, T., and Herminghaus, S. (2012). Droplet based microfluidics. *Rep. Prog. Phys.* 75: 016601.

71 Barbulovic-Nad, I., Yang, H., Park, P.S., and Wheeler, A.R. (2008). Digital microfluidics for cell-based assays. *Lab Chip* 8: 519–526.

72 Abdelgawad, M. and Wheeler, A.R. (2009). The digital revolution: a new paradigm for microfluidics. *Adv. Mater.* 21: 920–925.

73 Yeo, L.Y. and Friend, J.R. (2014). Surface acoustic wave microfluidics. *Annu. Rev. Fluid Mech.* 46: 379–406.

74 Ding, X., Li, P., Lin, S.C. et al. (2013). Surface acoustic wave microfluidics. *Lab Chip* 13: 3626–3649.

75 Choong, C.L., Milne, W.I., and Teo, K.B.K. (2008). Review: carbon nanotube for microfluidic lab-on-a-chip application Int. *J. Mater. Form.* 1: 117–125.

76 Crevillen, A.G., Pumera, M., Gonzalez, M.C., and Escarpa, A. (2009). Towards lab-on-a-chip approaches in real analytical domains based on microfluidic chips/electrochemical multi-walled carbon nanotube platforms. *Lab Chip* 9: 346–353.

77 Maehashi, K. and Matsumoto, K. (2009). Label-free electrical detection using carbon nanotube-based biosensors. *Sensors* 9: 5368–5378.

78 Tey, J.N., Wijaya, I.P.M., Wei, J. et al. (2010). Nanotubes-/nanowires-based, microfluidic-integrated transistors for detecting biomolecules. *Microfluid. Nanofluid.* 9: 1185–1214.

79 Okuno, J., Maehashi, K., Kerman, K. et al. (2007). Label-free immunosensor for prostate-specific antigen based on single-walled carbon nanotube array-modified microelectrodes. *Biosens. Bioelectron.* 22: 2377–2381.

80 Javey, A., Guo, J., Farmer, D.B. et al. (2004). Self-aligned ballistic molecular transistors and electrically parallel nanotube arrays. *Nano Lett.* 4 (3): 1319–1322.

81 Tey, J.N., Wijaya, I.P.M., Wang, Z. et al. (2009). Laminated, microfluidic-integrated carbon nanotube based biosensors. *Appl. Phys. Lett.* 94: 073704-1–073704-3.

82 Vlandas, A., Kurkina, T., Ahmad, A. et al. (2010). Enzyme-free sugar sensing in microfluidic channels with an affinity-based single-wall carbon nanotube sensor. *Anal. Chem.* 82: 6090–6097.

83 Hummers, W.S. and Offeman, R.E. (1958). Preparation of graphitic oxide. *J. Am. Chem. Soc.* 80 (6): 1339–1339.

84 Chua, C.K., Ambrosi, A., and Pumera, M. (2011). Graphene based nanomaterials as electrochemical detectors in lab-on-a-chip devices. *Electrochem. Commun.* 13: 517–519.

85 He, Q., Wu, S., Gao, S. et al. (2011). Transparent, flexible, all-reduced graphene oxide thin film transistors. *ACS Nano* 5 (6): 5038–5044.

86 Plowman, B.J., Bhargava, S.K., and O'Mullane, A.P. (2011). Electrochemical fabrication of metallic nanostructured electrodes for electroanalytical applications. *Analyst* 136: 5107–5119.

87 Zou, Z., Kai, J., Rust, M. et al. (2007). Functionalized nano interdigitated electrodes arrays on polymer with integrated microfluidics for direct bio-affinity sensing using impedimetric measurement. *Sens. Actuators, A* 136: 518–526.

88 Lee, D., Ondrake, J., and Cui, T. (2011). A conductometric indium oxide semiconducting nanoparticle enzymatic biosensor array. *Sensors* 11: 9300–9312.

89 Gamby, J., Rudolf, A., Abid, M. et al. (2009). Polycarbonate microchannel network with carpet of Gold NanoWires as SERS-active device. *Lab Chip* 9: 1806–1808.

90 Taylor, L.C., Kirchner, T.B., Lavrik, N.V., and Sepaniak, M.J. (2012). Surface enhanced Raman spectroscopy for microfluidic pillar arrayed separation chips. *Analyst* 137: 1005–1013.

91 Canpean, V. and Astilean, S. (2009). Multifunctional plasmonic sensors on low-cost subwavelength metallic nanoholes arrays. *Lab Chip* 9: 3574–3579.

92 Tavares, A.J., Noor, M.O., Vannoy, C.H. et al. (2012). On-chip transduction of nucleic acid hybridization using spatial profiles of immobilized quantum dots and fluorescence resonance energy transfer. *Anal. Chem.* 84 (1): 312–319.

93 Luo, Y., Sun, W., Liu, C. et al. (2011). Superlocalization of single molecules and nanoparticles in high-fidelity optical imaging microfluidic devices. *Anal. Chem.* 83: 5073–5077.

94 Demming, S., Hahn, A., Edlich, A. et al. (2010). Soft lithographic partial integration of surface-active nanoparticles in a PDMS matrix for microfluidic biodevices. *Phys. Status Solidi A* 207: 898–903.

95 Prabhakar, A. and Mukherji, S. (2010). A novel C-shaped, gold nanoparticle coated, embedded polymer waveguide for localized surface plasmon resonance based detection. *Lab Chip* 10: 3422–3425.

96 Qu, L.L., Li, D.W., Xue, J.Q. et al. (2012). Batch fabrication of disposable screen printed SERS arrays. *Lab Chip* 12: 876–881.

97 Cui, B., Wu, C., Chen, L. et al. (2007). One at a time, live tracking of NGF axonal transport using quantum dots. *Proc. Natl. Acad. Sci. U. S. A.* 104 (34): 13666–13671.

98 Kuyper, C.L., Budzinski, K.L., Lorenz, R.M., and Chiu, D.T. (2006). Real-time sizing of nanoparticles in microfluidic channels using confocal correlation spectroscopy. *J. Am. Chem. Soc.* 128: 730–731.

99 Wang, K., Liu, Q., Wu, X.Y. et al. (2010). Graphene enhanced electrochemiluminescence of CdS nanocrystal for H_2O_2 sensing. *Talanta* 82: 372–376.

100 Waggoner, P.S., Tan, C.P., and Craighead, H.G. (2010). Microfluidic integration of nanomechanical resonators for protein analysis in serum. *Sensors Actuators B Chem.* 150: 550–555.

101 Hwang, K.S., Lee, S.M., Kim, S.K. et al. (2009). Micro- and nanocantilever devices and systems for biomolecule detection. *Annu. Rev. Anal. Chem.* 2: 77–98.

102 Wagner, M.K., Li, F., Li, J. et al. (2010). Use of quantum dots in the development of assays for cancer biomarkers. *Anal. Bioanal. Chem.* 397: 3213–3224.

103 Ho, J.A., Lin, Y.C., Wang, L.S. et al. (2009). Carbon nanoparticle-enhanced immunoelectrochemical detection for protein tumor marker with cadmium sulfide biotracers. *Anal. Chem.* 81: 1340–1346.

104 Wang, S., Mamedova, N., Kotov, N.A. et al. (2002). Antigen/antibody immunocomplex from CdTe nanoparticle bioconjugates. *Nano Lett.* 2: 817–822.

105 Jokerst, J.V., Raamanathan, A., Christodoulides, N. et al. (2009). Nano-bio-chips for high performance multiplexed protein detection: determinations of cancer biomarkers in serum and saliva using quantum dot bioconjugate labels. *Biosens. Bioelectron.* 24: 3622–3629.

106 Zhang, H., Xu, T., Li, C.W., and Yang, M. (2010). A microfluidic device with microbead array for sensitive virus detection and genotyping using quantum dots as fluorescence labels. *Biosens. Bioelectron.* 25: 2402–2407.

107 Liu, W.T., Zhu, L., Qin, Q.W. et al. (2005). Microfluidic device as a new platform for immunofluorescent detection of viruses. *Lab Chip* 5: 1327–1330.

108 Nian, H., Wang, J., Wu, H. et al. (2012). Electrochemical immunoassay of cotinine in serum based on nanoparticle probe and immunochromatographic strip. *Anal. Chim. Acta* 713: 50–55.

109 Goluch, E.D., Stoeva, S.I., Lee, J.-S. et al. (2009). A microfluidic detection system based upon a surface immobilized biobarcode assay. *Biosens. Bioelectron.* 24 (8): 2397–2403.

110 Lei, K.F. (2011). Quantitative electrical detection of immobilized protein using gold nanoparticles and gold enhancement on a biochip. *Meas. Sci. Technol.* 22 (105802): 1–7.

111 Zhi, X., Liu, Q., Zhang, X. et al. (2012). Quick genotyping detection of HBV by giant magnetoresistive biochip combined with PCR and line probe assay. *Lab Chip* 12: 741.

112 Freitas, P.P., Ferreira, H.A., Graham, D.L. et al. (2004). Magnetoelectronics, Chapter 7. In: *Magnetoresistive DNA Chips* (ed. M. Johnson), 331–374. Oxford, UK: Elsevier.

113 Germano, J., Martins, V.C., Cardoso, F.A. et al. (2009). Autonomous magnetic detection platform for biosensing. *Sensors* 9: 4119–4137.

114 Martins, V.C., Germano, J., Cardoso, F. et al. (2010). *J. Magn. Magn. Mater.* 322: 1655–1663.

115 Chen, B., Hu, B., Jiang, P. et al. (2011). Nanoparticle labelling-based magnetic immunoassay on chip combined with electrothermal vaporization – inductively coupled plasma mass spectrometry for the determination of carcinoembryonic antigen in human serum. *Analyst* 136: 3934–3942.

116 Wang, A.J., Xu, J.J., Zhang, Q., and Chen, H.Y. (2006). The use of poly(dimethylsiloxane) surface modification with gold nanoparticles for the microchip electrophoresis. *Talanta* 69 (1): 210–215.

117 Guo, P., Hall, E.W., Schirhagl, R. et al. (2012). Microfluidic capture and release of bacteria in a conical nanopore array. *Lab Chip* 12 (3): 558–561.
118 Chen, G.D., Fachin, F., Fernandez-Suarez, M. et al. (2011). Nanoporous elements in microfluidics for multiscale manipulation of bioparticles. *Small* 7: 1061–1067.
119 Johansson, L., Gunnarsson, K., Bijelovic, S. et al. (2010). A magnetic microchip for controlled transport of attomole levels of proteins. *Lab Chip* 10: 654–661.
120 Khaderi, S.N., Craus, C.B., Hussong, J. et al. (2011). Magnetically-actuated artificial cilia for microfluidic propulsion. *Lab Chip* 11: 2002–2010.

10

Toxicity and Risk Assessment of Nanomaterials

10.1 Introduction

With the rapid advancements in nanotechnology, the design and production of engineered nanomaterials (NMs) as well as incidental NMs dramatically increase. Despite the excellent properties and many great applications of the NMs, there are still some questions, challenges, drawbacks, and concerns about the impacts of these nanomaterials on human health and environment [1–7]. Some of the potential concerns of NMs are their toxicity and safety. Most of the NMs will end up in the environment (air, water, and soil). Because many of the NMs are not biodegradable, they may remain in the environment for a long time. Biological organisms such as microorganisms, plants, and animals could take up NMs in the ecosystems and bring nanomaterials into the food chain. For example, nanoparticles (NPs) can easily enter human body via respiratory system. Once NPs enter the bloodstream after passing through the respiratory system, they can accumulate in the certain regions of the different organs [8]. Therefore, the negative impacts due to the NMs on human health should be carefully evaluated. For this purpose, a new field called "nanoecotoxicology" is gaining public attention [9].

Furthermore, future applications of engineered NMs are highly promising. Therefore, great efforts and evaluations on the safety of NMs on the environment and animal health will continue to be a research priority for the advances in nanotechnology. Consequently, sophisticated statistical understanding and methodologies are carefully needed in the design of toxicity studies and in analysis of toxicity data. Prior to the use of NMs especially for in vivo biological applications, various important physiological factors (i.e. distribution, absorption, and toxicity) of these NMs should be considered and carefully evaluated according to the international regulatory guidelines, which are still under development [10].

This chapter provides an overview of toxicity and risk assessment of NMs. The chapter starts with a brief description of NMs and their toxic impacts on human health and environment. In Section 10.2, hazard assessment of NMs is described. Then, toxicity mechanism of NMs is also evaluated. In Section 10.4, the traditional risk assessment paradigm is presented. Finally, various strategies for improving specific risk assessment are discussed.

Sample Preparation with Nanomaterials: Next Generation Techniques and Applications,
First Edition. Chaudhery Mustansar Hussain, Rüstem Keçili, and Chaudhery Ghazanfar Hussain.

10.2 Hazard Assessment of Nanomaterials

NMs may gain access to the body via different routes including through inhalation, ingestion, or dermal exposure. Once inside the body, NMs may interact with cells and tissues, including organs distal from the portal of entry, potentially leading to adverse effects on multiple organs [11]. In Sections 10.2.1–10.2.4, the impact of three main classes of NMs, i.e. carbon-based, metal, and metal oxide NMs on key organ systems including the pulmonary system, cardiovascular system, skin, gastrointestinal system, immune system, and central nervous system is presented. The possible carcinogenicity of NMs is also evaluated.

10.2.1 Dermal Toxicity of Nanomaterials

Dermal toxicity of NPs involves the penetration of NPs into the skin and the possible systemic impacts resulting from such exposure and the local impact of sensitization, irritation, or corrosion of particles on the skin surface. Skin absorption or dermal penetration of chemicals is a process of diffusion from the outer surface of the skin, through the skin, and finally into systemic circulation. In this regard, methods that utilize the entire (intact) skin must be used. The skin consists of three major layers: epidermis, dermis, and subcutis. The outermost layer of the epidermis consists of dead flattened corneocytes, and the intercellular space consists of a multilayer of lipids, making up the stratum corneum, which is the first barrier to pass for any chemical [12]. NMs have been effectively shown to penetrate through the skin in various routes, which are (i) through hair follicles, (ii) intracellularly through corneocytes, or (iii) intercellularly around the corneocytes [13, 14].

ZnO NPs are extensively used in the production of sunscreens. Thus, many studies investigated the penetration of NPs into the skin and absorption of NPs through the skin in human volunteers. These studies have shown that Zn can be absorbed through the skin after application of sunscreen for five consecutive days. However, it could not be concluded whether Zn was absorbed as NPs or as soluble Zn [15, 16]. In another research [17], Leite-Silva et al. showed that ZnO NPs were able to penetrate human skin and reach the viable epidermis, but most particles were located in the stratum corneum and in skin furrows.

Metal NPs (i.e. TiO_2, ZnO, Ag, and Au NPs) are extensively used in cosmetics, e.g. as UV filters and to achieve antibacterial/anti-inflammatory properties, accelerate wound healing, increase collagen production, etc. [18]. At the same time, studies on cultured skin cells often indicate toxic properties for these NPs. For TiO_2 NPs, one concern is their photocatalytic activity leading to reactive oxygen species (ROS) production. The EC Scientific Committee on Consumer Safety (SCCS) concludes in a recent report that it is possible that trace amount of NPs remains embedded in stratum corneum, in hair follicles, and/or sweat glands, over several days after skin application of a product containing TiO_2 NPs. If the NPs have photocatalytic activity, there is a possibility that they may generate ROS upon exposure to sunlight. Furthermore, such ROS production in a close proximity of living cells raises a concern over the possibility of harmful effects. Therefore, SCCS does not recommend the use of NMs that have high photocatalytic activity in dermal formulations (often anatase NPs). Cellular studies have reported on generation of ROS, mitochondrial DNA damage, and genotoxic effects in human keratinocytes following exposure of TiO_2 NPs [18]. It should be noted, however, that for TiO_2 such effects are often observed at relatively high doses (over 100 μg ml^{-1}), although phototoxicity may be observed also at lower concentration. In another interesting work [19], Horie et al. showed that rutile NPs used in sunscreen did not exhibit phototoxic activity following exposure of HaCat cells. Despite the strong phototoxic activity of anatase NPs in the cell cultures, no phototoxicity using a 3D skin model was observed. ZnO NPs has shown a wide range of toxic effects in HaCaT cells including impairment of the mitochondrial function, leakage of lactate dehydrogenase (LDH), ROS generation, membrane lipid peroxidation (LPO), genotoxicity, and apoptosis.

The most well-studied carbon-based NM concerning potential effects on skin cells are the fullerenes, carbon-based, nano-sized hollow structures also known as "buckyballs." Fullerenes have been reported to possess exceptional antioxidant properties, which has made them an attractive ingredient in many skin care products [20]. Recently, various effects of fullerenes on skin cells were summarized, and it was reported that several in vitro and some ex vivo and in vivo studies have shown protective effects of fullerenes against UV-induced skin damages [20]. On the other hand, photocytotoxicity has also been reported [21]. A study carried out by Sayes et al. [22] reported an LC50 of C60 of 20 ppb, i.e. 0.02 μg/ml, a concentration that must be considered to be very low. In contrast, Monterio-Riviere et al. reported only minor cytotoxicity of C60 in human embryonic kidney (HEK) cells [23]. However, it is important to also consider the dispersion protocol that is used since this may affect the toxicity. One study investigated toxicity of a high dose (100 μg/ml) of multiwalled carbon nanotubes (CNTs) in keratinocytes and reconstructed human skin. The results showed that water-suspended, non-sonicated CNTs caused only minor effects in keratinocytes, while the addition of hydroxypropylcellulose or Pluronic F108, which improved the dispersion, masked the toxic effects of sonicated CNTs [24].

10.2.2 Inhalational Toxicity of Nanomaterials

Nanotechnology paved the way for numerous technological advances for a variety of industrial techniques. This overwhelming exploration of nanotechnology in various fields led to the intervention of toxicologists to assess the ecological impact of these NMs. Among the various routes of exposure, inhalation has a special relevance owing to the fact that the presence of environmental pollutants may achieve direct access to the pulmonary intima causing respiratory hazards. The available literature suggests significant pulmonary toxicology concerns associated with the use of various NPs of organic, inorganic, and metal origin. The rise in respiratory disorders among the urban population is one of the proofs of the inhalational toxicology of suspended particulate matter. Owing to their high surface area and ultrafine size, inhaled NPs have a higher probability of deeply sitting inside the lungs compared with their conventional counterpart. This may lead to initiation of inflammatory responses in acute phase and may lead to fibrosis upon chronic exposure.

On the other hand, lungs are the primary target organ of inhaled NMs, and various types of NMs have shown the propensity for pulmonary toxicity. Moreover, recent reports suggest a significant extra pulmonary impact of inhaled NPs like CNTs. Among various metal NPs, AgNPs hold a significant place with multiple applications in the biomedical and industrial fields. They possess excellent antimicrobial properties. Various inhalational toxicology studies indicate the potential health hazards associated with the use of AgNPs. The results of a 28-day inhalational toxicity study of AgNPs indicated nonsignificant alterations according to the American Conference of Governmental Industrial Hygienists' silver dust limit (100 $\mu g/m^3$) indicating the relative safety of AgNPs [25].

Sung et al. carried out a subchronic inhalational toxicity study of AgNPs (18–19 nm). Inhalational exposure of animals to AgNPs for 13 weeks was carried out in a whole-body inhalation chamber. The NPs were found to elicit dose-dependent toxicity in both genders. The target organs for the toxicity were found to be the liver and lungs causing chronic alveolar inflammation, small granulomatous lesions, and inflammatory cells infiltrating the lungs. However, no significant alterations were observed at the maximal permissible silver dust limit (100 $\mu g/m^3$) [26].

Silica and mesoporous silica NPs are some of the most widely used NPs in chromatographic techniques [27–29]. Compared to the conventional silica NPs, mesoporous silica offers high surface area, which is higher than 900 m^2/g, and large pore volume, which is higher than 0.9 cm^3/g due to its novel honeycomb-like porous structure with hundreds of empty channels (mesopores). Various toxicological studies have been carried out with these unique NPs and indicate inflammation and oxidative stress triggering properties of SiNPs. In one of the studies, it was found to cause acute inflammatory response as evident from the increased neutrophil infiltration. Moreover, immunostaining of bronchoalveolar lavage fluid for Nrf_2 indicated higher nuclear colocalization showing acute inflammatory response [30].

CNTs are another widely used NMs for multiple chromatographic applications and have shown significant pulmonary and extra pulmonary toxicity concerns. CNTs have been shown to initiate inflammatory response by inducing cytokine expression. Upon chronic exposure, they cause upregulation of fibrogenic signaling and lead to pulmonary fibrosis [31].

Superparamagnetic iron oxide nanoparticles (SPIONs) have received wide attention due to their unique physicochemical properties [32–37] and hence were evaluated for toxicological effects in various animal species. Aerosol exposure of SPIONs (12.8 nm) was carried out in BALB/c mice. Detailed histological analysis was carried out along with target organ doses. In addition, the biomarkers of toxicological response were assessed by global transcriptomics. Application of target tissue dosimetry revealed good correlation between inflammatory response triggered in vitro and in vivo in the alveolar macrophage population. This study gave a new dimension for the utility of target organ dosimetry as a better means to assess the toxicological impact of NPs [38].

10.2.3 Carcinogenicity and Genotoxicity of Nanomaterials

Carcinogenicity is another toxicological dimension of the health hazards associated with the use of NMs. Logically, a material that causes an aberrant change in the chromosome may trigger DNA damage and the cells may eventually lose control over the normal cell cycle response. Although not many papers exist to establish the safety of NMs in terms of carcinogenicity, the available literature does confirm the carcinogenic propensity of various NMs. Titanium dioxide NPs are one of the most widely used metal-based materials in various industrial and cosmetics applications. The research reported by Lee and colleagues showed accumulation of massive TiO_2 NPs dust upon inhalational exposure at various doses for two years. Their findings showed bronchioloalveolar adenomas and cystic keratinizing squamous cell carcinomas at 250 mg/m^3 exposure, whereas no compound-related tumors were observed at lower doses [39]. Trouiller et al. deciphered the mechanism behind the genotoxic and carcinogenic potential of TiO_2 NPs. The studied NPs were found to delete DNA, upregulate micronuclei (MN), and induce 8-hydroxy-2′-deoxyguanosine and γ-H2AX foci suggesting potential molecular damage associated with the use of TiO_2 NPs [40].

To assess the skin carcinogenesis potential of TiO_2 NPs, Furukawa et al. exposed CD1 mice to uncoated and coated TiO_2 NPs after pre-exposure to known carcinogens, 7,12-dimethylbenz[*a*]anthracene and 12-*o*-tetradecanoylphorbol 13-acetate. The results of the histology and tumor incidence revealed the safety of the studied NPs as there was no evidence of morphometric changes in the tumor or the incidence [41]. In addition, NPs of metals such as nickel, cobalt, and copper also pose significant carcinogenic and genotoxic threat as numerous studies confirm their carcinogenic and genotoxic potential [42].

On the other hand, carbon-based NMs such as CNTs, fullerenes, and graphene oxide have been extensively used for a variety of industrial applications [43–48]. The small size and low density of CNTs make the respiratory exposure frequent at occupational sites. Various reports indicate the genotoxic potential of these carbon-based NMs [49–53].

On the other hand, genotoxicity deals with the toxicological properties of test substance at the genetic level. Genotoxicity is a potential health hazard as it may cause lethal mutations, DNA damage, and abnormalities in the chromosomes. The genotoxic response may become carcinogenic as well in certain cases forming malignant tumors.

The mechanisms of NP-induced genotoxicity are not completely understood. Furthermore, it is not clear if there are any "nano-specific" effects [54]. Genotoxicity can be classified as either "primary genotoxicity" or "secondary genotoxicity." The primary genotoxicity refers to genotoxicity from the NPs themselves (direct or indirect), whereas secondary genotoxicity refers to the induction of genotoxicity via ROS generated during particle-elicited inflammation [55]. Direct DNA interaction could represent a more nano-specific mechanism due to the fact that small NPs may reach the nucleus through nuclear pore complexes (NPCs) [56]. Even though NPCs consist of a tube with a diameter of approximately 30 nm, larger NPs have also been observed in the nucleus. One possibility is that these may gain access to the DNA in dividing cells when the nuclear membrane disassembles. In one of the few studies claiming a size-dependent interaction with DNA, gold NPs with a distinct particle size of 1.4 nm were shown to interact in a unique manner with the major grooves of DNA, which could account for the toxicity of these small NPs [57].

However, most studies suggest that the main mechanisms are not direct interaction with DNA, but rather an indirect cause of damage due to, e.g. oxidative stress as a result of either oxidant-generating properties of the NPs themselves or ROS formation due to interaction with mitochondria. Another possibility is that NPs interact with the mitotic spindle apparatus, centrioles, or their associated proteins and thereby cause aneugenic effects, i.e. loss or gain of chromosomes in daughter cells. As discussed by Sargent et al. [58], the long thin tubular-shaped CNTs show a striking similarity to cellular microtubules, suggesting a potential to interact with the mitotic spindle as well as the motor proteins that separate the chromosomes during cell division.

The most commonly used assays for assessing genotoxicity of NPs is the comet assay and analysis of MN in vitro [59, 60]. The general understanding is that the assays can be used for the evaluation of NMs, but with some caution. Regarding the comet assay, for example, photocatalytically active NPs, i.e. TiO_2 NPs, may cause false positives in the presence of light [61].

A recent study showed a positive response in comet assay for gold NPs, but the authors concluded that the response observed may be "false positive" due to additional DNA damage formed during the assay performance as a result of direct interaction between Au NPs and nucleoid DNA [62]. This has also been suggested in other studies [61], and such interactions may, indeed, result in additional DNA breaks. To what extent such damage formed during the assay performance cause false positives, rather than slightly exaggerated damage, remains to be elucidated. For analysis of micronuclei, the use of cytochalasin B (used to score MN specifically in once-divided binucleated cells) can inhibit uptake of the NPs [63]. Furthermore, bacterial cells are sometimes used for mutagenicity testing in the Ames test, but these tests are not recommended for mutagenicity testing of NMs.

It is very tedious to draw correlations between the genotoxicity and carcinogenicity propensity of NMs based on the available literature. A variety of physicochemical factors impact the physiological response elicited by NMs. Various NPs have been reported for their potential DNA-damaging potential. Metal NPs such as gold, iron, silica, silver, titanium, and zinc have been reported for their DNA-damaging potential [64–69].

Silica is a widely used additive and commonly used for manufacturing various chromatographic columns. It is deemed to possess low toxicity and has been approved by the FDA as a food additive. Barnes et al. assessed the genotoxicity of commercial and laboratory-synthesized silica NPs by Comet assay. The study showed nongenotoxic effects of SiNPs in 3T3 fibroblasts. Moreover, the results were reproducible in different laboratories indicating the validity of the Comet assay for the genotoxic assessment of NMs [70].

Demir et al. reported safety-indicating results in the *Drosophila melanogaster* model by using synthetic amorphous silica with sizes ranging from 6 to 55 nm. The results of wing spot assay showed the absence of somatic or recombination mutations upon treatment with SiNPs. However, a significant dose-dependent DNA damage was evident from the results of the Comet assay with maximum damage at 5 and 10 mM [71].

The biological effects of NPs might be modulated by various factors, and the interaction with serum proteins around NPs has been shown to significantly alter the toxicokinetics of NPs. In one of the studies, monodisperse SiNPs were tested on A549 human lung carcinoma cells in the presence or absence of serum. The results revealed potential interference of serum with the outcomes of flow-cytometry-based MN assay. Thus, consideration of the biological environment is of due importance in the selection of the right method of risk assessment of NMs [72].

In another study, the genotoxic- and ROS-inducing effects of various SiNPs were investigated. The results of the study revealed no DNA or oxidative damage by two pyrogenic (NM-202 and 203) and two precipitated forms (NM-200 and 201) of nano-sized silica upon oral administration to rats. In addition, the studied particles did not show any significant chromosome damage showing the relative safety of SiNPs [73]. In continuation of this study, genotoxic effects of amorphous silica NPs were further studied upon intratracheal and intravenous administration. The results revealed that although these NPs possess potential hepatotoxic and thrombocytopenic effect, these NPs still do not cause any accountable genotoxic response, thus giving a green light for its commercial acceptability and relatively safer genotoxic profile with the plausible routes of exposure [74].

The wide-scale production and use of various nano-species of carbon-based NMs are of significant concern. A significant number of reports indicate potential genotoxic concerns associated with the use of carbon-based NMs. For example, in a study reported by Chernova et al. [75], long, multiwalled CNTs replicated asbestos-induced mesothelioma following instillation (not inhalation) into the pleural cavity of mice Was reported. The authors presented evidence that hypermethylation of p16/Ink4a and p19/Arf in CNT and asbestos-induced inflammatory lesions preceded mesothelioma, consistent with epigenetic alterations playing a role in cancer.

In one of the studies, the cytotoxic and genotoxic potential of carbon nanofibers (CNFs) was compared with asbestos fibers. The study of cytotoxic effects revealed significant reduction in V79 cell viability in the following order: asbestos > CNF > SWCNT. Comet assay, MN assay, and fluorescence in situ hybridization were used to assess the genotoxic effects, and the results indicated significant genotoxic responses as evident from the increased frequency of centromere-positive micronuclei. Moreover, cellular uptake and generation of ROS in murine macrophages (Raw 264.7) cells showed increased oxidative stress with CNF and asbestos, whereas single-walled carbon nanotubes (SWCNTs) did not affect redox balance [76].

10.2.4 Neurotoxicity of Nanomaterials

Neurotoxicity refers to any reversible or irreversible adverse effect on the structure, function, or chemistry of the nervous system, during development or at maturity, produced by physical or chemical causes. The main mechanisms for neurotoxicity involve the excessive production of ROS leading to oxidative stress, the release of cytokines causing neuroinflammation, and dysregulations of apoptosis leading to neuronal death [77]. However, there is a considerable lack of information regarding the neurotoxicity of NMs, which complicates risk assessments following exposure. Therefore, the development of rapid, accurate, and effective strategies for determining the neurotoxic effects induced by NMs is achieved [78].

To date, because NMs can act as Trojan horses, they have been used as carriers for the delivery of drugs and imaging probes to specific cells and tissues [79]. Tremendous efforts have been made in developing nanomedicines for the diagnosis and therapy for the effective treatment of various brain diseases, including brain cancer, Alzheimer's disease, Parkinson's disease, amyotrophic lateral sclerosis, multiple sclerosis, and ischemic stroke [80–82]. A variety of NMs have been applied for brain diseases therapy, including polymeric NPs, liposomes, dendrimers, metallic NPs, CNTs, and quantum dots [83, 84]. However, in addition to the intentional application of NMs for brain disorders therapy, they can unintentionally enter the brain through blood circulation from peripheral disorders therapy or after inhalation from the air. Therefore, in spite of the impressive progress in the field, the application of NMs has brought about a new cause for concern, which is the neurotoxicity of NMs [85]. As they can enter the brain by penetrating the blood–brain barrier through a series of transporters or receptors expressed on the endothelial cells of brain capillaries, through adsorptive-mediated transcytosis, or by bypassing the blood–brain barrier through the intranasal route, neurotoxicity must be taken into consideration in the manufacturing process [86]. The common neurotoxicity mechanisms involve oxidative stress–induced cell apoptosis and autophagy, and immune response and inflammation, activating specific signaling pathways, which will subsequently affect the blood–brain barrier functions [87]. Moreover, a neurotoxic effect can directly alter the neuronal structure or activity, or it can result in a cascade of effects due to glial activation and glial–neuronal interactions. It can manifest immediately or after longer periods of time. The consequences of neurotoxicity can be reversible or permanent, affecting parts of the central nervous system or the whole system [88].

Conventional methods and techniques used to analyze the toxicity of NMs are not sufficiently accurate and reliable, and it is not safe to draw conclusions regarding the potential dangers and mechanisms of the toxicity of NMs [89]. Hence, there is a need to improve the novel discipline that combines nanotoxicology and neuroscience by dedicating original and specific experimental models and tools, as it is essential for studying the interactions of NMs with the brain [90]. Elucidating and understanding neurotoxicity processes represents a way to design safer nanocarrier systems and reduce their side effects [91].

10.3 Toxicity Mechanism of Nanomaterials

The rise of nanotechnology has changed our perspective toward the use of science and technology in various fields. The widespread use of NPs has been extended to chromatographic techniques with promising results. However, the issues related to the safety of NPs have not yet been resolved, which caused the beginning of a new branch called "nanotoxicology," which deals with the negative effects of NPs on human health and the environment. The toxicity of NPs such as Fe_3O_4, silica, fullerenes, and graphene, which are found to be useful in different chromatographic techniques, has already been reported [92–94]. Although mechanisms have been proposed for the toxicity of well-studied NPs, our knowledge of NP toxicity mechanisms is still limited.

The main concern for potential NM toxicity is their small size. Although a critical threshold number of NMs is required for the induction of any toxicological effect, some size-related toxicological properties are better understood by considering the behavior of a single NM, whereas for others it is better to consider the behavior of several NMs. If we consider the journey of a single NM inside the human body, it appears as a sort of fantastic voyage compared with that of the bulk form. It may in fact reach biological sanctuaries such as the intracellular environment and neural axons [95], usually interdicted to the bulk material. Furthermore, it may cross biological barriers such as the blood–brain barrier and the blood–testis barrier and may therefore reach highly protected organs such as the brain and the testes [96]. For pregnant women, it may translocate through the placenta and come in contact with fetal tissues, as recently reviewed by Campagnolo et al. [97]. Furthermore, it has been reported that the removal of NMs from tissues is slower than that of bulk material, because NMs are poorly taken over by macrophages [98]. Thus, for a given level of toxicity, NMs may cause injury to biological structures spared by the bulk form and may reside longer in these compartments. A brief description of various features of nanomaterials relevant for toxicology is given in Table 10.1.

On the other hand, there is a strong tendency to aggregation/agglomeration in the presence of a large number of NMs [100]. However, it is uncertain if this property confers specific toxicological attributes. The aggregation status probably influences cellular uptake in a specific manner according to different cell types [101] and may thus indirectly drive the target of toxic feature of NM.

The elucidation of the mechanism of NP toxicity is quite a daunting task as a plethora of mechanisms was proposed with most known mechanisms to be the generation of ROS. In addition to the involvement of ROS, different mechanisms of NP toxicity involve the release of proinflammatory cytokines; damage to cell membrane; DNA; altered cell cycle; activation of nuclear factor–kappa light chain enhancer of activated B cells (NF-kB); mitogen-activated protein kinase (MAPK); transforming growth factor beta signaling; and damage to organelles such as nucleus, mitochondria, and lysosome [102–105]. However, there is no single reason

Table 10.1 A brief description of various features of nanomaterials relevant for toxicology.

Property	Biological consequences	Relevance for toxicity
Size-related properties: single nanomaterial		
Dimension < 100 nm	Intracellular access	High
	Poor clearance by macrophages	Intermediate
	Transport through neural axons	Intermediate
	Crossing of biological barriers	High
Corona formation	Altered biodistribution	High
	Corona-mediated damage	Intermediate
High number of surface atoms	Increased damage of contact biological surfaces (oxidative stress)	High
Size-related properties: collection of nanomaterials		
Large surface area	Extensive contact with biological surfaces	High
Aggregation/agglomeration	Possible increase in cellular uptake	Intermediate
Non-size-related properties		
Shape	Possible fiber-like behavior	High
Surface charge	Possible damage to biological surfaces with opposite charge	Low
Biopersistence	Damage amplification	High
Loss of metal ions	Possible additional biological damage	Intermediate

Source: A. Pietroiusti et al., Chapter 2 – Mechanisms of nanomaterial toxicity, Editor(s): James Njuguna, Krzysztof Pielichowski, Huijun Zhu, Health and Environmental Safety of Nanomaterials, Woodhead Publishing, 2014, Pages 28–43, ISBN 9780857096555. © 2014, Elsevier.

that explains the cause of NP toxicity in a complete sense. In fact, NP toxicity is believed to be a consequence of the interplay between the mentioned mechanisms. NP exposure causes the generation of ROS, which can damage the membrane lipids, proteins, nucleus, and mitochondria followed by apoptosis/cell death. NPs due to reactive surface, surface defects, presence of metals, or by influencing intracellular production can increase the levels of ROS. Certain NPs including metal oxides, CNTs, fullerenes, and graphene, which are being used in the different steps of analytical techniques including sample preparation, separation, derivatization, and detection, have been reported to induce the generation of oxidative stress. Shvedova et al. reported that incubation of high doses of SWCNTs produced ROS, LPO, oxidative stress, and mitochondrial dysfunction [106]. Farcal et al. showed that silica NPs induced ROS production and proinflammatory cytokines such as TNF-α and IL-6 [107]. In another study, the role of oxidative stress and p38 MAPK pathway was investigated, and increased expression of p38, Nrf-2, and NF-kB was observed after exposure to silver NPs [108]. Apart from that, the study also showed increased DNA damage and cell death after exposure to silver NPs. In addition to the inherent pro-oxidant properties, NPs can also induce ROS generation from intracellular sources such as the mitochondria. High levels of ROS in the mitochondria further aggravate ROS production and can result in damage to membrane phospholipids altering mitochondrial membrane depolarization [109]. Internalization of NPs results in increased immune cell activation followed by inflammatory pathway activation. NPs, after entering the body, encounter different cells of the immune system such as monocytes, platelets, dendritic cells, and macrophages. The interaction of NPs with cells of immune component results in the release of various cytokines such as IL-6, IL-1β, and TNF-α, which causes inflammatory reactions [110]. Moreover, increased levels of ROS also trigger a cascade of proinflammatory cytokines and mediators via activation of redox-sensitive MAPK and NF-kB signaling pathways that control transcription of inflammatory genes such as IL-1β, IL-6, IL-8, and TNF-α [111]. The presence of metals in NPs further enhances the toxicity of NPs by inducing different reactions such as Fenton's reaction and Haber–Weiss reaction, which ultimately exaggerate the oxidative stress [112, 113]. However, there is a paradox on the role of oxidative stress in NP toxicity. Although the central role in the mechanism of toxicity is believed to be played by ROS, there are reports suggesting that certain toxicity incidents cannot be explained by the involvement of ROS. For example, Tabet et al. observed the toxicity of SWCNTs by interference with actin cytoskeleton without any increase in ROS levels [114]. Although a number of mechanisms involved in NP toxicity have been proposed, the exact mechanism of NP toxicity is still an unfathomable mystery. The elucidation of clearly defined underlying mechanisms will help to increase our understanding and will be useful in tackling the toxicity concerns leading to the safe applications of NPs.

10.4 The Traditional Risk Assessment Paradigm

The classical risk assessment is a well-established and formalized process intended to "calculate or estimate the risk to a given target organism, system or (sub)population, including the identification of uncertainties, following exposure to a particular agent, taking into account the inherent characteristics of the agent of concern as well as the characteristics of the specific target system" [115]. While the hazard of a substance is its potential to cause harm, whereas risk is the likelihood of that harm occurring, taking into account wider considerations of exposure and uncertainty. The process of the risk assessment consists of four steps: hazard identification, hazard characterization including dose–response assessment, exposure assessment, and risk characterization.

According to the Risk Commission [116], the scientific risk assessment process mainly deals with consequences of the effects of noxious agents to human health quantifying the probability of a harmful effect to individuals or populations. The framework was developed for conventional chemicals as an information and decision supporting tool for regulations in the context of safety and sustainability.

Associated uncertainty in the progress is managed by the application of safety factors. In general, the traditional approach is a chemical-by-chemical approach, focusing on a single substance, a single media, and a single toxic endpoint. In cases where the risk of adverse effects at exposures below a safe level is expected to be practically zero, the margin of exposure can be calculated to describe the quotient between expected exposure and the no observed adverse effect level. Thus, the output of risk assessment is based on a substance specific quotient produced on an absolute basis.

Triggers for chemical risk assessment are in general the quantities of chemical properties such as persistence or bioaccumulation and the effects such as acute and chronic toxicity, sensitization, genotoxicity, reproductive toxicity, or carcinogenicity. Thus, information requirements include physicochemical properties, toxicological and ecotoxicological information. Chemical risk assessment has been the standard approach for assessing the potential health and environmental risks of bulk chemicals and subsequently has been applied to NMs in recent years [117–119]. In Europe, this process is defined for chemicals by the EU chemicals policy Regulation on Registration, Evaluation, Authorization and Restriction of Chemicals (REACH) [120] and the detailed guidance documents provided by the European Chemicals Agency (ECHA).

As a decision support tool risk assessment uses the weight of evidence approach: This process is based on the totality of data in a holistic manner and facilitates transparency and quantification in decision-making. The data provided by toxicology are ranked and scored considering the varying quality, relevance, and consistency. This results in the so-called strength of evidence, which is inversely related to the degree of uncertainty. In addition, individual data are clustered in evidence lines. Complementary and contradictory evidence of varying strengths and evidence lines are considered in the weight of evidence process. The Scientific Committee on Emerging and Newly Identified Health Risks (SCENIHR) declared that this weighing of evidence is important as a guideline for preliminary risk assessment and for determining the priority for further assessments [121].

10.5 Strategies for Improving Specific Risk Assessment

10.5.1 Combining Life Cycle Methodology with the Risk Assessment Approach

Life cycle assessment is a systematic assessment of the potential environmental impacts of a defined good or service throughout all life cycle stages, contributing to the production of the product under investigation, its use, and disposal [122]. Life cycle assessment is a most extensively developed and standardized methodology. The assessment typically covers a broad range of environmental impacts, such as climate change, resource depletion, and toxicity on human health caused by releases of chemical agents. In contrast to risk assessment, the results of the life cycle assessment are normally presented on a relative or comparative basis, such as the comparison of two alternative solutions for product development, to make informed decisions regarding the relative environmental sustainability of products. The life cycle assessment inventory is a mass-based accounting system that does not take into account spatial, temporal, dose–response, and threshold information. Hence, this assessment tool could be helpful to quantify potential impacts of nanoproducts across different life cycle stages. In excess of the traditional chemical risk assessment, the manufacturing aspect but also consumption and disposal aspects are addressed.

The life cycle assessment has its own challenges and limitations. There is a lack of robust data regarding emissions and exposures related to the production, use, and disposal of NMs and products containing them and subsequently a need to establish a database that includes the most common production pathways. As discussed already, for the life cycle assessment, knowledge about the environmental fate, behavior, and toxicological impacts of NMs is generally limited. The unclear metrics used in these assessments, the inconsistencies in viable data sources, and the challenge of proper identification and characterization of NMs result in serious methodological uncertainties.

Formulating strategies to use life cycle assessment and risk assessment together for the assessment of chemicals has been pursued in a range of research efforts for more than a decade and remains a topic of ongoing work. Some proposed approaches range from the use of the same data, the complementary use of the results as well as efforts to integrate these two tools into one [123]. The proposed integrative approaches could be allocated to a more conventional risk assessment approach applied in a life cycle perspective or to the more conventional life cycle assessment supplemented by risk assessment in stages of specific life cycle [123]. However, the full integration into one tool is considered neither desirable nor meaningful as the two tools answer different questions and have different system boundaries. For example, the two methods have differing scopes, aims, and end results: while life cycle assessment assesses a wide variety of environmental impacts of a product or system related to the functional unit, risk assessment deals with the health and environmental risks of a single substance at a particular point in the chemical's life cycle. As these different systems have their own meanings of terms such as "life cycle" or "characterization," it may be more beneficial to use these two technical tools in a complementary manner. The following approaches were proposed:

- The performance of risk assessment (RA) at selected or each step of the life cycle of a product
- The careful identification of potential health concerns on a system-wide and hypothetical basis by life cycle assessment (LCA) that is followed by a detailed RA.
- LCA as a strategic tool to prioritize data that is needed to complete by RA.

10.5.2 The Support of Risk-Based Classification Systems

Despite of advances in frameworks and assessment methodologies, a comprehensive characterization of predictive properties for all nanomaterials is far from complete. Thus, some authors proposed a preliminary relative risk ranking or risk classification of nanomaterials. Zuin et al. [124] used physicochemical and toxicity data for a relative ranking of the hazard potential of quantum dots, fullerenes, SWCNTs and carbon black. The classification system of Olson and Gurian [125] is based on the properties persistence and toxicity. They recommended that persistence should be characterized first because this determines the burden of proof for nontoxicity. Tervonen et al. [126] described a classification system for NMs based on a set of performance metrics that measure both the toxicity and physicochemical characteristics as well as the expected environmental impacts through the product life cycle. The parameters included properties of NMs such as agglomeration, aggregation, reactivity, charge, critical functional groups, size, bioavailability, bioaccumulation, and the toxic potential. In addition, a relative ranking of the exposure potential of selected NPs was proposed using different transport and persistence properties in the aquatic environment [127].

10.6 Conclusions

The aim of this chapter is to summarize the general principles and theories of risk perception and communication, exploring how these might be applied for workers and employers working with NMs. Relevant research, studies, and surveys that have been done to date have been considered, with the aim of highlighting risk communication approaches that would be applicable to workplaces involved with NMs and in the process, discussing the challenges and difficulties that can and have been encountered.

Risk assessment of NMs is challenging. Efficient protection in a work environment can be achieved by avoiding exposure ("prevention by design"). Different tools and approaches for tiered testing and grouping have been developed in recent years. Case studies need to be performed to better understand the usefulness of these approaches. NPs that immediately dissolve are unlikely to pose nano-specific risks. Certain properties of NPs are, however, clearly considered to be of concern, such as high surface area and reactivity. Furthermore, biopersistent, high-aspect-ratio materials are considered to pose a risk. Occupational exposure limits (OELs) for NMs should also be developed.

References

1 Hussain, C.M. and Keçili, R. (2020). *Modern Environmental Analysis Techniques for Pollutants*. Amsterdam, The Netherlands: Elsevier.

2 Büyüktiryaki, S., Keçili, R., and Hussain, C.M. (2020). Functionalized nanomaterials in dispersive solid phase extraction: advances and prospects. *Trends Anal. Chem.* https://doi.org/10.1016/j.trac.2020.115893.

3 Keçili, R., Büyüktiryaki, S., and Hussain, C.M. (2019). Advancement in bioanalytical science through nanotechnology: past, present and future. *Trends Anal. Chem.* 110: 259–276.

4 Hussain, C.M. and Kharisov, B. (2017). *Advanced Environmental Analysis-Application of Nanomaterials*. The Royal Society of Chemistry.
5 Hussain, C.M. (2017). Magnetic nanomaterials for environmental analysis. In: *Advanced Environmental Analysis-Application of Nanomaterials* (eds. C.M. Hussain and B. Kharisov). The Royal Society of Chemistry.
6 Hussain, C.M. (2019). *Handbook of Nanomaterials in Analytical Chemistry: Modern Trends in Analysis*. Elsevier.
7 Büyüktiryaki, S., Sümbelli, Y., Keçili, R., and Hussain, C.M. (2019). Lab-on-chip platforms for environmental analysis. In: *Encyclopedia of Analytical Science*, 3e (eds. P. Worsfold, C. Poole, A. Townshend and M. Miró), 267–273. Academic Press.
8 Nikodinovska, V.V., Mladenovska, K., and Grozdanov, A. (2015). Risks and health effects from exposure to engineered nanostructures: a critical review. *J. Chem. Technol. Metall.* 50: 117–134.
9 Oberdörster, G., Oberdörster, E., and Oberdörster, J. (2005). Nanotoxicology: an emerging discipline evolving from studies of ultrafine particles. *Environ. Health Perspect.* 113: 823–839.
10 Holl, M.M.B. (2009). Nanotoxicology: a personal perspective. *WIREs Nanomed. Nanobiotechnol.* 1: 353–359.
11 Fadeel, B., Pietroiusti, A., and Shvedova, A. (eds.) (2017). *Adverse Effects of Engineered Nanomaterials: Exposure, Toxicology, and Impact on Human Health*, 2e. Elsevier-Academic Press.
12 WHO (2014). *Environmental Health Criteria 242*. IPCS: Dermal exposure http://www.who.int/ipcs/publications/ehc/ehc_242.pdf.
13 Baroli, B., Ennas, M.G., Loffredo, F. et al. (2007). Penetration of metallic nanoparticles in human full-thickness skin. *J. Invest. Dermatol.* 127 (7): 1701–1712.
14 Palmer, B.C. and DeLouise, L.A. (2016). Nanoparticle-enabled transdermal drug delivery systems for enhanced dose control and tissue targeting. *Molecules* (Basel, Switzerland) 21 (12): 1–17.
15 Gulson, B., McCall, M., Korsch, M. et al. (2010). Small amounts of zinc from zinc oxide particles in sunscreens applied outdoors are absorbed through human skin. *Toxicol. Sci.* 118 (1): 140–149.
16 Gulson, B., Wong, H., Korsch, M. et al. (2012). Comparison of dermal absorption of zinc from different sunscreen formulations and differing UV exposure based on stable isotope tracing. *Sci. Total Environ.* 420: 313–318.
17 Leite-Silva, V.R. et al. (2013). The effect of formulation on the penetration of coated and uncoated zinc oxide nanoparticles into the viable epidermis of human skin in vivo. *Eur. J. Pharm. Biopharm.* 84 (2): 297–308.
18 Niska, K., Zielinska, E., Radomski, M.W., and Inkielewicz-Stepniak, I. (2017). Metal nanoparticles in dermatology and cosmetology: interactions with human skin cells. *Chem. Biol. Interact.* 16: 30629-9.
19 Horie, M., Sugino, S., Kato, H. et al. (2016). Does photocatalytic activity of TiO_2 nanoparticles correspond to photo-cytotoxicity? Cellular uptake of TiO_2

nanoparticles is important in their photo-cytotoxicity. *Toxicol. Mech. Methods* 26 (4): 284–294.
20 Mousavi, S.Z., Nafisi, S., and Maibach, H.I. (2017). Fullerene nanoparticle in dermatological and cosmetic applications. *Nanomedicine: NBM* 13: 1071–1087.
21 Zhao, B., He, Y.Y., Bilski, P.J., and Chignell, C.F. (2008). Pristine (C60) and hydroxylated [C60(OH)24] fullerene phototoxicity towards HaCaT keratinocytes: type I vs type II mechanisms. *Chem. Res. Toxicol.* 21 (5): 1056–1063.
22 Sayes, C.M., Gobin, A.M., Ausman, K.D. et al. (2005). Nano-C60 cytotoxicity is due to lipid peroxidation. *Biomaterials* 26 (36): 7587–7595.
23 Monteiro-Riviere, N.A., Inman, A.O., and Zhang, L.W. (2009). Limitations and relative utility of screening assays to assess engineered nanoparticle toxicity in a human cell line. *Toxicol. Appl. Pharmacol.* 234 (2): 222–235.
24 Vankoningsloo, S., Piret, J.P., Saout, C. et al. (2010). Cytotoxicity of multi-walled carbon nanotubes in three skin cellular models: effects of sonication, dispersive agents and corneous layer of reconstructed epidermis. *Nanotoxicology* 4 (1): 84–97.
25 Ji, J.H., Jung, J.H., Kim, S.S. et al. (2007). Twenty-eight-day inhalation toxicity study of silver nanoparticles in Sprague-Dawley rats. *Inhal. Toxicol.* 19 (10): 857–871.
26 Sung, J.H., Ji, J.H., Park, J.D. et al. (2009). Subchronic inhalation toxicity of silver nanoparticles. *Toxicol. Sci.* 108 (2): 452–461.
27 Yıldırım, D., Kip, Ç., Tsogtbaatar, K. et al. (2020). Microfluidic immobilized metal affinity chromatography based on Ti(IV)-decorated silica microspheres for purification of phosphoproteins. *J. Chromatogr. B* 1140: 122010.
28 González, L., Morante-Zarcero, S., Pérez-Quintanilla, D., and Sierra, I. (2020). Hydroxymethylfurfural determination in cereal and insect bars by high-performance liquid chromatography-mass spectrometry employing a functionalized mesostructured silica as sorbent in solid-phase extraction. *J. Chromatogr. A* https://doi.org/10.1016/j.chroma.2020.461124.
29 Na, N., Cui, X., De Beer, T. et al. (2011). The use of silica nanoparticles for gas chromatographic separation. *J. Chromatogr. A* 1218 (28): 4552–4558.
30 Brown, D.M., Kanase, N., Gaiser, B. et al. (2014). Inflammation and gene expression in the rat lung after instillation of silica nanoparticles: effect of size, dispersion medium and particle surface charge. *Toxicol. Lett.* 224 (1): 147–156.
31 Shvedova, A.A., Kisin, E.R., Mercer, R. et al. (2005). Unusual inflammatory and fibrogenic pulmonary responses to single-walled carbon nanotubes in mice. *Am. J. Phys.* 289 (5): L698–L708.
32 Zhi, D., Yang, T., Yang, J. et al. (2020). Targeting strategies for superparamagnetic iron oxide nanoparticles in cancer therapy. *Acta Biomater.* 102: 13–34.
33 Alphandéry, E. (2020). Iron oxide nanoparticles for therapeutic applications. *Drug Discov. Today* 25 (1): 141–149.
34 Musielak, M., Piotrowski, I., and Suchorska, W.M. (2019). Superparamagnetic iron oxide nanoparticles (SPIONs) as a multifunctional tool in various cancer therapies. *Rep. Pract. Oncol. Radiother.* 24 (4): 307–314.

35 Roca, A.G., Gutiérrez, L., Gavilán, H. et al. (2019). Design strategies for shape-controlled magnetic iron oxide nanoparticles. *Adv. Drug Deliv. Rev.* 138: 68–104.

36 Tong, S., Zhu, H., and Bao, G. (2019). Magnetic iron oxide nanoparticles for disease detection and therapy. *Mater. Today* 31: 86–99.

37 Song, C., Sun, W., Xiao, Y., and Shi, X. (2019). Ultrasmall iron oxide nanoparticles: synthesis, surface modification, assembly, and biomedical applications. *Drug Discov. Today* 24 (3): 835–844.

38 Teeguarden, J.G., Mikheev, V.B., Minard, K.R. et al. (2014). Comparative iron oxide nanoparticle cellular dosimetry and response in mice by the inhalation and liquid cell culture exposure routes. *Part. Fibre Toxicol.* 11 (1): 46–63.

39 Lee, K., Trochimowicz, H., and Reinhardt, C. (1985). Pulmonary response of rats exposed to titanium dioxide (TiO_2) by inhalation for two years. *Toxicol. Appl. Pharmacol.* 79 (2): 179–192.

40 Trouiller, B., Reliene, R., Westbrook, A. et al. (2009). Titanium dioxide nanoparticles induce DNA damage and genetic instability in vivo in mice. *Cancer Res.* 69 (22): 8784–8789.

41 Furukawa, F., Doi, Y., Suguro, M. et al. (2011). Lack of skin carcinogenicity of topically applied titanium dioxide nanoparticles in the mouse. *Food Chem. Toxicol.* 49 (4): 744–749.

42 Magaye, R., Zhao, J., Bowman, L., and Ding, M. (2012). Genotoxicity and carcinogenicity of cobalt-, nickel-and copper-based nanoparticles (review). *Exp. Ther. Med.* 4 (4): 551–561.

43 Anzar, N., Hasan, R., Tyagi, M. et al. (2020). Carbon nanotube – a review on synthesis, properties and plethora of applications in the field of biomedical science. *Sens. Int.* 1: 100003.

44 Verma, B. and Balomajumder, C. (2020). Surface modification of one-dimensional carbon nanotubes: a review for the management of heavy metals in wastewater. *Environ. Technol. Innov.* 17: 100596.

45 Tiwari, S., Purabgola, A., and Kandasubramanian, B. (2020). Functionalised graphene as flexible electrodes for polymer photovoltaics. *J. Alloys Compd.* 825: 153954.

46 Nandanapalli, K.R., Mudusu, D., and Lee, S. (2019). Functionalization of graphene layers and advancements in device applications. *Carbon* 152: 954–985.

47 Korkmaz, S. and Afşin Kariper, İ.A. (2020). Graphene and graphene oxide based aerogels: synthesis, characteristics and supercapacitor applications. *J. Energy Storage* 27: 101038.

48 Pan, Y., Liu, X., Zhang, W. et al. (2020). Advances in photocatalysis based on fullerene C60 and its derivatives: properties, mechanism, synthesis, and applications. *Appl. Catal. B Environ.* 265: 118579.

49 Toyokuni, S. (2013). Genotoxicity and carcinogenicity risk of carbon nanotubes. *Adv. Drug Deliv. Rev.* 66 (15): 2098–2110.

50 Samadian, H., Salami, M.S., Jaymand, M. et al. (2020). Genotoxicity assessment of carbon-based nanomaterials: have their unique physicochemical properties made them double-edged swords? *Mutat. Res. Rev. Mutat.* 783: 108296.

51 Roller, M. (2011). In vitro genotoxicity data of nanomaterials compared to carcinogenic potency of inorganic substances after inhalational exposure. *Mutat. Res. Rev. Mutat.* 727 (3): 72–85.

52 Guo, X. and Mei, N. (2014). Assessment of the toxic potential of graphene family nanomaterials. *J. Food Drug Anal.* 22 (1): 105–115.

53 Lalwani, G., D'Agati, M., Khan, A.M., and Sitharaman, B. (2016). Toxicology of graphene-based nanomaterials. *Adv. Drug Deliv. Rev.* 105: 109–144.

54 Donaldson, K. and Poland, C.A. (2013). Nanotoxicity: challenging the myth of nanospecific toxicity. *Curr. Opin. Biotechnol.* 24 (4): 724–734.

55 Schins, R.P. and Knaapen, A.M. (2007). Genotoxicity of poorly soluble particles. *Inhal. Toxicol.* 19: 189–198.

56 Nabiev, I., Mitchell, S., Davies, A. et al. (2007). Nonfunctionalized nanocrystals can exploit a cell's active transport machinery delivering them to specific nuclear and cytoplasmic compartments. *Nano Lett.* 11: 3452–3461.

57 Tsoli, M., Kuhn, H., Brandau, W. et al. (2005). Cellular uptake and toxicity of Au55 clusters. *Small* (Weinheim an der Bergstrasse, Germany) 1 (8–9): 841–844.

58 Sargent, L.M., Reynolds, S.H., and Castranova, V. (2010). Potential pulmonary effects of engineered carbon nanotubes: in vitro genotoxic effects. *Nanotoxicology* 4: 396–408.

59 Golbamaki, N., Rasulev, B., Cassano, A. et al. (2015). Genotoxicity of metal oxide nanomaterials: review of recent data and discussion of possible mechanisms. *Nanoscale* 7 (6): 2154–2198.

60 Magdolenova, Z., Collins, A., Kumar, A. et al. (2014). Mechanisms of genotoxicity. A review of in vitro and in vivo studies with engineered nanoparticles. *Nanotoxicology* 8 (3): 233–278.

61 Karlsson, H.L., Di Bucchianico, S., Collins, A.R., and Dusinska, M. (2015). Can the comet assay be used reliably to detect nanoparticle-induced genotoxicity? *Environ. Mol. Mutagen.* 56 (2): 82–96.

62 George, J.M., Magogotya, M., Vetten, M.A. et al. (2017). An investigation of the genotoxicity and interference of gold nanoparticles in commonly used in vitro mutagenicity and genotoxicity assays. *Toxicol. Sci.* 156 (1): 149–166.

63 Swedish Chemical Agnecy (KEMI) (2016). *Nanomaterials and Genotoxicity – A Literature Review*. Report 13/16., 1–57.

64 Li, J.J., Zou, L., Hartono, D. et al. (2008). Gold nanoparticles induce oxidative damage in lung fibroblasts in vitro. *Adv. Mater.* 20 (1): 138–142.

65 Xu, A., Chai, Y., Nohmi, T. et al. (2009). Genotoxic responses to titanium dioxide nanoparticles and fullerene in gpt delta transgenic MEF cells. *Part. Fibre Toxicol.* 6 (1): 3.

66 AshaRani, P., Kah Mun, G.L., Hande, M.P. et al. (2008). Cytotoxicity and genotoxicity of silver nanoparticles in human cells. *ACS Nano* 3 (2): 279–290.

67 Singh, N., Jenkins, G.J.S., Nelson, B.C. et al. (2012). The role of iron redox state in the genotoxicity of ultrafine superparamagnetic iron oxide nanoparticles. *Biomaterials* 33 (1): 163–170.

68 Wang, J.J., Sanderson, B.J., and Wang, H. (2007). Cytotoxicity and genotoxicity of ultrafine crystalline SiO_2 particulate in cultured human lymphoblastoid cells. *Environ. Mol. Mutagen.* 48 (2): 151–157.
69 Sharma, V., Anderson, D., and Dhawan, A. (2011). Zinc oxide nanoparticles induce oxidative stress and genotoxicity in human liver cells (HepG2). *J. Biomed. Nanotechnol.* 7 (1): 98–99.
70 Barnes, C.A., Elsaesser, A., Arkusz, J. et al. (2008). Reproducible comet assay of amorphous silica nanoparticles detects no genotoxicity. *Nano Lett.* 8 (9): 3069–3074.
71 Demir, E., Aksakal, S., Turna, F. et al. (2015). In vivo genotoxic effects of four different nano-sizes forms of silica nanoparticles in *Drosophila melanogaster*. *J. Hazard. Mater.* 283: 260–266.
72 Gonzalez, L., Lukamowicz-Rajska, M., Thomassen, L.C.J. et al. (2014). Co-assessment of cell cycle and micronucleus frequencies demonstrates the influence of serum on the in vitro genotoxic response to amorphous monodisperse silica nanoparticles of varying sizes. *Nanotoxicology* 8 (8): 876–884.
73 Tarantini, A., Huet, S., Jarry, G. et al. (2015). Genotoxicity of synthetic amorphous silica nanoparticles in rats following short-term exposure. Part 1: oral route. *Environ. Mol. Mutagen.* 56 (2): 218–227.
74 Guichard, Y., Maire, M.A., Sébillaud, S. et al. (2015). Genotoxicity of synthetic amorphous silica nanoparticles in rats following short-term exposure. Part 2: intratracheal instillation and intravenous injection. *Environ. Mol. Mutagen.* 56 (2): 228–244.
75 Chernova, T., Murphy, F.A., Galavotti, S. et al. (2017). Long-fiber carbon nanotubes replicate asbestos-induced mesothelioma with disruption of the tumor suppressor gene Cdkn2a (Ink4a/Arf). *Curr. Biol.* 27 (21): 3302–3314.
76 Kisin, E., Murray, A.R., Sargent, L. et al. (2011). Genotoxicity of carbon nanofibers: are they potentially more or less dangerous than carbon nanotubes or asbestos? *Toxicol. Appl. Pharmacol.* 252 (1): 1–10.
77 Teleanu, D., Chircov, C., Grumezescu, A. et al. (2018). Impact of nanoparticles on brain health: an up to date overview. *J. Clin. Med.* 7: 490.
78 Maurizi, L., Papa, A.-L., Boudon, J. et al. (2018). Toxicological risk assessment of emerging nanomaterials: cytotoxicity, cellular uptake, effects on biogenesis and cell organelle activity, acute toxicity and biodistribution of oxide nanoparticle. In: *Unraveling the Safety Profile of Nanoscale Particles and Materials* (eds. A.C. Gomes and M.P. Sárria). London: IntechOpen.
79 Gao, H. and Jiang, X. (2017). The medical applications of nanomaterials in the central nervous system, Chapter 1. In: *Neurotoxicity of Nanomaterials and Nanomedicine* (eds. X. Jiang and H. Gao), 1–31. Cambridge, MA: Academic Press.
80 Siddiqi, K.S., Husen, A., Sohrab, S.S., and Yassin, M.O. (2018). Recent status of nanomaterial fabrication and their potential applications in neurological disease management. *Nanoscale Res. Lett.* 13: 231.
81 Veloz-Castillo, M.F., West, R.M., Cordero-Arreola, J. et al. (2016). Nanomaterials for neurology: state-of-the-art. *CNS Neurol. Disord. Drug Targets* 15: 1306–1324.

82 Omrani, M.M., Ansari, M., and Kiaie, N. (2016). Therapeutic effect of stem cells and nano-biomaterials on alzheimer's disease. *Biointerface Res. Appl. Chem.* 6: 1814–1820.

83 Huang, L., Hu, J., Huang, S. et al. (2017). Nanomaterial applications for neurological diseases and central nervous system injury. *Prog. Neurobiol.* 157: 29–48.

84 Shafiee, M.R.M. and Kargar, M. (2016). Preparation of aryl sulfonamides using cuo nanoparticles prepared in extractive rosmarinus officinalis leaves media. *Biointerface Res. Appl. Chem.* 6: 1257–1262.

85 Xiong, J. and Gao, H. (2017). Neurotoxicity of nanomaterials: where are we and what can we do? *EC Pharmacol. Toxicol.* 4: 93.

86 Liu, Y. and He, Q. (2017). The route of nanomaterials entering brain. In: *Neurotoxicity of Nanomaterials and Nanomedicine* (eds. X. Jiang and H. Gao), 33–57. Cambridge, MA: Academic Press.

87 Gao, H. and Jiang, X. (2017). Introduction and overview. In: *Neurotoxicity of Nanomaterials and Nanomedicine* (eds. X. Jiang and H. Gao). Cambridge, MA: Academic Press.

88 Li, J. and Martin, F.L. (2017). Current perspective on nanomaterial-induced adverse effects: neurotoxicity as a case example, Chapter 4. In: *Neurotoxicity of Nanomaterials and Nanomedicine* (eds. X. Jiang and H. Gao), 75–98. Cambridge, MA: Academic Press.

89 Xiaoli, F. and Longquan, S. (2018). Neurotoxicity of nanomaterials, Chapter 20. In: *Emerging Nanotechnologies in Dentistry*, 2e (eds. K. Subramani and W. Ahmed), 421–444. Norwich, NY: William Andrew Publishing.

90 Bencsik, A., Lestaevel, P., and Guseva Canu, I. (2018). Nano- and neurotoxicology: an emerging discipline. *Prog. Neurobiol.* 160: 45–63.

91 Karmakar, A., Zhang, Q., and Zhang, Y. (2014). Neurotoxicity of nanoscale materials. *J. Food Drug Anal.* 22: 147–160.

92 Jia, G., Wang, H., Yan, L. et al. (2005). Cytotoxicity of carbon nanomaterials: single-wall nanotube, multi-wall nanotube, and fullerene. *Environ. Sci. Technol.* 39 (5): 1378–1383.

93 Yang, K., Li, Y., Tan, X. et al. (2013). Behavior and toxicity of graphene and its functionalized derivatives in biological systems. *Small* 9 (9–10): 1492–1503.

94 Kim, J.S., Yoon, T.-J., Yu, K.-N. et al. (2006). Toxicity and tissue distribution of magnetic nanoparticles in mice. *Toxicol. Sci.* 89 (1): 338–347.

95 Elder, A., Gelein, R., Silva, V. et al. (2006). Translocation of inhaled ultrafi ne manganese oxide particles to the central nervous system. *Environ. Health Perspect.* 114: 1172–1178.

96 Pietroiusti, A. (2012). Health implications of engineered nanomaterials. *Nanoscale* 4: 1231–1247.

97 Campagnolo, L., Massimiani, M., Magrini, A. et al. (2012). Physico-chemical properties mediating reproductive and developmental toxicity of engineered nanomaterials. *Curr. Med. Chem.* 19: 4488–4494.

98 Kreyling, W.G., Semmler-Behnke, M., Takenaka, S., and Möller, W. (2012). Differences in the biokinetics of inhaled nano-versus micrometer-sized particles. *Acc. Chem. Res.* 46 (3): 714–722.

99 Pietroiusti, A., Magrini, A., and Campagnolo, L. (2014). Mechanisms of nanomaterial toxicity, Chapter 2. In: *Health and Environmental Safety of Nanomaterials* (eds. J. Njuguna, K. Pielichowski and H. Zhu), 28–43. Woodhead Publishing.

100 Hotze, E.M., Phenrat, T., and Lowry, G.V. (2010). Nanoparticle aggregation: challenges to understanding transport and reactivity in the environment. *J. Environ. Qual.* 39: 1909–1924.

101 Albanese, A. and Chan, W.C.W. (2011). Effect of gold nanoparticle aggregation on cell uptake and toxicity. *ACS Nano* 5: 5478–5489.

102 Fu, P.P., Xia, Q., Hwang, H.M. et al. (2014). Mechanisms of nanotoxicity: generation of reactive oxygen species. *J. Food Drug Anal.* 22 (1): 64–75.

103 Manke, A., Wang, L., and Rojanasakul, Y. (2013). Mechanisms of nanoparticle-induced oxidative stress and toxicity. *Biomed. Res. Int.* 2013 (1): 1–15.

104 Nel, A., Xia, T., Mädler, L. et al. (2006). Toxic potential of materials at the nanolevel. *Science* 311 (5761): 622–627.

105 Shvedova, A.A., Pietroiusti, A., Fadeel, B. et al. (2012). Mechanisms of carbon nanotube-induced toxicity: focus on oxidative stress. *Toxicol. Appl. Pharmacol.* 261 (2): 121–133.

106 Shvedova, A., Castranova, V., Kisin, E.R. et al. (2003). Exposure to carbon nanotube material: assessment of nanotube cytotoxicity using human keratinocyte cells. *J. Toxicol. Environ. Health A* 66 (20): 1909–1926.

107 Farcal, L.R., Uboldi, C., Mehn, D. et al. (2012). Mechanisms of toxicity induced by SiO_2 nanoparticles of in vitro human alveolar barrier: effects on cytokine production, oxidative stress induction, surfactant proteins A mRNA expression and nanoparticles uptake. *Nanotoxicology* 7 (6): 1095–1110.

108 Eom, H.J. and Choi, J. (2010). p38 MAPK activation, DNA damage, cell cycle arrest and apoptosis as mechanisms of toxicity of silver nanoparticles in Jurkat T cells. *Environ. Sci. Technol.* 44 (21): 8337–8342.

109 Lenaz, G. (2001). The mitochondrial production of reactive oxygen species: mechanisms and implications in human pathology. *IUBMB Life* 52 (3–5): 159–164.

110 Elsabahy, M. and Wooley, K.L. (2013). Cytokines as biomarkers of nanoparticle immunotoxicity. *Chem. Soc. Rev.* 42 (12): 5552–5576.

111 Thannickal, V.J. and Fanburg, B.L. (2000). Reactive oxygen species in cell signaling. *Am. J. Phys.* 279 (6): L1005–L1028.

112 Huang, Y.W., Wu, C., and Aronstam, R.S. (2010). Toxicity of transition metal oxide nanoparticles: recent insights from in vitro studies. *Materials* 3 (10): 4842–4859.

113 Valko, M., Rhodes, C.J., Moncol, J. et al. (2006). Free radicals, metals and antioxidants in oxidative stress-induced cancer. *Chem. Biol. Interact.* 160 (1): 1–40.

114 Tabet, L., Bussy, C., Amara, N. et al. (2008). Adverse effects of industrial multi-walled carbon nanotubes on human pulmonary cells. *J. Toxicol. Environ. Health A* 72 (2): 60–73.

115 OECD (2003). *OECD Series on Testing and Assessment 44 ENV/JM/MONO 15.* Paris: OECD.

116 Risk Commission. (2003). Ad hoc commission on "revision of risk analysis procedures and structures as well as of standard setting in the field of environmental health in the Federal Republic of Germany". Final Report (Berlin).

117 Rocks, S., Pollard, S., Dorey, R. et al. (2008). *Comparison of Risk Assessment Approaches for Manufactured Nanomaterials. Final Report*. London: Defra.

118 SCENIHR (2009). *Opinion on Risk Assessment of Products of Nanotechnologies*. European Commission: Brussels http://ec.europa.eu/health/ph_risk/committees/04_scenihr/docs/scenihr_o_023.pdf.

119 Aschberger, K., Micheletti, C., Sokull-Klüttgen, B., and Christensen, F.M. (2011). Analysis of currently available data for characterising the risk of engineered nanomaterials to the environment and human health-lessons learned from four case studies. *Environ. Int.* 37 (6): 1143–1156.

120 Regulation EC/1907/2006 of the European Parliament and the Council of 18 December 2006 concerning the Registration, Evaluation, Authorization and Restriction of Chemicals (REACH). OJ L 396/1, 30 December 2006. https://eur-lex.europa.eu/legal-content/EN/TXT/PDF/?uri=CELEX:32006R1907&from=EN.

121 SCENIHR (2012). *Opinion on the Memorandum on the Use of the Scientific Literature for Human Health Risk Assessment Purposes – Weighing of Evidence and Expression of Uncertainty*. Brussels: European Commission http://ec.europa.eu/health/scientific_committees/emerging/docs/scenihr_s_001.pdf.

122 Hauschild, M., Jeswiet, J., and Alting, L. (2005). From life cycle assessment to sustainable production: status and perspectives. *CIRP Ann. Manuf. Technol.* 54 (2): 1–21.

123 Grieger, K.D., Laurent, A., Miseljic, M. et al. (2012). Analysis of current research addressing complementary use of life-cycle assessment and risk assessment for engineered nanomaterials: have lessons been learned from previous experience with chemicals? *J. Nanopart. Res.* 14: 958–981.

124 Zuin, S., Micheletti, C., Critto, A. et al. (2011). Weight of evidence approach for the relative hazard ranking of nanomaterials. *Nanotoxicology* 5 (3): 445–458.

125 Olson, M.S. and Gurian, P.L. (2012). Risk assessment strategies as nanomaterials transition into commercial applications. *J. Nanopart. Res.* 14 (4): 1–7.

126 Tervonen, I., Linkov, J., Figueira, J. et al. (2009). Risk-based classification system of nanomaterials. *J. Nanopart. Res.* 11 (4): 757–766.

127 O'Brien, N.J. and Cummins, E.J. (2011). A risk assessment framework for assessing metallic nanomaterials of environmental concern: aquatic exposure and behavior. *Risk Anal.* 31 (5): 706–726.

11

Economic Aspects of Nanomaterials (NMs) for Sample Preparation

11.1 Introduction

Nanotechnology is the design and engineering of systems at the molecular or atomic scale. These systems usually exhibit sizes of less than 100 nm. Richard Feynman was one of the first scientists to mention the possibilities of a nano-world [1]. Since then, nanotechnology has rapidly spread and conquered almost all the industries all over the world including electronics [2], construction [3], cosmetics [4], agriculture [5], and medicine [6].

The attractiveness of nanomaterials (NMs) is explained by their unique properties such as mechanical, electrical, magnetic, and optical. At the nanoscale, electromagnetic forces are dominant and gravitational forces are negligible. Considerably smaller than bulk solids but still larger than individual atoms or molecules, NM properties do not obey to the classical physics either to the absolute quantum mechanical laws but lie in between. Furthermore, the ratio of surface to core atoms is much higher than that at the macroscale. Consequently, a slight change in diameter has important impact on the properties.

For example, optical properties of gold nanoparticles (Au NPs) are considerably different from those of bulk gold as Au colloidal solutions appear red, whereas bulk Au is yellow because of the influence of electron confinement on the absorption properties. Second, the melting point for Au NPs is considerably lower than that for bulk Au (e.g. 300–400 °C vs. 1064 °C) because of surface melting [7]. Third, bulk gold is inert; however, Au NPs are efficient catalysts in mild conditions [8–10].

Their remarkable features have efficiently been applied to many applications in today's world [11–28] and new opportunities are still appearing. The most common examples of NM applications exploit their (i) high surface-to-volume ratio, in particular for catalytic purposes, (ii) narrow size distribution, which provides them uniform properties, and (iii) ability to assemble into organized structures via molecular bonding for high-performance films for instance. Customers of the cosmetic industry are familiar with NPs including liposomes, spherical vesicles made of one or several phospholipid layers, being included in their formulation as delivery carriers of active ingredients such as pharmaceuticals.

On the other hand, TiO_2 and ZnO_2 NPs are usually incorporated in sun creams for their enhanced absorption of UV light but also for their interesting reflection and

Sample Preparation with Nanomaterials: Next Generation Techniques and Applications,
First Edition. Chaudhery Mustansar Hussain, Rüstem Keçili, and Chaudhery Ghazanfar Hussain.

scattering properties compared to the microscale because they offer transparency to the final formulation [29]. Ag NPs are successfully employed in water treatment systems, wound dressings, medical devices, and textiles because of their excellent antimicrobial activities [30–32].

As can be seen from the examples mentioned above, NMs have found so many application fields because of their unique features. However, in recent years, there has been a considerable shift in extensive studies on human potential and environmental exposures of NM-based consumer products. Therefore, this chapter aims to describe the toxicity concerns and economic aspects of NMs.

11.2 Toxicity Concerns of NMs

NMs exhibit distinguished toxicological features in comparison to the same substance in the larger micron-sized (bulk) form [33]. The toxicity is a function of particle number and surface area rather than mass. The decrease in size can attenuate to the toxicity of the NMs because they become electrochemically reactive and their structure in nanoscale renders the ability to enter the bloodstream and translocate in cells, tissues, and functional organelles [34]. This toxic behavior facilitates the higher possibility of interacting with biological surroundings [35] and having toxic features including inflammation, genotoxicity [36], autophagy, neurotoxicity, and ultimately cell death [37]. Figure 11.1 shows the schematic depiction of the interactions of NMs with the endocrine system.

The shape and size of the NMs directly impact the antibacterial activity feature. This was confirmed by a research carried out by Agnihotri and Mukherji [39]. In their research, size-controlled Ag NPs in the range between 5 and 100 nm were successfully synthesized and the achieved results showed that the effect was significantly enhanced as the size of the Ag NPs approached the sub-10 nm range, and the 5 nm Ag NPs demonstrated the fastest bactericidal activity compared to the Ag NPs

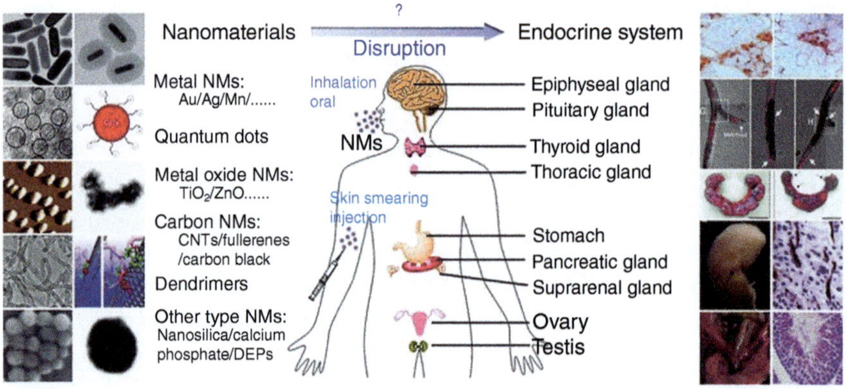

Figure 11.1 The schematic depiction of the interactions of NMs with the endocrine system. Source: Republished with permission from Lu et al. [38]. © 2013, John Wiley & Sons.

in the range from 7 to 10 nm. A number of researchers synthesized Ag NPs with different sizes and shapes and reported the anisotropic shapes of Ag particles, such as nanoplates or triangular nanoprisms [40]. The smaller size will provide the rapid penetration, whereas the larger sized particles (i.e. 1000 nm) were not capable of translocating through this barrier [41]. It is very crucial to have a basic knowledge to recognize that not all the NMs exhibit toxic features [42], which mainly depend on the chemical composition, size, and shape of the related NMs. Consequently, nanotoxicology research is now gaining much more attention, and its importance is gradually being recognized in recent years [38].

Extensive effort on toxicological research confirmed that NMs are potentially harmful and can lead to a change and inhibition of cell growth causing to different pathophysiological states in and aquatic organisms, animals, and humans [43].

Several works indicated that the enhanced toxic features of NPs mainly depend on the decrease in their size. In vitro and in vivo research studies conducted by exposure to NMs exhibited that smaller particles display greater toxicological impacts. Smaller sizes of the particles attribute to larger surface areas. Shilo et al. demonstrated the toxic effect of Au NPs. They synthesized a series of the Au NPs of sizes varying between 20 and 110 nm [44]. Among the different-sized particles, the 70 nm Au NPs show the highest ability of penetration in an in vitro blood–brain barrier model, resulting in a maximum concentration in brain cells.

The evaluation of the risks exerted by NMs is a challenge for scientists, people in the industry, and regulatory agencies all over the world. Experimental evidence points toward significant toxicity for a range of NMs both in vivo and in vitro. Extensive efforts aim at uncovering the underlying mechanisms of this toxicity. In the Trojan horse mechanism, NMs are usually employed as Trojan horses [45] that are loaded with pharmaceutical compounds, genes, or proteins that are aimed at targeting and destroying the infected cells internalized within the cells and then efficiently release high levels of toxic ions, as particle degradation is strongly promoted by the lysosomal environment.

11.3 Global Market for NM-Based Products

As of May 2020, according to Statnano [46], the number of products using NMs is 9006 manufactured by 2445 different companies in 61 countries, mainly in the cosmetic, textile, and construction industries with United States being the country with the largest quantity and a variety of products containing nanotechnology. Besides United States, China, Germany, Switzerland, Japan, United Kingdom, South Korea, Iran, Russia, and Taiwan are the other top countries that produce a large quantity and a variety of NM-based products.

On the other hand, according to the data provided by The Nanodatabase [47], most NM-based products increased over years as can be seen from Figure 11.2.

On the other hand, the data provided by The Nanodatabase indicated that Ag is the most extensively used NM in most commercial products (Figure 11.3). Thus, nanotechnology is everywhere in our life including cosmetics, medicines, clothing,

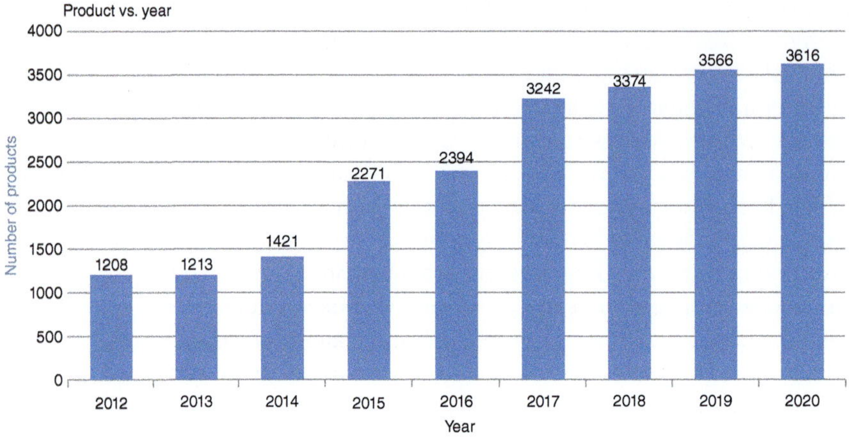

Figure 11.2 Number of products using NMs in the period from 2012 to 2020. Source: The Nanodatabase, CONSUMER PRODUCTS, Dashboard: Product vs Year. © 2020, The Nanodatabase. Retrived from: https://nanodb.dk/en/analysis/consumer-products/#chartHashsection

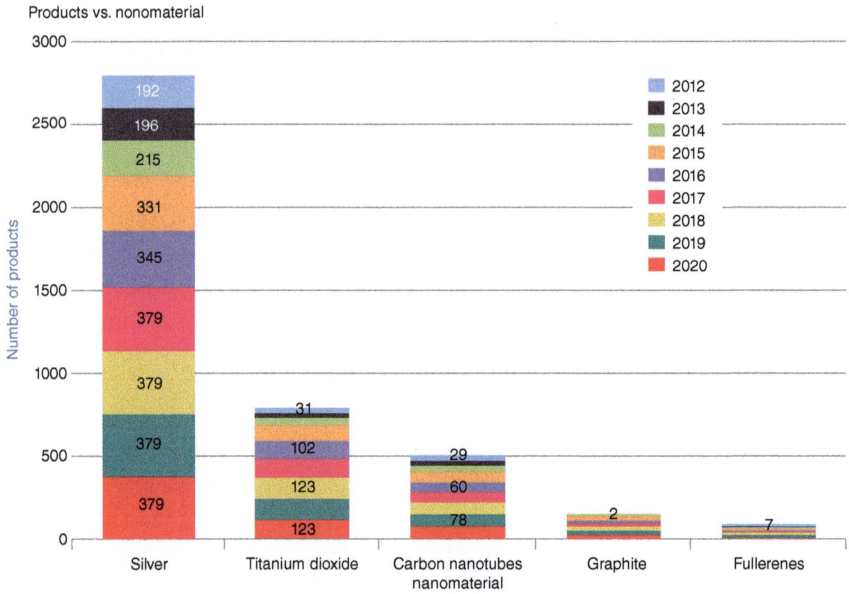

Figure 11.3 Number of products and types of NMs produced in the years between 2012 and 2020. Source: The Nanodatabase, CONSUMER PRODUCTS, Dashboard: Product vs Nanomaterial. © 2020, The Nanodatabase. Retrived from: https://nanodb.dk/en/analysis/consumer-products/#chartHashsection

automobiles, computers, and monitors, adding new functionalities, intelligence, integrations, portability, and networking capability in many new products with high market potential [48].

It is estimated that commercial production of NMs is expected to reach 100 000 t over the next decade. In spite of the unique features, superiorities, and great application of NMs, it is also known that these NMs can potentially lead to deleterious impacts to environment and humans. Thus, it is an absolutely urgent need for directives for their safe usage.

According to the market analysis for nanotechnology reported by Adıgüzel [49], the nanotechnology market all over the world is expected to exceed a net value of approximately US$ 124 billion by 2024. Because nanotechnological applications are rapidly growing within the biomedicine, cosmetic, energy, electronic, automotive, defense, and agricultural sectors, the increase in the support provided by the governments and private sectors to develop nanotechnology is also expected to gradually increase. In the United States, the National Nanotechnology Initiative was founded in 2001, which is a collaboration of a number of agencies and federal departments for the development and commercialization of nanotechnology-based products.

The National Nanotechnology Initiative not only supported the extensive efforts on the studies and advancement of nanotechnology in the United States but also provided excellent collection of articles in many prestigious journals having a high impact factor on the global stage, especially over the past 10 years. One of the most important challenges which the community in science faces with regard to nanotechnology progresses is the ability to integrate NMs into novel systems following the discovery of their unique features at nanoscale. There are also several agencies in the National Nanotechnology Initiative including the Nano Manufacturing and Small Business Innovation Research that are specifically focused on the development of the production technologies that are not urgently needed to transfer newly designed and developed NMs into novel products for commercial use.

On the other hand, the European Union invested approximately €896 million in NM-based research in the years from 2007 to 2011. The global investment in NMs is estimated to be approximately a quarter of a trillion USD, with both China and the United States investing more than US$ 2 billion. With so much effort and investment, it is understandable that nanotechnology is continuing its rapid growth. A large number of NM-based products including cosmetics, coatings (paints and lacquers), antibacterial clothing, and food products are commercially available in the European market and the market is expanding day by day [49].

11.4 Conclusions

Nanotechnology is a highly multidisciplinary area drawing attention from various scientific fields including materials science, physics, chemistry, biology, and engineering. Instruments produced by using NMs are going to be high-quality products providing high efficiency with low cost in the future. Nanotechnology can be seen as an extension of the existing sciences into the nanoscale or as a recasting of the existing sciences using a new, innovative, and more modern term that has a crucial role in a variety of industrial applications. In the future, extensive efforts should be made for the development of unique features of new NMs for innovative applications. Particularly, the current status and trends of studies on NMs should be updated, and industrial development related to the application of novel NMs in various sectors such as cosmetic, engineering, biomedical, and environmental fields should also be made. In addition, the most interesting and crucial scientific papers, expert opinions by regulatory agencies, and patent literature from different countries concerning the application of novel NMs in dermatological, dental, and hair care products should be intended for improving the studies on this field. Also, the novel NMs including biopolymeric materials and nanofibrous scaffolds from a variety of materials would be possibly designed for effluent treatment, food and cosmetic industries, and hard and soft tissue engineering.

However, as nanotechnology revolutionizes the society by designing and developing novel NM-based products and processes, it also leads to significant concerns about the risks and safety issues of NMs associated with their different applications. In this sense, it is also needed to learn more about how NMs are changing the route of creating products in the modern industry. One of the crucial issue emphasize to note that is the difficulty in comparing the toxicity data of NMs reported in the current literature, because not only to the large spectrum of approaches and techniques of for the design and synthesis of NMs but also to the lack of systematic research reporting an enough physical/chemical characterization experiments of the sample investigated in the researches.

References

1 Feynman, R.P. (1960). There's plenty of room at the bottom. *Eng. Sci.* 23 (5): 22–36.

2 Dahman, Y. (2017). Electronic and electro-optic nanotechnology, Chapter 9. In: *Micro and Nano Technologies, Nanotechnology and Functional Materials for Engineers* (eds. Y. Dahman, D. Le, N. Niznik and N. Sadyathasan), 191–206. Elsevier.

3 Oke, A.E., Aigbavboa, C.O., and Semenya, K. (2017). Energy savings and sustainable construction: examining the advantages of nanotechnology. *Energy Procedia* 142: 3839–3843.

4 Nafisi, S. and Maibach, H.I. (2017). Nanotechnology in cosmetics. In: *Cosmetic Science and Technology* (eds. K. Sakamoto, R.Y. Lochhead, H.I. Maibach and Y. Yamashita), 337–369. Elsevier.

5 Chhipa, H. (2019). Applications of nanotechnology in agriculture, Chapter 6. In: *Methods in Microbiology*, vol. 46 (eds. V. Gurtler, A.S. Ball and S. Soni), 115–142. Academic Press.

6 Davidovits, P. (2019). Nanotechnology in biology and medicine, Chapter 18. In: *Physics in Biology and Medicine*, 5e (ed. P. Davidovits), 293–305. Academic Press.

7 Daniel, M.-C. and Astruc, D. (2004). Gold nanoparticles: assembly, supramolecular chemistry, quantum-sized related properties and applications towards biology catalysis and nanotechnology. *Chem. Rev.* 104: 296–346.

8 Haruta, M., Kobayashi, T., Sano, H., and Yamada, N. (1987). Novel gold catalysts for the oxidation of carbon monoxide at a temperature far below 0°C. *Chem. Lett.* 16: 405–408.

9 Hvolbæk, B., Janssens, T.V.W., Clausen, B.S. et al. (2007). Catalytic activity of Au nanoparticles. *Nano Today* 2: 14–18.

10 Perks, B. (2010). Gold fever. *Chemistry World* 3: 48–50.

11 Palit, S. and Hussain, C.M. (2020). Green nanomaterials: a sustainable perspective. In: *Green Nanomaterials. Advanced Structured Materials*, vol. 126 (eds. S. Ahmed and W. Ali), 23–41. Singapore: Springer.

12 Singh, S.B. and Hussain, C.M. (2020). Functionalized nanographene for catalysis, Chapter 5. In: *Micro and Nano Technologies, Handbook of Functionalized Nanomaterials for Industrial Applications* (ed. C.M. Hussain), 111–129. Elsevier.

13 Palit, S. and Hussain, C.M. (2020). Functionalization of nanomaterials for industrial applications: recent and future perspectives, Chapter 1. In: *Micro and Nano Technologies, Handbook of Functionalized Nanomaterials for Industrial Applications* (ed. C.M. Hussain), 3–14. Elsevier.

14 Hussain, C.M. and Keçili, R. (2020). *Modern Environmental Analysis Techniques for Pollutants*. Amsterdam, The Netherlands: Elsevier.

15 Büyüktiryaki, S., Keçili, R., and Hussain, C.M. (2020). Functionalized nanomaterials in dispersive solid phase extraction: advances and prospects. *Trends Anal. Chem.* https://doi.org/10.1016/j.trac.2020.115893.

16 Keçili, R., Büyüktiryaki, S., and Hussain, C.M. (2019). Advancement in bioanalytical science through nanotechnology: past, present and future. *Trends Anal. Chem.* 110: 259–276.

17 Hussain, C.M. and Kharisov, B. (2017). *Advanced Environmental Analysis-Application of Nanomaterials*. The Royal Society of Chemistry.

18 Hussain, C.M. (2017). Magnetic nanomaterials for environmental analysis. In: *Advanced Environmental Analysis-Application of Nanomaterials* (eds. C.M. Hussain and B. Kharisov). The Royal Society of Chemistry.

19 Keçili, R., Büyüktiryaki, S., Dolak, İ., and Hussain, C.M. (2020). The use of magnetic nanoparticles in sample preparation devices and tools, Chapter 5. In: *Handbook of Nanomaterials in Analytical Chemistry* (ed. C.M. Hussain), 75–95. Elsevier.

20 Keçili, R., Büyüktiryaki, S., and Hussain, C.M. (2020). Membrane applications of nanomaterials, Chapter 7. In: *Handbook of Nanomaterials in Analytical Chemistry* (ed. C.M. Hussain), 159–182. Elsevier.

21 Büyüktiryaki, S., Keçili, R., and Hussain, C.M. (2020). Modern age of analytical chemistry: nanomaterials, Chater 2. In: *Handbook of Nanomaterials in Analytical Chemistry* (ed. C.M. Hussain), 29–40. Elsevier.

22 Keçili, R., Büyüktiryaki, S., and Hussain, C.M. (2020). Micro total analysis systems with nanomaterials, Chapter 8. In: *Handbook of Nanomaterials in Analytical Chemistry* (ed. C.M. Hussain), 185–198. Elsevier.

23 Büyüktiryaki, S., Sümbelli, Y., Keçili, R., and Hussain, C.M. (2019). Lab-on-chip platforms for environmental analysis. In: *Encyclopedia of Analytical Science*, 3e (eds. P. Worsfold, C. Poole, A. Townshend and M. Miró), 267–273. Academic Press.

24 Keçili, R. and Hussain, C.M. (2018). Recent progress of imprinted nanomaterials in analytical chemistry. *Int J. Anal. Chem.*: 8503853.

25 Keçili, R. and Hussain, C.M. (2018). *Mechanism of Adsorption on Nanomaterials Nanomaterials Chromatography*, Chapter 4 (ed. C.M. Hussain), 89–115. Elsevier.

26 Keçili, R., Sibel Büyüktiryaki, S., and Hussain, C.M. (2018). Engineered nanosensors based on molecular imprinting technology, Chapter 57. In: *Chaudhery Mustansar Hussain, In Micro and Nano Technologies, Handbook of Nanomaterials for Industrial Applications*, vol. 2018, 1031–1046. Elsevier.

27 Keçili, R., Özcan, A.A., Ersöz, A. et al. (2011). Superparamagnetic nanotraps containing MIP based mimic lipase for biotransformations uses. *J. Nanopart. Res.* 13 (5): 2073–2079.

28 Say, R., Keçili, R., and Ersöz, A. (2016). Molecularly imprinted polymer-based micro- and nanotraps for solid-phase extraction. In: *Advanced Molecularly Imprinting Materials* (eds. A. Tiwari and L. Uzun). Wiley.

29 Smijs, T.G. and Pavel, S. (2011). Titanium dioxide and zinc oxide nanoparticles in sunscreens: focus on their safety and effectiveness. *Nanotechnol. Sci. Appl.* 4: 95–112.

30 Le Ouay, B. and Stellacci, F. (2015). Antibacterial activity of silver nanoparticles: a surface science insight. *Nano Today* 10: 339–354.

31 Abou El-Nour, K.M.M., Eftaiha, A.A., Al-Warthan, A., and Ammar, R. (2010). Synthesis and applications of silver nanoparticles. *Arab. J. Chem.* 3: 135–140.

32 Mirjalili, M., Yaghmaei, N., and Mirjalili, M. (2013). Antibacterial properties of nano silver finish cellulose fabric. *J. Nanostructure Chem.* 3: 43.

33 Yong, K., Roy, I., Law, W., and Hu, R. (2010). Synthesis of cRGD-peptide conjugated near-infrared CdTe/ZnSe core–shell quantum dots for in vivo cancer targeting and imaging. *Chem. Commun.* 46 (38): 7136.

34 Tucci, P., Porta, G., Agostini, M. et al. (2013). Metabolic effects of TiO_2 nanoparticles, a common component of sunscreens and cosmetics, on human keratinocytes. *Cell Death Dis.* 4 (3): e549.

35 García, A., Delgado, L., Torà, J. et al. (2012). Effect of cerium dioxide, titanium dioxide, silver, and gold nanoparticles on the activity of microbial communities intended in wastewater treatment. *J. Hazard. Mater.* 199–200: 64–72.

36 Shukla, R., Sharma, V., Pandey, A. et al. (2011). ROS-mediated genotoxicity induced by titanium dioxide nanoparticles in human epidermal cells. *Toxicol. In Vitro* 25 (1): 231–241.

37 Zhang, L., Jiang, Y., Ding, Y. et al. (2006). Investigation into the antibacterial behaviour of suspensions of ZnO nanoparticles (ZnO nanofluids). *J. Nanopart. Res.* 9 (3): 479–489.

38 Lu, X., Liu, Y., Kong, X. et al. (2013). Nanotoxicity: a growing need for study in the endocrine system. *Small* 9: 1654–1671.

39 Agnihotri, S. and Mukherji, S. (2014). Size-controlled silver nanoparticles synthesized over the range 5-100 nm using the same protocol and their antibacterial efficacy. *RSC Adv.* 4 (8): 3974–3983.

40 Chernousova, S. and Epple, M. (2012). Silver as antibacterial agent: ion, nanoparticle, and metal. *Angew. Chem. Int. Ed.* 52 (6): 1636–1653.

41 Jani, P., Halbert, G., Langridge, J., and Florence, A. (1989). The uptake and translocation of latex nanospheres and microspheres after oral administration to rats. *J. Pharm. Pharmacol.* 41 (12): 809–812.

42 Oberdörster, G., Maynard, A., Donaldson, K. et al. (2005). Principles for characterizing the potential human health effects from exposure to nanomaterials: elements of a screening strategy. *Part. Fibre Toxicol.* 2 (1): 8.

43 Lai, M.B., Jandhyam, S., Dukhande, V.V. et al. (2008). Differential cytotoxicity of metallic oxide nanoparticles in mammalian cells. *TechConnect Briefs* 2: 130–133.

44 Shilo, M., Sharon, A., Baranes, K. et al. (2015). The effect of nanoparticle size on the probability to cross the blood-brain barrier: an in-vitro endothelial cell model. *J. Nanobiotechnol.* 13 (1): 19.

45 Niu, Q., Huang, X., Zhang, L. et al. (2010). Trojan horse mechanism of bacterial pathogenesis against nematodes. *Proc. Natl. Acad. Sci. U. S. A.* 107 (38): 16631–16636.

46 https://product.statnano.com (accessed 7 May 2020).

47 http://nanodb.dk/en/analysis/consumer-products/#chartHashsection (accessed 7 May 2020).

48 Nanofutures (2019). European initiative for sustainable development by Nanotechnologies. http://nanofutures.eu (accessed 7 May 2020).

49 Adıguzel, O. (2019). Nanotechnology 2020 market analysis. *Biomater. Med. Appl.* 3: 1.

12

Legal Aspects of Nanomaterials (NMs) for Sample Preparation

12.1 Introduction

Because of the limited performance of conventional bulk-scale materials, the development of novel nanomaterials (NMs) with enhanced properties has attracted increasing great attention of research from different fields of science. Significant developments were made in nanotechnology over the past years. Compared with the traditional materials, NMs have large surface area-to-volume ratio and great superiorities including electrical, mechanical, optical, and morphological features. These unique features of NMs make them great candidates in many application areas [1–18].

On the other hand, although NMs are employed in attractive and promising application fields, there are still some issues and challenges about their negative impacts on the environment and human health. One of the crucial issues is their toxicity. The ongoing research studies are focused on the investigation of toxic features and development of the international regulatory guidelines, which is vital for the advancements of further technology. There are still various drawbacks that need to be considered and overcome before the NM-based products are commercially available in near future. Some of the main concerns are the design and fabrication of biocompatible, nontoxic, environmentally friendly, and low-cost NMs. Some other issues of these unique materials are their manufacturing in large-scale and commercialization processes.

The aim of this chapter is to describe and highlight various concerns, legal and regulatory issues, and challenges associated with NMs. It starts with the safety issues of NMs. Then, regulatory issues of NMs (i.e. ethical concerns in environmental effects of NMs, ethical concerns in occupational health and safety of workers, ethical concerns of NMs in food, and ethical concerns of NMs in drugs, cosmetics, and human health) are discussed. Finally, conclusions and future perspectives are presented.

12.2 Safety Issues of NMs

With the rapid advancements in nanotechnology, the design and production of engineered NMs dramatically increase, as well as incidental NMs. Despite the excellent

Sample Preparation with Nanomaterials: Next Generation Techniques and Applications,
First Edition. Chaudhery Mustansar Hussain, Rüstem Keçili, and Chaudhery Ghazanfar Hussain.

properties and many great applications of the NMs, there are still some questions, challenges, drawbacks, and concerns about the impacts of these NMs on human health and environment. One of the potential concerns of NMs is their toxicity and safety. Most of the NMs will end up in the environment (air, water, and soil). Because most of the NMs are not biodegradable, they may remain in the environment for a long time. Biological organisms such as microorganisms, plants, and animals could take up NMs in the ecosystems and bring NMs into the food chain. For example, nanoparticles (NPs) can easily enter the human body via the respiratory system. Once the NPs enter the bloodstream after passing through the respiratory system, they can accumulate in certain regions of different organs [19]. Therefore, the negative impacts caused by the NMs on human health should be carefully evaluated. For this purpose, a new field called "nanoecotoxicology" is gaining public attention [20].

On the other hand, the other issue that should always be considered is biomolecular coronas. The term "corona," i.e. the biomolecular layer that surrounds the NMs, was first introduced by Cedervall et al. [21]. The biomolecular corona mainly interacts with biological systems and thereby constitutes a main element of the biological identity of the NMs affecting multiple interactions in molecular scale [22, 23].

The formation of a biomolecular corona (i.e. protein corona) decreases the surface energy of the NMs, endows them with new biological identities, and affects not only the targeting and delivery features of NMs but also their toxic features [24–26].

As for the targeting of NMs, the physical and chemical characteristics of NMs play a crucial role in the potential toxic or nontoxic impacts exerted by the biomolecular corona. Moreover, even in the case of NMs having the same physical and chemical features, it is not easy to compare the toxicological profile of the biomolecular corona formed in different conditions because the relative quantities of the adsorbed biomolecules can significantly change [27].

The biomolecular coronas of functionalized NMs are far from being an already resolved issue in basic as well as in nanotoxicology. Even regulatory authorities begin to accept that corona signatures may be relevant for risk assessment and prediction of functionalized NMs. Although the relationship between NM design and physiological responses was extensively investigated, only some general principles emerged. Fortunately, the biomolecule corona is no longer ignored as an unknown parameter and is acknowledged as a potential but yet a fully unexploited opportunity to understand and predict the effects of functionalized NMs on the physiological medium.

12.3 Regulatory Aspects of NMs

The economic potential of the produced novel NMs is threatened by a limited understanding of the related environment, health, and safety concerns. Although toxicity data of NMs are continuously becoming available, the relevance to regulators is usually unproven and unclear. The shrinking time to market innovative NMs drives the need for urgent action by the international regulatory agencies. As an example, NANoREG is the first FP7 project to provide the answers needed by regulators and legislators on environment, health, and safety concerns by linking them to a scientific evaluation of the achieved data and test techniques [28, 29]. In addition, NANoREG aims to provide legislators with a set of tools for risk assessment and decision-making tools in the short term and medium term by evaluating the data and risk assessment, including exposure, monitoring, and control, to a selected number of NMs already used in products; to develop, in the long term, new testing strategies, adapted to a large number of NMs, where many factors can affect their environmental and health impact; to establish a close collaboration between governments and industries on the knowledge needed to properly manage risks; and to provide the basis for conventional approaches, mutually acceptable data sets, and risk management practices.

Besides, under the Certification, Labeling and Packaging (CLP) Regulation [30] compounds including NMs exist in the European market are required to be notified to the European Chemicals Agency (ECHA) according to their hazard categorization, regardless of their tonnage [31]. In this sense, careful regulation for NMs is advanced in the European Union (EU) throughout the community and in each EU member state.

For the regulation of NMs, the US Environmental Protection Agency (USEPA) asks manufacturers to provide scientific evidence that the use of those NMs does not lead to environmental damage and public health risks [32]. Therefore, specific investigations on the behavior are required to evaluate which type of NM exists.

12.3.1 Ethical Concerns in the Environmental Effects of NMs

The Parma Declaration on Environment and Health issued by the Fifth Ministerial Conference on Environment and Health [33] considered the implications of nanotechnology to be an important issue equivalent to bioaccumulating and endocrine disrupting compounds. NPs were specifically mentioned as pollutants responsible for respiratory diseases, whereas the existence of NPs and NMs in products was considered to be an issue that needs careful investigation. The European Code of Conduct [34] stated that the studies on nanotechnology should contribute to the Millennium (Sustainable) Development Goals and should do no harm to animals, plants, and the environment. A portion of this book has dealt with the analysis of risks and impacts of nanotechnologies. The full potential benefits that nanotechnologies could bring with them in regard to reaching the goals are still unclear, and similarly, it remains difficult to estimate all possible impacts [35]. One possible positive impact can be seen in the nanoformulations of pesticides and fertilizers: if these materials were more effective than conventional counterparts, this would decrease the amount of chemicals used and reduce the risks to workers.

The detectability and traceability of NMs in the environment, including air, water, and soil, are critical issues that need significant additional research. Accurate and detailed exposure assessment is still elusive in many instances, and concern is great where children and other vulnerable subgroups are involved [36]. It is also highly likely that exposure will not be equally distributed for all countries, populations, or individuals; this creates an obvious concern with regard to the principles of equity.

From the perspective of toxicity relevant to environmental effects, the EU approach categorizes NMs based on their similarities in physicochemical, toxicological, ecotoxicological, and environmental fate properties [37]. A specific procedure for assessment of environmental impacts should likely be required for nanomedicines or nanopharmaceuticals, which can reach water and soil after transfer through or metabolism in the human body.

The USEPA has issued a "science in ACTION" leaflet to explain to the lay community about the research being done on NMs and nanotechnologies in order to protect the environment and human health. Applications of NMs to environmental problems, such as pollutant remediation, are described to highlight the benefits of these new technologies [38].

12.3.2 Ethical Concerns in Occupational Health and Safety of Workers

Working environments are the places where the risk of exposure to high concentrations of engineered NMs is the greatest. The precautionary principle mandates controls in the work environment, and when there is a lack of adequate information about nanotechnology hazards and risks, employers need to implement additional control measures. This situation may result in difficult choices for employers. As noted earlier, the risk assessment for nanotechnologies is performed on a case-by-case basis. There are added difficulties of effective risk communication, as well as how to obtain acceptance by workers: there are difficulties in obtaining the timeliness of communication, and there are requirements to include workers in decision-making.

A common problem encountered in work environments is that workers often have little control over their working conditions because of the pressures from economic and social conditions. Regarding the need for medical surveillance and screening of workers, the usual ethical problems also apply to nanotechnologies: how to stimulate voluntary participation of workers, how to plan actions in the case of positive results, and how to protect the privacy of the results [39].

A final consideration on possible consequences involves the economics of nanotechnology production: the development of new production approaches based on nanoscience could displace traditional markets, affecting the employment of workers in traditional sectors [40].

On the subject of worker safety, the World Health Organization (WHO) released standard guidelines on protecting workers from potential risks of manufactured NMs [41] that are based on the precautionary principle and advocate the involvement of workers in all phases of risk assessment and control. The Organisation for Economic Co-operation and Development has also issued documents on NM exposure in workplaces. Documents prepared by the European Agency for Safety and Health at Work (EU-OSHA) advocate the use of "control banding" procedures for risk assessment and management to assist in minimizing the exposure of workers and provide employers with accessible tools to ensure compliance and safety. However, these documents do not provide guidance relative to situations in which "susceptible" workers might be exposed.

12.3.3 Ethical Concerns of NMs in Food

NMs existing in food were recently included in the EU Directive on Novel Foods [42] that also includes the currently valid description of NMs. Official documents dealing with NMs in food are usually released by the EU by the European Food Safety Agency (EFSA) [43]. Issues concerning NMs are carefully considered in several aspects of food, feed, and flavorings. NMs in food can be intentionally added or derived from manufacturing processes. The documents issued by EFSA do not explain the benefits derived from inclusion of NMs in food processing or manufacturing. The documents mainly address the evaluation of potential hazards because of the specific features of materials in nanoscale that can lead to a significant change in the toxicokinetic behavior of compounds.

Labeling should inform consumers about the existence of NMs in food products, but it seems that currently no food product is correctly labeled. Perceived benefits of nanotechnologies can be envisaged in the design and fabrication of sensitive sensor platforms and measures for improving the safety of food.

On the other hand, U.S. Food and Drug Administration (FDA) [44] released a guidance concerning NMs in food ingredients, additives, and food contact compounds. The existence of poisonous or deleterious compounds makes the food adulterated and unsafe. In addition, it is a responsibility of the producer and of the end user to ensure the safe use of food.

Nanotechnology can be seen as a potential source of altered features in food products, which requires special testing methodologies to assess the safety of food. In this case, FDA does not consider nanoengineered food to be intrinsically harmful or benign.

WHO and Food and Agriculture Organization (FAO) of the United Nations together released a document [45] that addresses all concerns of NMs in agriculture and food.

12.3.4 Ethical Concerns of NMs in Drugs, Cosmetics, and Human Health

One of the crucial reference documents for any application to humans and health is the Convention on Human Rights and Biomedicine also called "Oviedo Convention" that protects individuals against exploitation [46].

According to Article 1, "Parties to this Convention shall protect the dignity and identity of all human beings and guarantee everyone, without discrimination, respect for their integrity and other rights and fundamental freedoms with regard to the application of biology and medicine." In addition, the European Charter of Fundamental Rights ensures human dignity, equality, and solidarity [47].

The application of nanotechnology in medicine and related fields should consider some crucial issues that can affect the rights of individuals. The first issue is to establish the collateral benefits and harmful impacts of nanomedicine. The benefits can be estimated by considering the current and future applications. Considering the extensive use of NMs in drug release or designing and fabrication of novel and sensitive sensor systems, a main concern is that knowledge on the possible impacts on the human body is still incomplete.

One of the most promising and crucial applications of nanomedicine is designing and development of novel nanotheranostics systems, which is the integration of diagnostic and therapeutic functions in one system using the benefits of nanotechnology. The combination of nanosensor systems and nanoenabled pharmaceutical compounds with communication technology can efficiently provide the acquisition of personal data in real time. Employing this approach, nanotechnology in personalized medicine will bring issues of privacy protection and patient communication to the forefront.

One additional consequence of nanomedicine that should be considered is the concept of distributive justice. There is potential unequal access to new nanotechnologies that are quite expensive for the diagnostic or therapeutic applications, causing inequity and potential discrimination among patients. Other ethical issues that can be touched by nanotechnology are gender equality, ethnic equality, and consideration for pluralism in lifestyles. These approaches were investigated by the European Medicines Authority for the EU and by the FDA for the United States. The FDA states in all the guidance documents that they do not consider NMs to be intrinsically harmful or benign. Thus, all concerns are product specific and evaluated on a case-by-case basis. Concerning pharmaceutical compounds, a specific mention is made of the fact that NMs can interact with biological systems through different routes depending on the different parameters including age or gender and this can introduce elements of inequality in medicinal treatment.

On the other hand, in the EU, a specific directive covers NMs present in cosmetic products [48], which require appropriate labeling with an indication of the nanoformulation of the ingredient. Importantly, the NMs may or may not have toxic features; the labeling indication should only ensure the freedom of preference by the customers. On the other hand, there are also long-standing ethical concerns, which are studies on animals. These studies are replaced with alternative approaches [49].

12.4 Conclusions

This chapter attempted to provide an overview of the concerns, legal and regulatory issues, and challenges associated with NMs. For the production of NMs, good manufacturing practices should be encouraged. On the other hand, the ethical concerns of NMs were previously addressed in various documents released by the regulatory agencies involved in their design, development, and regulation. However, some undefined fields still exist, and there is an urgent need for a more targeted effort especially to address eventual inequalities in approaches and unfairness toward classification of workers and end users.

The future applications of functionalized NMs are highly promising. Therefore, great efforts and evaluations on the safety of NMs on the environment and animal health will continue to be a research priority for the advances in nanotechnology. Consequently, sophisticated statistical understanding and methodologies are carefully needed in the designing of toxicity studies and in analyzing toxicity data. Prior to the use of functionalized NMs especially for in vivo biological applications, various important physiological factors (i.e. distribution, absorption and toxicity) of these NMs should be considered and carefully evaluated according to the international regulatory guidelines that are still under development [50].

References

1 Palit, S. and Hussain, C.M. (2020). Green nanomaterials: a sustainable perspective. In: *Green Nanomaterials: Advanced Structured Materials*, vol. 126 (eds. S. Ahmed and W. Ali). Singapore: Springer.

2 Singh, S.B. and Hussain, C.M. (2020). Functionalized nanographene for catalysis, Chapter 5. In: *Micro and Nano Technologies, Handbook of Functionalized Nanomaterials for Industrial Applications* (ed. C.M. Hussain), 111–129. Elsevier.

3 Palit, S. and Hussain, C.M. (2020). Functionalization of nanomaterials for industrial applications: recent and future perspectives, Chapter 1. In: *In Micro and Nano Technologies, Handbook of Functionalized Nanomaterials for Industrial Applications* (ed. C.M. Hussain), 3–14. Elsevier.

4 Hussain, C.M. and Keçili, R. (2020). *Modern Environmental Analysis Techniques for Pollutants*. The Netherlands: Elsevier.

5 Büyüktiryaki, S., Keçili, R., and Hussain, C.M. (2020). Functionalized nanomaterials in dispersive solid phase extraction: advances and prospects. *Trends Anal. Chem.* https://doi.org/10.1016/j.trac.2020.115893.

6 Keçili, R., Büyüktiryaki, S., and Hussain, C.M. (2019). Advancement in bioanalytical science through nanotechnology: past, present and future. *Trends Anal. Chem.* 110: 259–276.

7 Hussain, C.M. and Kharisov, B. (2017). *Advanced Environmental Analysis-Application of Nanomaterials*. The Royal Society of Chemistry.

8 Hussain, C.M. (2017). Magnetic nanomaterials for environmental analysis. In: *Advanced Environmental Analysis-Application of Nanomaterials* (eds. C.M. Hussain and B. Kharisov). The Royal Society of Chemistry.

9 Keçili, R., Büyüktiryaki, S., Dolak, İ., and Hussain, C.M. (2020). The use of magnetic nanoparticles in sample preparation devices and tools, Chapter 5. In: *Handbook of Nanomaterials in Analytical Chemistry* (ed. C.M. Hussain), 75–95. Elsevier.

10 Keçili, R., Büyüktiryaki, S., and Hussain, C.M. (2020). Membrane applications of nanomaterials, Chapter 7. In: *Handbook of Nanomaterials in Analytical Chemistry* (ed. C.M. Hussain), 159–182. Elsevier.

11 Büyüktiryaki, S., Keçili, R., and Hussain, C.M. (2020). Modern age of analytical chemistry: nanomaterials, Chater 2. In: *Handbook of Nanomaterials in Analytical Chemistry* (ed. C.M. Hussain), 29–40. Elsevier.

12 Keçili, R., Büyüktiryaki, S., and Hussain, C.M. (2020). Micro total analysis systems with nanomaterials, Chapter 8. In: *Handbook of Nanomaterials in Analytical Chemistry* (ed. C.M. Hussain), 185–198. Elsevier.

13 Büyüktiryaki, S., Sümbelli, Y., Keçili, R., and Hussain, C.M. (2019). Lab-on-chip platforms for environmental analysis. In: *Encyclopedia of Analytical Science*, 3e

(eds. P. Worsfold, C. Poole, A. Townshend and M. Miró), 267–273. Academic Press.

14 Keçili, R. and Hussain, C.M. (2018). Recent progress of imprinted nanomaterials in analytical chemistry. *Int. J. Anal. Chem.*: 8503853.

15 Keçili, R. and Hussain, C.M. (2018). Mechanism of adsorption on nanomaterials. In: *Nanomaterials in Chromatography* (ed. C.M. Hussain), 89–115. Elsevier.

16 Keçili, R., Sibel Büyüktiryaki, S., and Chaudhery Mustansar Hussain, C.M. (2018). Engineered nanosensors based on molecular imprinting technology, Chapter 57. In: *Micro and Nano Technologies, Handbook of Nanomaterials for Industrial Applications* (ed. C.M. Hussain), 1031–1046. Elsevier.

17 Keçili, R., Özcan, A.A., Ersöz, A. et al. (2011). Superparamagnetic nanotraps containing MIP based mimic lipase for biotransformations uses. *J. Nanopart. Res.* 13 (5): 2073–2079.

18 Say, R., Keçili, R., and Ersöz, A. (2016). Molecularly imprinted polymer-based micro- and nanotraps for solid-phase extraction. In: *Advanced Molecularly Imprinting Materials* (eds. A. Tiwari and L. Uzun). Wiley.

19 Nikodinovska, V.V., Mladenovska, K., and Grozdanov, A. (2015). Risks and health effects from exposure to engineered nanostructures: a critical review. *J. Chem. Technol. Metall.* 50: 117–134.

20 Oberdörster, G., Oberdörster, E., and Oberdörster, J. (2005). Nanotoxicology: an emerging discipline evolving from studies of ultrafine particles. *Environ. Health Perspect.* 113: 823–839.

21 Cedervall, T., Lynch, I., Lindman, S. et al. (2007). Understanding the nanoparticle-protein corona using methods to quantify exchange rates and affinities of proteins for nanoparticles. *Proc. Natl. Acad. Sci. U. S. A.* 104: 2050–2055.

22 Treuel, L., Brandholt, S., Maffre, P. et al. (2014). Impact of protein modification on the protein corona on nanoparticles and nanoparticle-cell interactions. *ACS Nano* 8: 503–513.

23 Treuel, L., Eslahian, K.A., Docter, D. et al. (2014). Physicochemical characterization of nanoparticles and their behavior in the biological environment. *Phys. Chem. Chem. Phys.* 16: 15053–15067.

24 Tenzer, S., Docter, D., Kuharev, J. et al. (2013). Rapid formation of plasma protein corona critically affects nanoparticle pathophysiology. *Nat. Nanotechnol.* 8 (10): 772–781.

25 Zeng, L., Gao, J., Liu, Y. et al. (2019). Role of protein corona in the biological effect of nanomaterials: investigating methods. *Trends Anal. Chem.* 118: 303–314.

26 Duan, Y., Coreas, R., Liu, Y. et al. (2020). Prediction of protein corona on nanomaterials by machine learning using novel descriptors. *NanoImpact* 17: 100207.

27 Corbo, C., Molinaro, R., Parodi, A. et al. (2016). The impact of nanoparticle protein corona on cytotoxicity, immunotoxicity and target drug delivery. *Nanomedicine (Lond.)* 11 (1): 81–100.

28 NANoREG (2019). A common European approach to the regulatory testing of manufactured nanomaterials. http://www.nanoreg.eu. (accessed 5 May 2020).

29 NANoREG (2017). NANoREG framework for the safety assessment of nanomaterials. https://ec.europa.eu/jrc/en/publication/eur-scientific-and-technical-

research-reports/nanoreg-framework-safety-assessment-nanomaterials (accessed 5 May 2020).
30 Occupational Safety and Health Administration (2008). CLP — Classification, Labelling and Packaging of substances and mixtures. http://publications.europa.eu/resource/cellar/a0da21bf-b795-4b30-983b-f146aa15e594.0006.04/DOC_1 (accessed 5 May 2020).
31 United States Environmental Protection Agency (US EPA) (2017). https://echa.europa.eu/en/home (accessed 5 May 2020).
32 U.S. Environmental Protection Agency (2017). Control of nanoscale materials under the toxic substances control act. https://www.epa.gov/reviewing-new-chemicals-under-toxic-substances-control-act-tsca/control-nanoscale-materials-under (accessed 5 May 2020).
33 Fifth Ministerial Conference on Environment and Health (2010). Parma declaration on environment and health. EUR/55934/5.1 Rev.2. http://www.euro.who.int/__data/assets/pdf_file/0011/78608/E93618.pdf (accessed 5 May 2020).
34 Commission of the European Communities (2008). Commission Recommendation of 07/02/2008 on a Code of Conduct for Responsible Nanosciences and Nanotechnologies Research, 424 final.
35 EEA (2015). *The European Environment-State and Outlook 2015: Synthesis Report*. Copenhagen: European Environment Agency.
36 WHO (2013). Nanotechnology and human health: scientific evidence and risk governance. In: *Report of the WHO Expert Meeting* (10–11 December 2012), Bonn, Germany. Copenhagen: WHO Regional Office for Europe.
37 Arts, J.H.E., Hadi, M., Keene, A.M. et al. (2014). A critical appraisal of existing concepts for the grouping of nanomaterials. *Regul. Toxicol. Pharmacol.* 70: 492–506.
38 U.S. Environmental Protection Agency, Office of Research and Development (2013). https://www.epa.gov/sites/production/files/2013-12/documents/nanotechnology-fact-sheet.pdf (accessed 5 May 2020).
39 Schulte, P.A. and Salamanca-Buentello, F. (2007). Ethical and scientific issues of nanotechnology in the workplace. *Environ. Health Perspect.* 115: 5–12.
40 Sodano, V. (2018). Nano-food regulatory issues in the European Union. *AIP Conf. Proc.* 1990: 020018.
41 WHO (2017). WHO guidelines on protecting workers from potential risks of manufactured nanomaterials. https://apps.who.int/iris/bitstream/handle/10665/259671/9789241550048-eng.pdf;jsessionid=709A4CECA53C9DC1154E8379FFCB89AB?sequence=1 (accessed 5 May 2020).
42 European Union (2015). Regulation (EU) 2015/2283 of the European Parliament and of the Council of 25 November 2015 on novel foods, amending Regulation (EU) no 1169/2011 of the European Parliament and of the Council and repealing Regulation (EC) no 258/97 of the European Parliament and of the Council and Commission Regulation (EC) no 1852/2001. *Off. J. Eur. Union*: L327/1–L327/22.
43 World Health Organization (WHO) (2017). Guidance on the risk assessment of the application of nanoscience and nanotechnologies in the food and feed chain.

https://efsa.onlinelibrary.wiley.com/doi/epdf/10.2903/j.efsa.2011.2140 (accessed 5 May 2020).

44 Food and Drug Administration (FDA) (2014). https://www.fda.gov/ScienceResearch/SpecialTopics/Nanotechnology/ucm602536.htm (accessed 5 May 2020).

45 FAO/WHO (Food and Agriculture Organization of the United Nations/World Health Organization) (2013). FAO/WHO Paper: State of the Art on the Initiatives and Activities Relevant to Risk Assessment and Risk Management of Nanotechnologies in the Food and Agriculture Sectors, Geneva.

46 Council of Europe (1997). Convention for the protection of human rights and dignity of the human being with regard to the application of biology and medicine: convention on human rights and biomedicine. https://www.coe.int/en/web/conventions/full-list/-/conventions/treaty/164 (accessed 5 May 2020).

47 The European Parliament, the Council and the Commission (2012). *Charter of Fundamental Rights of the European Union*. Official Journal of the European Union. C326/391.

48 European Commission (2009). *Council Regulation (EC) 1223/2009 of the European Parliament and of the Council of 30 November 2009 on Cosmetic Products (Recast)*. Official Journal of the European Union, L342/59.

49 SCCS (Scientific Committee on Consumer Safety) (2012). Guidance on the safety assessment of nanomaterials in cosmetics. SCCS/1481/12.

50 Holl, M.M.B. (2009). Nanotoxicology: a personal perspective. *WIREs Nanomedi. Nanobiotechnol.* 1: 353–359.

13

Monitoring of Nanomaterials (NMs) in the Environment

13.1 Introduction

Nanotechnology refers to the study, application, and characterization of materials, structures, and devices with a size on the nanoscale. By definition, nanotechnology involves the manipulation of matter in a way that produces structures with at least one dimension ranging from 1 to 100 nm in size [1–10]. Nanomaterials (NMs) have unique physicochemical properties that differentiate them from their bulk counterparts [11–18].

NMs can be categorized into different types according to their origin, dimensions, and elemental composition. Based on their origin, NMs are divided into naturally occurring and engineered nanomaterials (ENMs). Naturally occurring NMs are those commonly produced by nature (e.g. volcanic ash, ocean spray, soot from forest fires, viruses, and protein molecules). ENMs refer to materials that are engineered by manipulation of matter at the nanoscale [19]. Silver NPs (Ag NPs) for antibacterial applications or carbon nanotubes (CNTs) for the strengthening of materials are examples of NMs. Furthermore, a subgroup of NMs exists and is recognized as incidentally occurring NMs that are the by-product of man-made processes such as welding fumes, diesel engine exhaust, cigarette smoke, and paint pigments. NMs are currently employed in a wide range of applications and fields, including the medical, chemical, pharmaceutical, and environmental industries [20]. The main objective of nanoscience is to explore and exploit unique chemical, physical, electrical, and mechanical properties that emerge when matter is structured at the nanoscale [21].

Two main parameters are responsible for the increased interest in NMs relative to their bulk material counterparts. First, nanoscale solids contain a higher fraction of atoms at their surfaces than bulk materials. This phenomenon causes an increase in the specific surface area, leading to a significant increase in the surface reactivity. Second, at the nanoscale, quantum confinement effects (electrons confined in small area) govern the properties and interactions of matter. The physicochemical and intrinsic properties of NMs can be significantly different from those of the same material (elemental composition) in bulk form [22].

Inside the unlimited diversity of these materials, some NMs may exhibit toxic features to biological species, others are relatively benign, and yet others are produced for their health benefits. It must be noted that the same features that make NMs

Sample Preparation with Nanomaterials: Next Generation Techniques and Applications,
First Edition. Chaudhery Mustansar Hussain, Rüstem Keçili, and Chaudhery Ghazanfar Hussain.

unique with beneficial technological applications may also endanger human beings directly through the potential induction of cyto- and genotoxic effects, inflammation, and cancer [23, 24]. In addition, the environmental health and hazard risks associated with applications of nanotechnology for industrial and commercial uses are not fully known [25]. Furthermore, the physicochemical features associated with NMs not only differ from the corresponding bulk matter but also between different forms of NMs with the same chemical composition [26, 27]. NMs containing the same chemical or elemental composition with different morphological characteristics (e.g. size, shape, and surface properties) can exhibit different physicochemical features [28]. This phenomenon further complicates the investigation of these materials as a plethora of NMs from several materials in many forms and sizes with a variety of surface coatings are possible. Therefore, the potential risks and benefits associated with the utilization of NMs should be carefully evaluated concentrating on the specific physicochemical features of each NM. Furthermore, potential alterations to the physicochemical features of NMs, caused by time or environmental conditions, may cause dissimilar environmental and public health implications and thus must be taken into careful consideration.

This chapter demonstrates a systematic approach to investigate NMs in the environment. For this purpose, toxicity and safety concerns of NMs are presented in Section 13.2. Then, the main sources and transport routes of NMs are described. In Section 13.4, requirements of analytical approaches are discussed. Then, sampling and separation of NMs in the environmental samples are also demonstrated. Finally, various detection techniques for the characterization of NMs are presented.

13.2 Toxicity and Safety Concerns of NMs

Nanotechnology and nanoscience involve the scientific research, design, preparation, characterization, and application of extremely small materials at nanoscale between 1 and 100 nm. Nanotechnology can successfully be applied in several branches of science such as engineering, chemistry, physics, biology, materials science, pharmaceutical science, and medicine. All of these branches of science involve the design, preparation, characterization, and application of NMs.

In recent years, the scientists put so much effort into the design and preparation of NMs to take advantage of their excellent features such as lighter weight, higher strength, large surface area, small size, enhanced control of light spectrum, and excellent chemical reactivity than their counterparts at the microscale [29]. The physicochemical features of these unique materials in the range between 1 and 100 nm are completely different compared to the same materials at the microscale. For example, Au NPs are red in color, whereas the bulk gold particles exhibit yellow color because of the interaction between light and the restricted electrons of the Au NPs. The surface area of the materials increases when its size decreases. The physicochemical features of the materials in the nanoscale can efficiently be controlled by structural design, incorporation of appropriate functional groups or compounds, and modification of the surface.

Despite the unique features and several applications of these NMs, there are still some questions, challenges, drawbacks, and concerns about the impacts of NMs on human health and environment. One of the potential concerns of NMs is their toxicity and safety. For example, NPs have the potential to enter the human body via the respiratory system. They may also come in contact with the skin and cause some health problems. The toxicity features of NMs are affected by many physicochemical properties such as chemical composition, reactivity, and particle size. [30]. Even small amounts of these NMs may have serious toxic effects because they have large surface area that is usually correlated with high reactivity by formation of free radicals. When they enter the bloodstream after passing through the respiratory system, they may accumulate at the specific sites of various organs. They may even directly penetrate into the brain [31]. Thus, the effects of NMs on human health should be carefully investigated and defined. Recently, a lot of efforts were put into the development of innovative techniques for the evaluation of toxic effects of these NMs. These intensive efforts led to birth of a new field of toxicology, which is called "nanotoxicology" [32].

The material safety and risks on health and environment need to be carefully investigated because each NM is unique [33]. Before using the NMs for in vivo biological applications, the crucial physiological factors such as distribution, absorption, and toxicity of NMs should be considered and evaluated. The potential toxicity of NMs should be investigated according to the international regulatory guidelines, which are still under development [34, 35].

The researchers should also be careful during the disposal step of the NMs. Disposal of the NMs may generate the new forms of environmental nanowastes, which is another challenge for the recycling efforts and waste management.

NMs have also potential toxic impacts on the environment. The release of these materials into the environment causes environmental pollution and ecotoxicity. For example, NPs are rapidly released into the environment as a natural consequence of the rapid increase in nanoparticle-containing consumer products and nanoparticle production facilities. This release results in the NPs being introduced into the air, water, or soil by various processes. The fate of NPs in the environment depends on the type, shape, physicochemical properties, and the receptive environment characteristics. The release of NPs into the environment can occur during the production process or the completion of the life cycle of a product. Environmental oscillations of NPs in the industrial production stage are inevitable. In addition to production, NPs that are separated from the products during transport, storage, and use stages are released into the environment by discharges and leaks and are involved in the ecosystem with diffusion and diffusion and come into contact with air, water, nutrients, and humans as a part of the ecosystem Considering the application areas of nanoparticle-based products during the production in industrial scale, it has been observed that it is quite easy to spread the NPs to the environment by both water and air [36].

On the other hand, the generated by-products in the form of NPs typically come from emissions because of incomplete combustion of diesel fuel or nanosized minerals resulting from acid drainage in mining operations. The gradually

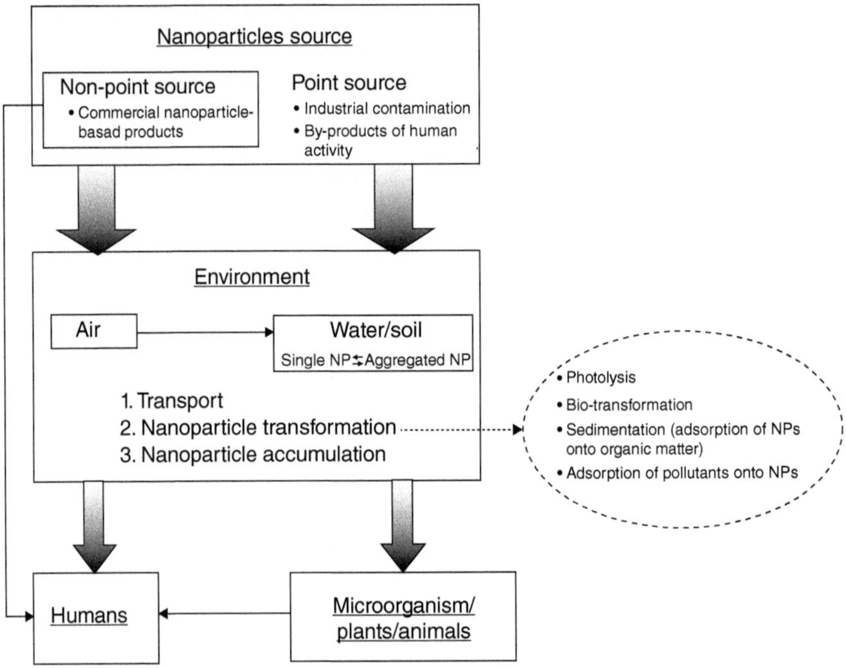

Figure 13.1 The schematic depiction of various routes and crucial transformation reactions of NPs in the environment. Source: Simonet and Valcárcel [39]. © 2008, Springer.

increasing manufacturing and extensive application of NMs will inevitably cause their accumulation in the environment that can have crucial implications for environmental and human health [37, 38]. Figure 13.1 shows the schematic depiction of various routes and crucial transformation reactions of NPs in the environment.

13.3 Main Sources and Transport Routes of Nanopollutants

The major types of environmental pollutants based on NMs (nanopollutants) that display toxic impacts on humans, animals, and plants are water pollutants, soil pollutants, and air pollutants. Water is one of the main transport routes of nanopollutants [40]. The transport of nanopollutants can occur under dissolved or particulate forms in the ground or surface water environment. The transport in a particulate form occurs for the nanopollutants in ground water, whereas the transport in a dissolved form in an aqueous environment is crucial because it is highly related to the more available forms of soil pollutants, which is a crucial issue. The nanopollutants can easily transport to the other environments from their main sources by this way. In surface water environments, the particles of the soil can be released into the water streams and transport under a particulate form by sliding and rolling and further accumulated in the downstream. This transport route of nanopollutants mainly depends on their shape, density, turbulence, and flow rate of the water. Most of the water pollutants are transported to seas and oceans through rivers. Pollutants transported through rivers in some regions of the world can also pollute the deltas and coasts miles away.

On the other hand, nanopollutants can easily enter the soil that significantly impacts the various organisms living in soil. However, the significant impact of these nanopollutants in soil on both animals and ecosystems is much more crucial because these nanopollutants can accumulate in food chains. Fossil fuels may also cause contamination of soil. Various sources of fossil fuel-based soil pollution involve petrochemical plants, refineries, and motor vehicles. On the other hand, indiscriminate application of various agricultural compound including herbicides and pesticides and improper disposal of industrial wastes are other causes of soil pollution [41].

The release of soil nanopollutants into the water or air can occur mainly via the changes in the status of water saturation in the soil, changes in the composition and chemistry of gaseous and water phase, and the changes in the properties of the particle surface. The release rate of nanopollutants may be relatively fast varying between minutes and hours or very slow (in years) depending on the type of the nanopollutant in the soil [42]. Transport of the nanopollutants in soil can also occur between biotic and abiotic environment [43]. Furthermore, the transport process in the abiotic environment has strong effect on the dispersion of the pollutants in soil and includes the migration of pollutants that are dissolved in the solution of soil and water. Wind is another route for the transport of soil pollutants. The transport of nanopollutants in soil can occur in aerosols or as windblown particles. The transport by wind is the most efficient way. Therefore, only nanopollutants can be transported through this route. The other parameters that affect the transportation of the soil pollutants by wind are the moisture content of the soil that promotes the aggregation of the pollutant particles and the vegetative layer that behaves as a barrier for the dispersion of pollutant particles by helping on their fixation. One of the crucial soil pollutants is the mixture of small particles generated during the mining processes where the small particle size fractions facilitate the dispersion and transport by wind to the far distances from the sources of soil pollutants [44, 45].

Air pollution can be defined as the existence of any solid, liquid, or gas compounds in the atmosphere at such concentrations which can directly or indirectly affect humans, animals, and plants. Air pollutants are categorized into two types that are natural resources and anthropogenic sources, which are related to human activity [46]. Natural sources of air pollutants are volcanic eruptions, forest fires, dust storms, ocean, seas, and plants. On the other hand, the main anthropogenic sources that caused air pollution are fossil fuels in the industry, mining, transportation, and construction of buildings. One of the major types of air pollutants are suspended materials (i.e. dust, smoke, and fume). Gas pollutants including carbon monoxide, nitrogen oxides, and sulfur dioxide are the other types of air pollutants.

Considering the water pollution concern that occurs especially in industrialized areas after industrial revolution, industrialization has a great impact on air pollution. In other words, the crucial part of this is to meet the energy needs that arise in order to increase the quality of life. The change in the proportion of gases in the air required for the continuation of life brings with it many problems in terms of living things. Any change in the rate of rapid rise and decline in these ratios disrupts the balance of the atmosphere. As a result of industrialization, urbanization, fast growth of population, heating, traffic, and improper land use, the amount of gas released into the atmospheric environment changes. As a result of this situation, not only the structure of the atmosphere will change but also dangerous environments will be created for the life of the living things [47].

13.4 Requirements of Analytical Approaches

The efficient monitoring of NPs in environmental samples entails meeting several requirements. One of these requirements is applying powerful analytical approaches that have ability to sensitively detect these NPs at relevant concentrations (i.e. ppm or ppb level) and the other requirement is avoiding the potential interferences of natural NPs usually existing in environmental samples [48, 49]. One crucial approach of meeting the previous requirements is by an effective combination of the existing approach with novel, developed, and innovative ones to provide efficient screening and sensitive detection of NPs. However, it should be noted that the toxic features of NPs depend on their morphology. The reactivity of NPs is usually dependent on their shape and size. This supports the applying efficient analytical techniques affording size measurements and being aware that different approaches can also provide the various size averages that depend on their particular basis. Therefore, the response from instrumental-based experiments can be achieved by employing a specific volume, mass, optical features, and number of NPs. The ensuing size average may be identical for spherical NPs having a narrow distribution in size that is rarely the case in practice. On the basis of the reported results, the analytical requirements for the efficient and accurate assessment of the toxic features of NPs are appropriate sampling, efficient sample treatment, separation NPs, and finally sensitive detection capabilities toward target NPs in the environmental samples.

13.5 Sampling of NMs in Environmental Samples

Sampling and sample preparation constitute crucial stages in the characterization of NMs including NPs. NPs in the environment exist in dynamic nonequilibrium systems that are usually sensitive to chemical and physical disturbances. Sampling of NPs and laboratory procedures can change the dispersion state of NPs in the environmental samples. In addition, the existence of natural NPs and organic matter in environmental samples can complicate the analytical processes. For example, studies on geological samples by employing transmission electron microscopy (TEM) revealed the existence of carbon NMs such as fullerenes in geological formations dating back almost 1.85 billion years [50]. In addition, fullerenes and CNTs were detected in a Greenland ice core dated as approximately 10 000 years old [51].

Because of the unstable dispersions of NPs in the environmental samples, the sensitive determination of these materials is preferably carried out in situ. However, this process is impossible because no portable instrument that has high sensitivity for this purpose is currently available. Therefore, alternative approaches and techniques leading to minimum perturbation from sampling step to analysis are applied instead. Various crucial issues including the adsorption of organic and inorganic NPs to sample bottles still require much investigation. The efficient analysis of NPs in soil and sediment samples is further complicated by the existence of increased amounts of natural solid particles of size similar to that of the target NPs. The application of dispersive techniques involving chemical dispersants such as surfactants or sonication can be efficient for the release of target NPs to the water phase. However, they can change the physicochemical state of the target NPs in the environmental sample. In fact, the effective separation of NPs from soil samples including various components is a challenging process. For this purpose, a centrifugation technique is usually suggested, but this process may induce some aggregation [52].

13.6 Separation of NMs in Environmental Samples

Separation of NMs in environmental samples is one of the most crucial and key stage in their analytical processing. The existing techniques for this purpose are still unrefined. For example, pristine fullerenes can be extracted in environmental samples such as water by exploiting their solubility in toluene [53]. However, CNTs are poorly soluble even in organic solvents that lead to various drawbacks and difficulties in their purification. In a reported approach, the extraction of carboxylic CNTs from surface water was successfully carried out by employing a filter modified with multiwalled CNTs as preconcentrators [54].

The efficient removal of the particles existing in environmental samples usually entails their prefractionation applying filtration of centrifugation techniques. Centrifugation approach is more effective in the removal of denser mineral particles. On the other hand, microfiltration process is extensively employed on account of its simplicity. NPs are deposited on the surface of a membrane via collision or electrostatic interactions. Cross-flow ultrafiltration process was extensively assessed for studying natural NPs. The achieved outcome of the fractionation process mainly depends on the nature of the membrane and the particular process conditions [55]. It is possible to use membrane materials having pore sizes smaller than 1 nm for the so-called nanofiltration. These membranes thus exhibit a high potential for the efficient analysis of NPs in environmental samples.

On the other hand, field-flow fractionation process is used for the separation of NPs according to particle size in terms of their diffusion coefficients in a very thin open channel. In this technique, separation process relies on the combination of an applied field and a longitudinal carrier flow. The field leads to transport of NPs to an accumulation wall. Smaller NPs will not be affected to the same extent as larger NPs. Monitoring size distribution entails using a coupled detector responding to the number of NPs or their mass concentration. One of the most commonly applied detection approaches here is inductively coupled plasma mass spectrometry (ICP-MS) technique that is particularly efficient and powerful for the characterization of metallic NPs [56]. Field-flow fractionation technique was efficiently employed for the analysis a large range of NPs such as SiO_2, CNTs, Au NPs, and Ag NPs. Size exclusion chromatography was also successfully used for the analysis of CNTs and fullerenes [57, 58]. In addition, efficient separation and purification of NMs including CNTs according to their size-to-length ratio was successfully performed by applying capillary electrophoresis technique [59–61].

13.7 Detection Techniques for the Characterization of NMs

Although excluded from real-time particulate detection suitability based on the need for sample preparation and subsequent lack of real-time information, techniques such as scanning electron microscopy (SEM), TEM, and atomic force microscopy (AFM) can offer very useful information on the morphology and topology of the target NPs. All the three techniques are extensively applied for the characterization of NPs in combination with other techniques. The AFM technique uses a cantilever with a very thin tip that oscillates over the surface of the sample. Also, a laser-based balance is used to measure both repulsive and attractive forces between the tip and the sample. The resulting image is an atomic force topogram. In contrast to electron microscopy, AFM can be applied to both conductive and nonconductive samples. In addition, AFM possesses spatial and vertical resolution. On the other hand, AFM provides partial about the chemical composition if the tip is coated with a material phobic or philic to the nanoparticle; it can also cause sample and tip damage and displacement of NPs and make data interpretation difficult.

TEM analysis may provide both qualitative and quantitative sample information with high resolution, leading to its regular use in biological studies. For example, Lasagna-Reeves and colleagues employed TEM to demonstrate the accumulation of Au NPs in mice and toxicity after repeated administration at different doses [62]. SEM analyses were also carried out in a number of studies. For example, SEM was applied to provide surface morphology information of the coatings added to polymeric NPs for the efficient release of anticancer drugs [63] and to investigate the antimicrobial features of Ag NPs [64].

There are significant limitations with regard to using microscopy to monitor airborne NPs. Prior to analysis, appropriate sampling methods and sample preparation procedures are required, making microscopy unsuitable for real-time detection of airborne nanoparticle samples. In addition, although quantitative information can be obtained, limited sample amounts can be visualized at any one time, thus providing a limited view with regard to exposure monitoring and uncertainty about the representativeness of the sample in addition to excessive time requirements.

One of the shortcomings of these high-resolution techniques is the use of a very small amount of NMs for the characterization studies, which can make it extremely difficult to provide representativeness of the achieved results. On the other hand, the information they provide can never be achieved by using the bulk analytical methods. One other common feature of these techniques is that almost all of them need some sample treatment steps such as evaporating a solvent that may lead to aggregation of NPs and precipitation of salts. Because of its subnanometer resolution, TEM is indispensable for accurately measuring NPs. In the TEM technique, an electron beam is transmitted through a conducting grid; for NPs, the grid frequently consists of copper embedded in a thin layer of resin. Once the beam is transmitted through the sample, non-adsorbed electrons are focused onto an imaging detector (i.e. a fluorescence screen). This has the advantage that it provides a direct image of the target NPs. TEM affords determinations for size distribution, particle size, and morphology of the target NPs. By applying suitable standards, the particle size values of the NPs can be successfully determined with an accuracy of $\pm 5\%$ [65]. This technique was effectively employed for the characterization of carbon NPs in ecotoxicological research studies [66].

Another efficient approach is scanning tunneling microscopy (STM) that relies on the monitoring of the tunneling current between a conductive or a semiconductive sample and a conductive tip. In addition to its high resolution (molecular to atomic scale), STM technique is quite sensitive to the chemical identity of the atoms at the surface [67]. Tip-enhanced Raman scattering is a related technique that enables molecular information on the nanometer scale.

The characterization of NMs can be conducted at the bulk-scale level as well. The techniques preferred for this purpose are employed to achieve the data about the chemical structure or composition of the target NMs. The average bulk chemical composition of a sample can be determined by using various spectroscopy-based approaches including inductively coupled plasma atomic emission spectroscopy (ICP-AES), atomic absorption spectroscopy, and ICP-MS. These techniques feature good limits of detection; also, the composition results they provide are sample averages and they afford multielement analysis. On the other hand, they are destructive and subject to matrix interferences. These techniques are especially recommended for the analysis of metallic NPs. When the number concentration is low (less than one particle per measurement), the signal gives spurious results because of slurry nebulization. Also, measurements may be unquantitative owing, for example, to incomplete atomization in elemental analyses. Particle size has proved to be the critical factor with a view to ensuring complete atomization for slurry nebulization in ICP-AES and ICP-MS. A particle size below 3–5 μm is recommended to ensure quantitative recovery [68, 69].

Fluorescence-based techniques can be an interesting choice for the analysis of NPs with a strong native fluorescence, which can be characterized from their adsorption or fluorescence spectra. Thus, the position of emission bands can be correlated with particle size and is commonly used to monitor size in the synthesis of quantum dots [70], for example. Also, particle concentrations can be quantified from adsorption or fluorescence measurements provided the optical constants for the NPs are known.

Vibrational spectroscopies, both Raman and infrared, are useful for the analysis of NPs (particularly carbon NPs). The two techniques differ in the excitation source used, the way signals are monitored, and the selection rules employed on vibrational modes. At present, Raman spectroscopy is a widely used, powerful tool for characterizing CNTs [71].

13.8 Conclusions

Environmental pollution is the main source of concerns to human health and sustainable development of society and economy all over the world. The presence of environmental pollutants including NMs in water, soil, and air is an important problem that drew great public concern and attention from many researchers in industry and scientists from different disciplines.

With the extensive use of NMs in commercial products, there is no doubt about the existence of various NMs such as nanoparticles in the environmental samples. Researchers and scientists should not only proclaim the benefits of nanotechnology to the society but they should also demonstrate the possible concerns, risk, and safety issues associated with its implementation.

Among the other environmental pollutants, NPs were efficiently characterized by employing various approaches and techniques. However, the accurate and sensitive analysis of NPs in the environmental samples still requires the design and development of innovative and powerful preconcentration approaches and the application of efficient, selective, and sensitive detection techniques.

References

1 Palit, S. and Hussain, C.M. (2020). Green nanomaterials: a sustainable perspective. In: *Green Nanomaterials. Advanced Structured Materials*, vol. 126 (eds. S. Ahmed and W. Ali), 23–41. Singapore: Springer.

2 Singh, S.B. and Hussain, C.M. (2020). Functionalized nanographene for catalysis, Chapter 5. In: *Micro and Nano Technologies, Handbook of Functionalized Nanomaterials for Industrial Applications* (ed. C.M. Hussain), 111–129. Elsevier.

3 Palit, S. and Hussain, C.M. (2020). Functionalization of nanomaterials for industrial applications: recent and future perspectives, Chapter 1. In: *In Micro and Nano Technologies, Handbook of Functionalized Nanomaterials for Industrial Applications* (ed. C.M. Hussain), 3–14. Elsevier.

4 Hussain, C.M. and Keçili, R. (2020). *Modern Environmental Analysis Techniques for Pollutants*. Netherlands: Elsevier.

5 Büyüktiryaki, S., Keçili, R., and Hussain, C.M. (2020). Functionalized nanomaterials in dispersive solid phase extraction: advances and prospects. *Trends Anal. Chem.* 127: 115893.

6 Keçili, R., Büyüktiryaki, S., and Hussain, C.M. (2019). Advancement in bioanalytical science through nanotechnology: past, present and future. *Trends Anal. Chem.* 110: 259–276.

7 Hussain, C.M. and Kharisov, B. (2017). *Advanced Environmental Analysis-Application of Nanomaterials*. The Royal Society of Chemistry.

8 Hussain, C.M. (2017). Magnetic nanomaterials for environmental analysis, Chapter 19. In: *Advanced Environmental Analysis-Application of Nanomaterials* (eds. C.M. Hussain and B. Kharisov). 1–13. The Royal Society of Chemistry.

9 Keçili, R., Büyüktiryaki, S., Dolak, İ., and Hussain, C.M. (2020). The use of magnetic nanoparticles in sample preparation devices and tools, Chapter 5. In: *Handbook of Nanomaterials in Analytical Chemistry* (ed. C.M. Hussain), 75–95. Elsevier.

10 Keçili, R., Büyüktiryaki, S., and Hussain, C.M. (2020). Membrane applications of nanomaterials, Chapter 7. In: *Handbook of Nanomaterials in Analytical Chemistry* (ed. C.M. Hussain), 159–182. Elsevier.

11 Büyüktiryaki, S., Keçili, R., and Hussain, C.M. (2020). Modern age of analytical chemistry: nanomaterials, Chapter 2. In: *Handbook of Nanomaterials in Analytical Chemistry* (ed. C.M. Hussain), 29–40. Elsevier.

12 Keçili, R., Büyüktiryaki, S., and Hussain, C.M. (2020). Micro total analysis systems with nanomaterials, Chapter 8. In: *Handbook of Nanomaterials in Analytical Chemistry* (ed. C.M. Hussain), 185–198. Elsevier.

13 Büyüktiryaki, S., Sümbelli, Y., Keçili, R., and Hussain, C.M. (2019). Lab-on-chip platforms for environmental analysis. In: *Encyclopedia of Analytical Science*, 3e (eds. P. Worsfold, C. Poole, A. Townshend and M. Miró), 267–273. Academic Press.

14 Keçili, R. and Hussain, C.M. (2018, 2018). Recent progress of imprinted nanomaterials in analytical chemistry. *Int. J. Anal. Chem.*: 8503853.

15 Keçili, R. and Hussain, C.M. (2018). Mechanism of adsorption on nanomaterials nanomaterials chromatography, Chapter 4. In: , 89–115. Elsevier.

16 Keçili, R., Sibel Büyüktiryaki, S., and Hussain, C.M. (2018). Engineered nanosensors based on molecular imprinting technology, Chapter 57. In: *In Micro and Nano Technologies, Handbook of Nanomaterials for Industrial Applications* (ed. C.M. Hussain), 1031–1046. Elsevier.

17 Keçili, R., Özcan, A.A., Ersöz, A. et al. (2011). Superparamagnetic nanotraps containing MIP based mimic lipase for biotransformations uses. *J. Nanopart. Res.* 13 (5): 2073–2079.

18 Say, R., Keçili, R., and Ersöz, A. (2016). Molecularly imprinted polymer-based micro- and nanotraps for solid-phase extraction. In: *Advanced Molecularly Imprinting Materials* (eds. A. Tiwari and L. Uzun). Wiley.

19 Colvin, V.L. (2003). The potential environmental impact of engineered nanomaterials. *Nat. Biotechnol.* 21: 1166.

20 Ju-Nam, Y. and Lead, J.R. (2008). Manufactured nanoparticles: an overview of their chemistry, interactions and potential environmental implications. *Sci. Total Environ.* 400: 396–414.

21 Buzea, C., Pacheco, I.I., and Robbie, K. (2007). Nanomaterials and nanoparticles: sources and toxicity. *Biointerphases* 2: MR17–MR71.

22 Roduncr, E. (2006). Size matters: why nanomaterials are different. *Chem. Soc. Rev.* 35: 583–592.

23 Borm, P.J., Robbins, D., Haubold, S. et al. (2006). The potential risks of nanomaterials: a review carried out for ECETOC. *Part. Fibre Toxicol.* 3: 11.

24 Yang, H., Liu, C., Yang, D. et al. (2009). Comparative study of cytotoxicity, oxidative stress and genotoxicity induced by four typical nanomaterials: the role of particle size, shape and composition. *J. Appl. Toxicol.* 29: 69–78.

25 Savolainen, K., Pylkkänen, L., Norppa, H. et al. (2010). Nanotechnologies, engineered nanomaterials and occupational health and safety – a review. *Saf. Sci.* 48: 957–963.

26 Nowack, B. and Bucheli, T.D. (2007). Occurrence, behavior and effects of nanoparticles in the environment. *Environ. Pollut.* 150: 5–22.

27 Schwirn, K., Tietjen, L., and Beer, I. (2014). Why are nanomaterials different and how can they be appropriately regulated under REACH? *Environ. Sci. Eur.* 26: 4.

28 Simon-Deckers, A., Loo, S., Mayne-L'hermite, M. et al. (2009). Size-, composition-and shape-dependent toxicological impact of metal oxide nanoparticles and carbon nanotubes toward bacteria. *Environ. Sci. Technol.* 43: 8423–8429.

29 Ahmadi, M., Elmongy, H., Madrakian, T., and Abdel-Rehim, M. (2017). Nanomaterials as sorbents for sample preparation in bioanalysis: a review. *Anal. Chim. Acta* 958: 1–21.

30 Nikodinovska, V.V., Mladenovska, K., and Grozdanov, A. (2015). Risks and health effects from exposure to engineered nanostructures: a critical review. *J. Chem. Technol. Metallurgy.* 50: 117–134.

31 Maynard, A.D., Aitken, R.J., Butz, T. et al. (2006). Safe handling of nanotechnology. *Nature* 444: 267–269.

32 Oberdörster, G., Oberdörster, E., and Oberdörster, J. (2005). Nanotoxicology: an emerging discipline evolving from studies of ultrafine particles. *Environ. Health Perspect.* 113: 823–839.

33 Ng, J.H. and Ilag, L.L. (2004). The dangers of generalization in nanotechnology. *Drug Discov. Today* 9: 60–61.

34 Oberdörster, G. (2010). Safety assessment for nanotechnology and nanomedicine: concepts of nanotoxicology. *J. Intern. Med.* 267: 89–105.

35 Holl, M.M.B. (2009). Nanotoxicology: a personal perspective. *WIREs Nanomed Nanobio.* 1: 353–359.

36 Kokdemir Ünşar, E. and Perendeci, A. (2016). Environmental fate of nanoparticles and their impacts on anaerobic digestion process. *Pamukkale Uni. J. Eng. Sci.* 22: 503–512.

37 Anastasio, C. and Martin, S.T. (2001). Atmospheric nanoparticles. *Rev. Mineral. Geochem.* 44: 293–349.

38 Wiesner, M.R., Lowry, G.V., Alvarez, P. et al. (2006). Assessing the risks of manufactured nanomaterials. *Environ. Sci. Technol.* 15: 4336–4345.

39 Simonet, B.M. and Valcárcel, M. (2009). Monitoring nanoparticles in the environment. *Anal. Bioanal. Chem.* 393: 17–21.

40 Geissen, V., Mol, H., Klumpp, E. et al. (2015). *Int. J. Soil Water Conserv. Res.* 3: 57–65.

41 Ashraf, M.A. and Hanfiah, M.H. (2017). Recent advances in assessment on clear water, soil and air. *Environ. Sci. Pollut. Res.* 24: 22753–22754.

42 National Research Council (2003). *Bioavailability of Pollutants in Soils and Sediments: Processes, Tools, and Applications*. Washington, DC: The National Academies Press.

43 Thibodeaux, L.J. (1996). *Environmental Chemodynamics: Movement of Chemicals in Air, Water, and Soil*, 2e. New York, NY: Wiley.

44 Ravi, S., D'Odorico, P., Over, T.M., and Zobeck, T.M. (2004). On the effect of air humidity on soil susceptibility to wind erosion: the case of air-dry soils. *Geophys. Res. Lett.* 31: 1–4.

45 Ravi, S., D'Odorico, P., Breshears, D.D. et al. (2011). Aeolian processes and the biosphere: interactions and feedback loops. *Rev. Geophys.* 49: RG3001.

46 Ghorani-Azam, A., Riahi-Zanjani, B., and Balali-Mood, M. (2016). Effects of air pollution on human health and practical measures for prevention in Iran. *J. Res. Med. Sci.* 21: 65.

47 Menteşe, S. (2017). Soil, water and air pollution in terms of environmental sustainability: theoretical review. *J. Int. Soc. Res.* 10 (53): 381–389.

48 Wigginton, N.S., Haus, K.L., and Hochella, M.F. (2007). Aquatic environmental nanoparticles. *J. Environ. Monit.* 9: 1306–1316.

49 Lead, J.R. and Wilkinson, K.J. (2006). Aquatic colloids and nanoparticles: current knowledge and future trends. *Environ. Chem.* 3: 159–171.

50 Becker, L., Bada, J.L., Winans, R.E. et al. (1994). Fullerenes in the 1.85-billion-year-old Sudbury impact structure. *Science* 265: 642–645.

51 Murr, L.E., Esquivel, E.V., Bang, J.J. et al. (2004). Chemistry and nanoparticulate compositions of a 10,000 year-old ice core melt water. *Water Res.* 38: 4282–4296.

52 Gimbert, L.J., Haygarth, P.M., Beckett, R., and Worsfold, P.J. (2006). The influence of sample preparation on observed particle size distributions for contrasting soil suspensions using flow field-flow fractionation. *Environ. Chem.* 3: 184–191.

53 Fortner, J.D., Lyon, D.Y., Sayes, C.M. et al. (2005). C60 in water: nanocrystal formation and microbial response. *Environ. Sci. Technol.* 39: 4370–4316.

54 Suárez, B., Moliner-Martínez, Y., Cárdenas, S. et al. (2008). Monitoring of carboxylic carbon nanotubes in surface water by using multiwalled carbon nanotube-modified filter as preconcentration unit. *Environ. Sci. Technol.* 42 (16): 6100–6104.

55 Liu, R. and Lead, J.R. (2006). Partial validation of cross flow ultrafiltration by atomic force microscopy. *Anal. Chem.* 78: 8105–8112.

56 Hu, J., Deng, D., Liu, R., and Lv, Y. (2018). Single nanoparticle analysis by ICPMS: a potential tool for bioassay. *J. Anal. At. Spectrom.* 33: 57–67.

57 Jackson, B.P., Ranville, J.F., Bertsch, P.M., and Sowder, A. (2005). Characterization of colloidal and humic-bound Ni and U in the "dissolved" fraction of contaminated sediment extracts. *Environ. Sci. Technol.* 39: 2478–2485.

58 Duesberg, G.S., Burghard, M., Muster, J. et al. (1998). Separation of carbon nanotubes by size exclusion chromatography. *Chem. Commun.* 3: 435–436.

59 López-Pastor, M., Domínguez-Vidal, A., Ayora-Cañada, M.J. et al. (2008). Separation of single-walled carbon nanotubes by use of ionic liquid-aided capillary electrophoresis. *Anal. Chem.* 80: 2672–2679.

60 Suárez, B., Simonet, B.M., Cárdenas, S., and Valcárcel, M. (2006). Separation of carbon nanotubes in aqueous medium by capillary electrophoresis. *J. Chromatogr. A* 1128: 282–289.

61 Doorn, S.K., Fields, R.E., Hu, H. et al. (2002). High resolution capillary electrophoresis of carbon nanotubes. *J. Am. Chem. Soc.* 124: 3169–3170.

62 Lasagna-Reeves, C., Gonzalez-Romero, D., Barria, M.A. et al. (2010). Bioaccumulation and toxicity of gold nanoparticles after repeated administration in mice. *Biochem. Biophys. Res. Commun.* 393: 649–655.

63 Yin Win, K. and Feng, S.S. (2005). Effects of particle size and surface coating on cellular uptake of polymeric nanoparticles for oral delivery of anticancer drugs. *Biomaterials* 26: 2713–2722.

64 Sondi, I. and Salopek-Sondi, B. (2004). Silver nanoparticles as antimicrobial agent: a case study on E. coli as a model for Gram-negative bacteria. *J. Colloid Interface Sci.* 275: 177–182.

65 Williams, D.B. and Carter, C.B. (1996). *Transmission Electron Microscopy: A Textbook for Materials Science*. New York, NY: Plenum.

66 Suzuki, Y., Kelly, S.D., Kemmer, K.M., and Banfield, J.F. (2002). Nanometre-size products of uranium bioreduction. *Nature* 419: 134–134.

67 Das, B.C. and Pal, A.J. (2008). Core-shell hybrid nanoparticles with functionalized quantum dots and ionic dyes: growth, monolayer formation, and electrical bistability. *ACS Nano* 2: 1930–1938.

68 Ebdon, L., Foulkes, M., and Sutton, K. (1997). Slurry nebulization in plasmas. *J. Anal. At. Spectrom.* 12: 213–229.

69 Santos, M.C. and Nobrega, J.A. (2006). Slurry nebulization in plasmas for analysis of inorganic materials. *Appl. Spectrosc. Rev.* 41: 427–448.

70 Yu, Q.Q., Qu, L., Guo, W., and Peng, X. (2003). Experimental determination of the extinction coefficient of CdTe, CdSe, and CdS nanocrystals. *Chem. Mater.* 15: 2854–2860.

71 Dresselhaus, M.S., Dresselhaus, G., and Jorio, A. (2007). Raman spectroscopy of carbon nanotubes in 1997 and 2007. *J. Phys. Chem. C* 111: 17887–17893.

14

Future Prospect of Sampling

14.1 Introduction

The proper choice of a measurement technique is the only one step in the development of a successful application. All of the steps leading up to the analyte measurement are equally important. The sampling and sample preparation processes begin at the point of collection and extends to the measurement step [1–4]. The proper collection of a sample during the sampling process also called the "primary sampling," the transport of this representative sample from the point of collection to the analytical laboratory, the proper selection of the laboratory sample itself also called the "secondary sampling," and the sample preparation method used to convert the sample into a suitable form for the measurement step can have a greater effect on the overall accuracy and reliability of the results than the measurement itself [5–9].

Although many of the chromatographic techniques have matured and automation is commonplace, sample preparation is still considered to be slow, labor intensive, and even a bottleneck in laboratory processes. Advances in analytical chemistry have led to the development of instruments with detection limits as low as one part per billion. Sample preparation techniques, however, have lagged behind in development. These antiquated techniques may take hours to days to complete and are greatly dependent on the skills of the operator. It is important to note here that sample preparation contributes as much as, if not more, toward the final results as analytical techniques. Some high-throughput laboratories, particularly in the pharmaceutical industry, take advantage of the latest automation equipment to process hundreds and sometimes thousands of samples a day, but many laboratories use techniques based on age-old methodologies with some degree of miniaturization or low levels of automation.

This chapter provides an overview of future prospect of sampling that is a crucial step in analytical processes. For this purpose, sampling and sample preparation are described in Sections 14.2 and 14.3, respectively. Then, green chemistry and its key principles are discussed. In the final section, miniaturization of analytical systems (i.e. miniaturized separation techniques and lab-on-valve (LOV) approach) is presented.

Sample Preparation with Nanomaterials: Next Generation Techniques and Applications,
First Edition. Chaudhery Mustansar Hussain, Rüstem Keçili, and Chaudhery Ghazanfar Hussain.

14.2 Sampling

Primary sampling is the process of selecting and collecting the sample to be analyzed. The objective of sampling is the reduction of mass or volume from the parent batch, which itself can be homogeneous or heterogeneous. If collected incorrectly, then all of the further stages in the analysis are meaningless and the resulting data are worthless. Therefore, sampling forms a very crucial start to the entire analytical process [7, 9, 10].

Once the primary sample is taken, it must be transported to the analytical laboratory without a physical or chemical change in its characteristics. When the system under investigation is a dynamic entity, such as samples containing volatile, unstable, or reactive materials, the act of transportation can present a challenge, especially if the laboratory is a long distance from the point of collection. Often, prepared laboratory standards, surrogate samples, and blanks are carried through the entire preservation, transport, and storage processes to ensure that sample integrity is maintained. Physical, chemical, and/or microbiological degradation are minimized by proper preservation techniques. Appropriate sampling containers, addition of chemical stabilizers such as antioxidants and antibacterial agents, freezing the sample to avoid degradation, etc., are examples of preservation techniques. Once the sample has been brought into the laboratory, storage conditions are equally important to maintain sample integrity before analysis. Often, prepared laboratory standards, surrogate samples, and blanks are carried through the entire preservation, transport, and storage processes to ensure that sample integrity is maintained.

On the other hand, secondary sampling is another crucial step in the analytical processes. After primary sampling, a representative subsample must be taken. The size or inhomogeneity of the sample may be a problem in secondary sampling. Statistically appropriate sampling procedures are applied to avoid discrimination, which can further degrade analytical data.

14.3 Sample Preparation

Sampling, sample preparation, analysis of the sample, and evaluation of the achieved results are the key steps that should be carefully followed in the analytical processes. Sample preparation is a critical step and is one of the most time-consuming and challenging tasks among these steps in the analytical processes. The sampling and sample preparation processes take almost two-third of the total time of the analyses [11]. Figure 14.1 shows the schematic depiction of the percentage share of the steps conducted in the analytical processes. These all stages are also the main source of uncertainties in analytical processes [12]. Thus, researchers put so much effort on the design and development of easy, rapid, cheap, powerful, sensitive, selective, and green sample preparation approaches.

Over the past years, considerable time has been devoted to improving analysis speed, resolution, and automation of analytical measurement techniques and developing and improving data handling and report generation software. In

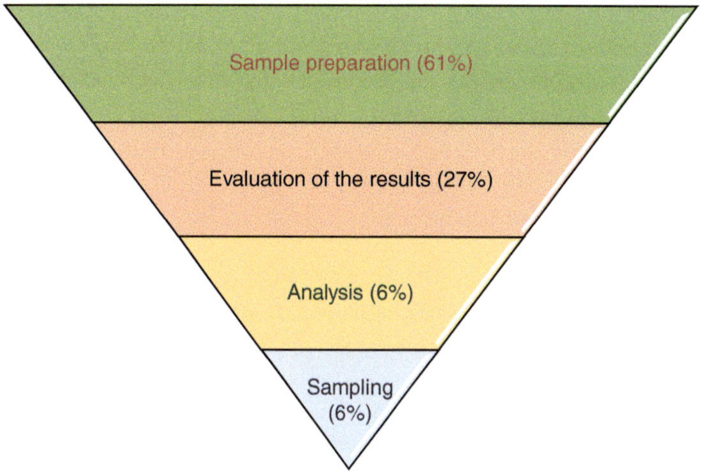

Figure 14.1 The schematic depiction of the percentage share of the steps conducted in the analytical processes. Source: Hussain et al. [57]. © 2020, Elsevier.

contrast, sample preparation, particularly its automation, has been neglected. Many analytical chemists use time-consuming manual methods that have been around for decades. For example, a liquid chromatography (LC)-mass spectrometry (MS)-based separation process can require a few minutes. However, preparation of the sample itself can take much more time. Speeding up or automating the sample preparation will certainly reduce the analysis time and improve the sample throughput.

With the rapid developments in the automation and computerization of analytical tools, the onus for the precision and accuracy lies on the sample preparation processes now more than ever before. Being a part of an analytical method, any variances resulting from sample preparation methods contribute to the total variance of the analytical method.

On the other hand, some of the most frequently encountered problems in sample preparation are time, cost, and lack of reproducibility (Figure 14.3). An often overlooked aspect of sample preparation is its effect on error generation. Each sample transfer and each stage in the analytical process represent a potential source of error because of sample loss or modification. Sample preparation accounts for almost one-third of the error generated during the performance of an analytical method; operator error is responsible for another 20%. Thus, improving and automating sample preparation can decrease error in a typical analytical method by as much as 50% [1]. It is very crucial that a clear sample preparation strategy should be outlined to minimize the number of steps [13]. Optimization is also another extremely crucial issue.

The traditional sample preparation techniques including liquid–liquid extraction display some disadvantages including time consumption, low selectivity for the target compound(s), and high consumption of toxic organic solvents. Various innovative and new sample preparation approaches techniques were designed and developed to overcome the drawbacks of those traditional approaches [14–21].

These sample preparation techniques exhibit a number of superiorities such as high selectivity for the target compound(s), facile automation, low cost, environmentally friendliness, rapid processing of samples, high reproducibility, and the need of low volumes of organic toxic solvents.

14.4 Green Chemistry

Green chemistry refers to the design and application of chemical processes and products that eliminate or decrease the generation and use of chemicals hazardous to the environment and human health [22]. The term "green chemistry" is commonly used as a synonym of the term "sustainable chemistry." However, these terms have not the same meaning. Sustainable chemistry is a much broader term that considers financial, social, and environmental effects that are related to design, manufacturing, and applications of new products. The term sustainability is a balance between financial, social, and environmental factors. The aspects of sustainability are schematically depicted in Figure 14.2.

As the market for nanotechnology continues to develop and increase, so do the chemical/physical production costs associated with synthesizing these materials. Moreover, the chemical synthesis method can produce toxic by-products that may affect the function of technological devices (i.e. solar devices), as well as have an impact on human health and the environment. In an effort to address these adverse effects, in the 1990s, the U.S. Environment Protection Agency (EPA) established the

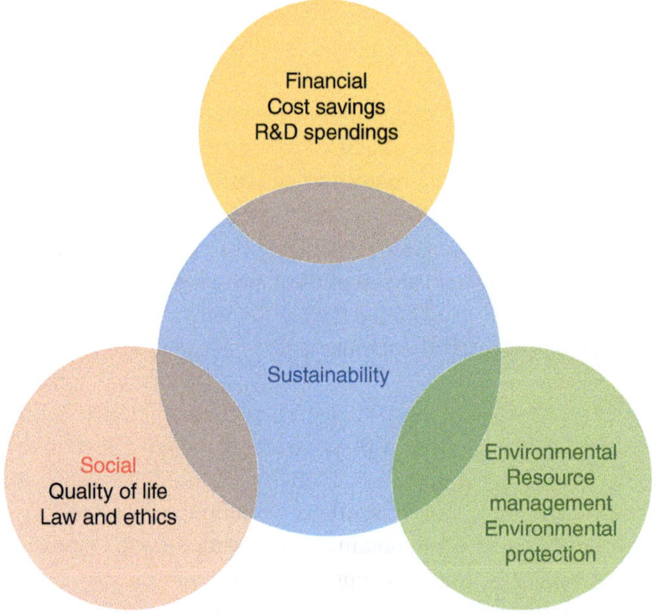

Figure 14.2 The schematic depiction of the aspects of sustainability. Source: Keçili et al. [58]. © 2020, Elsevier.

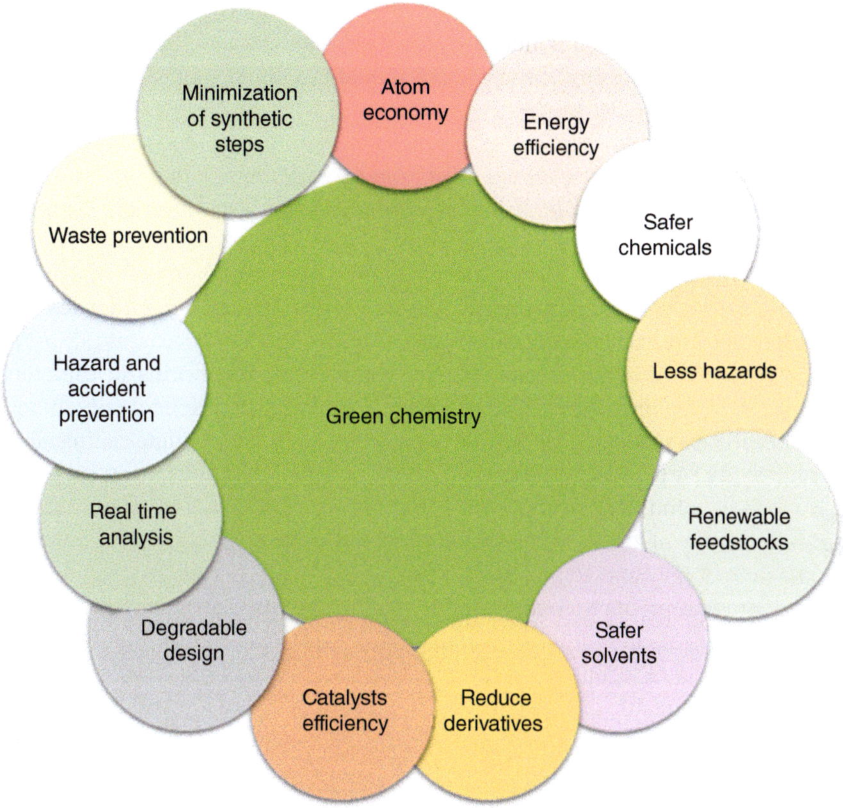

Figure 14.3 The aspects of green chemistry.

field of green chemistry as a legitimate scientific field [23, 24]. In doing so, the EPA defined green chemistry as "the utilization of a set of principles that reduces or eliminates the use or generation of hazardous substances in the design manufacture and application of chemical products" [25, 26]. Green chemistry is the core term for innovative techniques, approaches, and new strategies for the development of sustainable chemistry. In 1998, Anastas and Warner proposed the key principles of green chemistry [26]. The schematic depiction of these key principles of green chemistry is given in Figure 14.3.

1. *Waste prevention*: Design and application of synthesis techniques to prevent waste components.
2. *Atom economy*: Design and application of synthesis techniques to increase the incorporation of all starting materials to obtain a high-yield product.
3. *Use of less hazardous chemicals*: Design and application of synthesis techniques to reduce environmental pollution by generating compounds with lower toxicity.
4. *Use of safer chemicals*: Design chemical products that exhibit minimum toxic effects to the environment.

5. *Use of safer solvents and process conditions*: Avoid using toxic solvents and chemicals.
6. *Energy efficiency*: Conduct all processes at mild conditions such as room temperature, lower pressure, etc.
7. *Use of renewable feedstocks*: Use of renewable starting materials also called as feedstocks such as materials obtained from agricultural products or waste.
8. *Reduce derivatives*: Avoid using any protective or derivatization agents if possible.
9. *Catalyst efficiency*: Use green catalysts to reduce the waste.
10. *Degradable design*: Design degradable chemicals and products that are not potentially hazardous to the environment.
11. *Real-time analysis*: Real-time analysis, control, and monitoring of the compounds during the process to eliminate and minimize undesired products.
12. *Minimize the potential hazards and accidents*: Choose appropriate chemicals to reduce their potential hazards and minimize potential risks for the accidents such as fires, explosions, and releases of chemicals into the environment.
13. *Minimization of synthetic steps*: Use of eco-effective synthetic routes by minimization of synthetic steps and isolations

Since the publication and dissemination of these principles, green synthetic techniques have gained considerable attention as they have provided innovative alternatives for chemical and physical synthetic methods through the employment of cost-effective, environmentally friendly, and scalable techniques in the absence of either high energy or toxic chemical by-products. The cost-effective nature of environmentally friendly synthetic techniques in addition to a proper control over unique properties of nanomaterials (NMs)s will aid in a more realistic practical translation from the lab scale to commercial applications.

14.5 Miniaturization of Analytical Systems

Miniaturization is a rapidly progressing trend with the innovative technologies and approaches in recent years. Like other fields of science, analytical chemistry was also influenced by this superior technology. A wide range of new techniques including molecular self-assembly and micromachining enable the decrease in dimensions of analytical devices and instruments with micro- to nanometer size. Miniaturization is considered to be the trend to downscale biological and chemical processes, offering various superiorities compared to the macroscopic systems [27–32]. The time for the mass transfer and heat are reduced in microfluidic channels because of the shorter distances in the miniaturized systems. On the other hand, the high surface-to-volume ratio also provides homogeneous heat distribution. Miniaturization of various functional units (i.e. reaction, separation, and detection) enables the serial processing of samples as well as the parallelization of the process for high throughput. The term "miniaturization" is not only associated with the decrease in size but also linked to the various challenges in analytical systems including automation, rapid analysis, operational cost, and portability.

The rapid progresses of the microelectronic industry since 1960s established semiconductor preparation techniques in order to enhance the density of transistors in integrated circuits. In 1980s, these techniques led to the first design and fabrication of the microelectromechanical systems (MEMS) [33]. Further refinement of the fabrication technologies of silicon-based MEMS made it straight forward to design and preparation of simple microfluidic platforms. The first demonstration of microfluidic systems was carried out by Terry et al. [34]. In their study, a microfluidic system was prepared by using a silicon wafer, and the prepared system was successfully used as a miniaturized gas analysis system in terms of gas chromatography (GC) with a thermal conductivity detector. The microfluidic system was fabricated by applying photolithography technique and chemical etching, and it consisted of a 1.5 m long spiral capillary column and an injection valve on a 5 cm diameter wafer. Being able to use the device as efficient as using an old-fashioned and generally accepted GC device indicated that the need of big equipment was cleared out. Although the system showed the possibility, researchers were not very keen to use the technology until the beginning of 1990s. The efforts on the design and fabrication of microfluidic platforms did not pick up momentum until the early 1990s, when Manz first introduced the concept of miniaturized or micro total analysis system (μTAS) [35]. According to Manz's proposed μTAS term, injection, separation, and detection systems were in the same interface. This research was not only defining the μTAS systems but it was a realization at the same time for scientists to use the technique in life sciences. Since then, the produced systems were used for different applications such as separation [36–38], detection [39–41], or used as a reactor [42–45]. Then, scientists realized the potential of these unique systems having great advantages in other fields of science besides analytical chemistry.

14.5.1 Miniaturization of Separation Techniques

Various separation techniques such as LC, GC, and capillary electrophoresis (CE) are widely applied for the efficient separation of target compound(s) prior to their sensitive detection. The main purpose of the miniaturization of the separation techniques is to enhance the speed and efficiency of the separation process, reduce the cost of analysis, and increase the portability as well as decrease the volume of sample, reagents, solvents, and wastes, which are consumed and produced during the analysis. Great efforts were put on the design and development of novel miniaturized analytical systems such as nano-LC, chip-based LC, capillary LC, and micro-LC. The decrease in the dimensions of the columns in LC systems has a crucial effect on the other components of the system such as the injector, tubings, and cell volume in detection. Specifically, the tubings that are used for the connection of the detector and column with the injector are downscaled in accordance with the reduction of the dimensions of the LC column. The decrease in the internal diameter of the LC column can also led to the increase in the sensitivity of the detection system [46].

On the other hand, one of the main superiorities of the miniaturized LC systems using columns having reduced dimensions is the use of a very low amount and volume of both stationary and mobile phases. This is very important when we consider the environmentally friendly analyses in green chemistry because it promotes the decrease in the consumption of both solvent and waste generated during the analysis. The decrease in the column dimensions also provides the investigation of elevated temperature values during the chromatographic analysis as well as conducting temperature gradient instead of using mobile-phase gradient for the modification of the resolution to optimize the retention factor for a given analytical separation.

Advancements toward the design and development of automated chip-based CE platforms for the remote-controlled analysis were also recently reported. These novel platforms ensure one to conduct each stage of the analytical separation process such as sampling, sample treatment, separation of the analytes, and sensitive detection without the need of expert user. In an important work conducted by Culbertson et al. [47], a novel, portable, automated chip-based CE platform that is capable of effectively separating the target compounds in a very short analysis time was developed.

In 1952, James and Martin invented the first GC system for the efficient separation of various fatty acid compounds using an 11-ft stainless steel packed GC column [48]. After this milestone work, capillary GC columns were introduced and this was the first stage toward the design and development of miniaturized analytical techniques in separation field. The dimensions of GC systems were significantly decreased in the past years. For example, a novel portable GC–MS system based on a miniaturized toroidal ion trap mass analyzer (TMS) and a low thermal mass GC was developed by Contreras et al. in 2008 [49]. The developed miniaturized GC-TMS system has a relatively low weight (lower than 13 kg) and includes a battery power source and carrier gas supply. In their work, a solid-phase microextraction fiber mounted in a syringe-style holder was efficiently used for the collection of the sample and injection into a specially designed low thermal mass GC injection port. System start-up took about three minutes and analysis of the sample including library matching took about five minutes including the time for column cooling.

On the other hand, the design and development of miniaturized GC systems is of great attention not only in analytical science but also in other technological and scientific areas. The first chip-based GC system was developed in 1979 when Terry et al. proposed a miniaturized GC system using a silicon wafer (5 cm diameter) by applying chemical etching and photolithography approaches [34]. In this study, the developed chip-based GC system composed of a sample injection port, a column having 1.5 m length, a carrier gas supply, and a thermal conductivity detection system. However, the prepared stationary phase in a capillary column exhibited relatively low separation performance compared to the columns in traditional GC systems. To overcome this problem, the geometry of capillary columns in micro gas chromatography systems is currently the focus of the researchers [50, 51]. In a study reported by Radadia et al., it was revealed that a microfabricated serpentine geometry of GC systems provides an enhanced separation performance compared to the square-spiral and circular-spiral channel configurations in miniaturized GC columns [52].

14.5.2 Lab-on-a-Valve (LOV) as a Powerful Tool to Meet Green Chemical Principles

Ever since the first flow injection (FI) analysis was reported by Ruzicka and Hansen in 1975 [53], the technique has been utilized in many research applications. FI is an analytical technique that allows precise and reproducible fluid handling and reaction monitoring. More recently, it has evolved to sequential injection [54] and bead injection [55] where more versatile liquid and even particulate manipulation has become possible. Although the fluidics is reproducible using the same setup, some changes were inevitable when the setup was built using different flow geometries, tubings, or fittings. The LOV approach was developed by Ruzicka in 2000 to overcome this problem [56]. The LOV is made from a block of material (typically plexiglass or polyvinyl chloride) in which approximately 1.6 mm diameter channels are drilled. This is placed atop the 6-position valve, and the fluidics is controlled by the positioning of the valve and the reversible motion of pumps. Because of the monolithic nature of LOV and the rigidity of the flow geometry, system-to-system reproducibility is readily attained. The additional benefit of LOV is the shorter channel distance from the selection valve to the detector or reaction chamber, which can be mounted in the LOV. This minimizes the dispersion of the sample in carrier solvent and reduces the amount of solvent.

Miniaturized LOV-based techniques have the ability to cope with most of the key principles of green chemistry for the decrease in the consumption of toxic chemicals and organic solvents and ultimately the design and development of reagent-free chemical assays. The efficient combination of automated flow with microsyringe pumps enables the accurate control and injection of reagent solutions at the microliter levels with a downscaling option to the nanoliter range. This is especially appropriate for enzymatic assays using expensive enzymes either as target analytes or as catalysts.

14.6 Conclusions

Sampling and sample preparation steps are critical challenging tasks in the analytical processes. Thus, it is very crucial to design and develop easy, effective, fast, cheap, and environmentally friendly sampling and sample preparation techniques before the analysis of target compound(s).

The developments in the extensive research on nanotechnology and miniaturized analytical systems have rapidly grown in parallel because of the rapid technological and scientific progresses including innovative design and fabrication technologies, novel nanomaterials, improved designs, and advanced auxiliary components for the miniaturized analytical systems. The decreased size in miniaturized analytical systems enables many advantages such as low consumption of sample and reagents, fast sample processing and analysis, high integration of multistep processes, precise fluid control, portability, and automation. However, the design and development of certain components or processes also need to be addressed and carefully developed.

In addition, various crucial issues such as sample pretreatment, matrix complexity, stability, and storage of reagents are still hindering the design, fabrication, and commercialization of these miniaturized analytical systems.

References

1 Majors, R.E. (1991). An overview of sample preparation. *LC-GC* 9: 16–20.
2 Majors, R.E. (1999). An overview of sample preparation methods for solids. *LC-GC* 17: S8–S13.
3 Majors, R.E. (2002). Trends in sample preparation. *LC-GC* 20 (12): 1098–1113.
4 Majors, R.E. (1999). An overview of sample introduction and sample preparation of volatile organic compounds. *LC-GC* 17 (9): S7–S13.
5 Lopez Avila, V. (1999). Sample preparation for environmental analysis. *Crit. Rev. Anal. Chem.* 29: 195–230.
6 Keçili, R., Büyüktiryaki, S., and Hussain, C.M. (2019). Advancement in bioanalytical science through nanotechnology: past, present and future. *Trends Anal. Chem.* 110: 259–276.
7 Hussain, C.M. and Kharisov, B. (2017). *Advanced Environmental Analysis-Application of Nanomaterials*. The Royal Society of Chemistry.
8 Hussain, C.M. (2017). Magnetic nanomaterials for environmental analysis. In: *Advanced Environmental Analysis-Application of Nanomaterials* (eds. C.M. Hussain and B. Kharisov). The Royal Society of Chemistry.
9 Keçili, R., Büyüktiryaki, S., Dolak, İ., and Hussain, C.M. (2020). The use of magnetic nanoparticles in sample preparation devices and tools, Chapter 5. In: *Handbook of Nanomaterials in Analytical Chemistry* (ed. C.M. Hussain), 75–95. Elsevier.
10 Hussain, C.M. (2017). Magnetic nanomaterials for environmental analysis. In: *Advanced Environmental Analysis-Application of Nanomaterials* (eds. C.M. Hussain and B. Kharisov), 1–13. The Royal Society of Chemistry.
11 Wasik, A., Kot-Wasik, A., and Namiesnik, J. (2016). New trends in sample preparation techniques for the analysis of the residues of pharmaceuticals in environmental samples. *Curr. Anal. Chem.* 12: 280–302.
12 Ribeiro, C., Ribeiro, A.R., Maia, A.S. et al. (2014). New trends in sample preparation techniques for environmental analysis. *Crit. Rev. Anal. Chem.* 44: 142–185.
13 Han, S.M. and Munro, A. (1999). Transfer from manual to automated sample preparation: a case study. *J. Pharm. Biomed. Anal.* 20: 785–790.
14 Calderilla, C., Maya, F., Leal, L.O., and Cerda, V. (2018). Recent advances in flow-based automated solid-phase extraction. *Trends Anal. Chem.* 108: 370–380.
15 Płotka-Wasylka, J., Szczepanska, N., de la Guardia, M., and Namiesnik, J. (2016). Modern trends in solid phase extraction: new sorbent media. *Trends Anal. Chem.* 77: 23–43.
16 Xu, J., Zheng, J., Tian, J. et al. (2013). New materials in solid-phase microextraction. *Trends Anal. Chem.* 47: 68–83.

17 Buszewski, B. and Szultka, M. (2012). Past, present, and future of solid phase extraction: a review. *Crit. Rev. Anal. Chem.* 42: 198–213.

18 Azzouz, A., Kailasa, S.K., Lee, S.S. et al. (2018). Review of nanomaterials as sorbents in solid-phase extraction for environmental samples. *Trends Anal. Chem.* 108: 347–369.

19 Reyes-Garcés, N., Gionfriddo, E., Gómez-Ríos, G.A. et al. (2018). Advances in solid phase microextraction and perspective on future directions. *Anal. Chem.* 90: 302–360.

20 Khezeli, T. and Daneshfar, A. (2017). Development of dispersive micro-solid phase extraction based on micro and nano sorbents. *Trends Anal. Chem.* 89: 99–118.

21 Wierucka, M. and Biziuk, M. (2014). Application of magnetic nanoparticles for magnetic solid-phase extraction in preparing biological, environmental and food samples. *Trends Anal. Chem.* 59: 50–58.

22 Palit, S. and Hussain, C.M. (2020). Green nanomaterials: a sustainable perspective. In: *Green Nanomaterials. Advanced Structured Materials*, vol. 126 (eds. S. Ahmed and W. Ali). Singapore: Springer.

23 Duan, H.H., Wang, D.S., and Li, Y.D. (2015). Green chemistry for nanoparticle synthesis. *Chem. Soc. Rev.* 2015 (44): 5778–5792.

24 Collins, T.J. (1997). Green chemistry. In: *Macmillan Encyclopedia of Chemistry*, vol. 2 (ed. J.J. Lagowsky). New York: Simon and Schuster Macmillan.

25 Varma, R.S. (2013). Greener routes to organics and nanomaterials: sustainable applications of nanocatalysts. *Pure Appl. Chem.* 85: 1703–1710.

26 Anastas, P.T. and Warner, J.C. (1998). *Green Chemistry: Theory and Practice.* New York, NY: Oxford University Press.

27 Lindström, S. and Andersson-Svahn, H. (2011). Miniaturization of biological assays – overview on microwell devices for single-cell analyses. *Biochim. Biophys. Acta* 1810: 308–316.

28 Ríos, Á. and Zougagh, M. (2015). Modern qualitative analysis by miniaturized and microfluidic systems. *Trends Anal. Chem.* 69: 105–113.

29 Mitsudo, K., Kurimoto, Y., Yoshioka, K., and Suga, S. (2018). Miniaturization and combinatorial approach in organic electrochemistry. *Chem. Rev.* 118: 5985–5999.

30 Dixit, C.K., Kadimisetty, K., and Rusling, J. (2018). 3D-printed miniaturized fluidic tools in chemistry and biology. *Trends Anal. Chem.* 106: 37–52.

31 Büyüktiryaki, S., Sümbelli, Y., Keçili, R., and Hussain, C.M. (2019). Lab-on-chip platforms for environmental analysis. In: *Encyclopedia of Analytical Science*, 3e (eds. P. Worsfold, C. Poole, A. Townshend and M. Miró), 267–273. Academic Press.

32 Keçili, R., Büyüktiryaki, S., and Hussain, C.M. (2020). Micro total analysis systems with nanomaterials, Chapter 8. In: *Handbook of Nanomaterials in Analytical Chemistry* (ed. C.M. Hussain), 185–198. Elsevier.

33 Bao, M. and Wang, W. (1996). Future of microelectromechanical systems (MEMS). *Sens. Actuators, A* 56: 135–141.

34 Terry, S.C., Hermanand, J.H., and Angell, J.B. (1979). A gas chromatographic air analyzer fabricated on a silicon wafer. *IEEE Trans. Electron Devices* 26 (12): 1880–1886.

35 Manz, A., Graber, N., and Widmer, H.M. (1990). Miniaturized total chemical analysis systems: a novel concept for chemical sensing. *Sens. Actuators, B* 1 (1–6): 244–248.

36 Seiler, K., Harrison, D.J., and Manz, A. (1993). Planar glass chips for capillary electrophoresis: repetitive sample injection, quantitation, and separation efficiency. *Anal. Chem.* 15: 1481–1488.

37 Woolley, A.T. and Mathies, R.A. (1994). Ultra-high-speed DNA fragment separations using microfabricated capillary array electrophoresis chips. *Proc. Natl. Acad. Sci. U. S. A.* 91: 11348–11352.

38 Adamski, K., Kubicki, W., and Walczak, R. (2016). 3D printed electrophoretic lab-on-chip for dna separation. *Procedia Eng.* 168: 1454–1457.

39 Figeys, D., Ning, Y., and Aebersold, R. (1997). A microfabricated device for rapid protein identification by microelectrospray ion trap mass spectrometry. *Anal. Chem.* 69 (16): 3153–3160.

40 Weigl, B.H. and Yager, P. (1997). Silicon-microfabricated diffusion-based optical chemical sensor. *Sens. Actuators, B* 38–39: 452–457.

41 Xie, P., Cao, X., Lin, Z., and Javanmard, M. (2017). Top-down fabrication meets bottom-up synthesis for nanoelectronic barcoding of microparticles. *Lab Chip* 17 (11): 1939–1947.

42 Harrison, D.J., Fluri, K., Chiem, N. et al. (1996). Micromachinng chemical and biochemical analysis and reaction systems on glass substrates. *Sens. Actuators, B* 33 (1–3): 105–109.

43 Daniel, J.H., Iqbal, S., Millingston, R.B. et al. (1998). Silicon microchambers for DNA amplification. *Sensor Actuat. A Phys.* 81: 81–88.

44 Pang, X., Lewis, A.C., and Ródenas-García, M. (2013). Microfluidic lab-on-a-chip derivatization for gaseous carbonyl analysis. *J. Chromatogr. A* 1296: 93–103.

45 Hür, D., Say, M.G., Diltemiz, S. et al. (2018). 3D micropatterned all-flexible microfluidic platform for microwave-assisted flow organic synthesis. *ChemPlusChem* 83 (1): 42–46.

46 Szumski, M. and Buszewski, B. (2002). State of the art in miniaturized separation techniques. *Crit. Rev. Anal. Chem.* 32: 1–46.

47 Culbertson, C.T., Tugnawat, Y., Meyer, A.R. et al. (2005). Microchip separations in reduced-gravity and hypergravity environments. *Anal. Chem.* 77: 7933–7940.

48 James, A. and Martin, A. (1952). Gas-liquid partition chromatography: the separation and microestimation of volatile fatty acids from formic acid to dodecanoic acid. *Biochem. J* 50: 679–690.

49 Contreras, J.A., Murray, J.A., Tolley, S.E. et al. (2008). Hand-portable gas chromatograph-toroidal ion trap mass spectrometer (GC-TMS) for detection of hazardous compounds. *J. Am. Soc. Mass. Spectrom.* 19: 1425–1434.

50 Mittermuller, M. and Volmer, D.A. (2012). Micro- and nanostructures and their application in gas chromatography. *Analyst* 137: 3195–3201.

51 Regmi, B.P. and Agah, M. (2018). Micro gas chromatography: an overview of critical components and their integration. *Anal. Chem.* 90 (22): 13133–13150.

52 Radadia, A., Salehi-Khojin, A., Masela, R., and Shannon, M. (2010). The effect of microcolumn geometry on the performance of micro-gas chromatography columns for chip scale gas analyzers. *Sens. Actuators, B* 150: 456–464.

53 Ruzicka, J. and Hansen, E.H. (1975). Flow injection analyses. I. New concept of fast continuous flow analysis. *Anal. Chim. Acta* 78: 145–157.

54 Gubeli, T., Christian, G.D., and Ruzicka, J. (1991). Fundamentals of sinusoidal flow sequential injection spectrophotometry. *Anal. Chem.* 63: 2407–2413.

55 Ruzicka, J. and Hansen, E.H. (2000). Flow injection analysis: from beaker to microfluidics. *Anal. Chem.* 72: 212A–217A.

56 Ruzicka, J. (2000). Lab-on-valve: universal microflow analyzer based on sequential and bead injection. *Analyst* 125: 1053–1060.

57 Hussain, C.M. and Keçili, R. (2020). Sampling and sample preparation techniques for environmental analysis, Chapter 4. In: *Chaudhery Mustansar Hussain, Rüstem Keçili, Modern Environmental Analysis Techniques for Pollutants*, 75–119. Elsevier.

58 Keçili, R., Yılmaz, E., Ersöz, A., and Say, R. (2020). Imprinted materials: from green chemistry to sustainable engineering, Chapter 12. In: *Sustainable Nanoscale Engineering* (eds. G. Szekely and A. Livingston), 317–350. Elsevier.

Index

Sample Preparation with Nanomaterials: Next Generation Techniques and Applications,
First Edition. Chaudhery Mustansar Hussain, Rüstem Keçili, and Chaudhery Ghazanfar Hussain.

l

m